PROLONGED WARS

A Post-Nuclear Challenge

Edited by

Karl P. Magyar, PhD
Constantine P. Danopoulos, PhD

Library of Congress Cataloging-in-Publication Data

Prolonged wars: the post-nuclear challenge / editors, Karl P. Magyar, Constantine P. Danopoulos.
p. cm.
Includes bibliographical references and index.
1. Prolonged war. 2. Protracted conflicts (Military science) 3. Africa—History, Military. 4. Asia—History, Military. 5. World politics—1945- I. Magyar, K. P. (Karl P.) II. Danopoulos, Constantine P. (Constantine Panos)
U243.P76 1994
355'.027'09045—dc20 94-32131
CIP

ISBN 1-58566-056-6

First Printing October 1994
Second Printing July 2001

Disclaimer

This publication was produced in the Department of Defense school environment in the interest of academic freedom and the advancement of national defense-related concepts. The views expressed in this publication are those of the authors and do not reflect the official policy or position of the Department of Defense or the United States government.

This publication has been reviewed by security and policy review authorities and is cleared for public release.

Contents

Essays

Preface

This book was conceived on the battlefields of Vietnam, where the term Vietnam became more than a geopolitical or cultural designation and came to denote a phenomenon. Vietnam is today a euphemism for getting mired in a war, for getting bogged down, for being drawn into a guagmire. Since that war, the US has not entered any military engagement without the fear of encountering another Vietnam. Nor are we alone. The Soviets met their Vietnam in Afghanistan; the South Africans experienced theirs in Angola; and the Nigerians encountered theirs in Liberia. In these cases, the problem concerned the usual expectations of a war of brief duration—the "short, sharp strike" and the realities of subsequent military involvement which came to be measured in terms of years.

Histories have been written of such long wars. Indeed, Thucydides offered the first masterful account of a prolonged war, recording the vicissitudes of battles as they changed the fortunes of societies locked in a struggle which could not be anticipated when the first spear pierced the last moment of peace. Since then, there have been short wars, but they are the exception. More often, wars have bogged down and produced results hardly anticipated by the conflicts' perpetrators. The general term for such conflicts has been *protracted war.*

We argue, however, that wars may be long for two reasons and that these reasons are so antithetical that to call both protracted wars is analytically misleading. Some wars are at the outset planned around a protracted war strategy, usually by an insurgent force which realizes that a quick victory against a superior enemy will not be gained on a conventional battlefield. Hence, protraction is preferred by one of the sides. The other long wars are those in which both protagonists expect quick victory, but for a variety of reasons, they are frustrated in their expectations. These should be termed *prolonged wars.* Understanding protracted war is easier than arriving at wide acceptance of why wars become prolonged.

Because so many wars are prolonged, they leave an unanticipated wake of death and destruction. Their results rarely allow either side to claim a victory that justifies the human and material expenses incurred. Yet, apart from the histories of such wars, few analytic energies have been expended to understand the generic phenomenon. Occasionally, certain dimensions, such as the expansion of war aims or of the number of participants, have been examined, but few have studied the wider, overarching concept. In view of the prevelance of prolonged wars, surely it is time to undertake systematic analytic efforts to learn about them.

Our objective is modest: to establish the distinction between protracted and prolonged wars, to present a number of case studies of prolonged wars drawn from mostly contemporary examples, and to offer the rudimentary outline of a proto theory of prolonged wars. The introductory chapter was sent to all contributors to this volume, along with a list of about 30 factors which we suggest may play a central role in prolonging wars. We encouraged the contributors to consider these factors as relevant variables but also to identify other items unique to their cases. We did not attempt to impose a straight jacket on the authors. The concluding chapter synthesizes the findings of the case studies and provides an outline of a theory of prolonged wars which subsequent researchers may use as a guide to probe the prolonged war phenomenom in greater depth. Certainly, our own speculations should be subjected to more rigorous scrutiny.

However, we are confident that ultimately analysts will offer more than theory to explain prolonged wars. They may aspire to develop guidelines for anticipating in the early stages of a new conflict those conditions which lend themselves to becoming prolonged wars. Additionally, by divining the character of prolonged wars, it might be possible—and productive—to know at which stage a particular war is. This new knowledge should enable them to prepare for strategic responses and determine whether to terminate a conflict or face another Vietnam.

This project has received the full support and official backing of Col Dennis Drew, then head of the Airpower Research Institute, Center for Aerospace Doctrine, Research, and Education, a unit at Air University, Maxwell Air Force Base. We appreciate his support. Later, at the Air University Press, we benefited greatly from the support of Mr Thomas E. Mackin and Mr John E. Jordan, Jr. We are most grateful also for the hard work and professional expertise of our editor, Dr Richard Bailey.

Karl P. Magyar, PhD
Constantine P. Danopoulos, PhD

About the Contributors

As'ad AbuKhali, born in Lebanon, holds a doctor of philosophy degree from Georgetown University. He has taught at Georgetown University, Tufts University, Randolph-Macon Woman's College, Colorado College, and George Washington University. He was scholar in residence at the Middle East Institute in 1992. He currently teaches political science at California State University, Stanislaus, and is research associate at the University of California, Berkeley.

Stephen Blank has been a national security affairs specialist at the US Army Strategic Studies Institute since 1989. Previously, he was associate professor of Soviet Studies at the Center for Aerospace Doctrine, Research, and Education at Air University. Dr Blank holds masters and doctoral degrees in Russian History from the University of Chicago. He has published numerous articles on Soviet military and foreign policies.

Constantine P. Danopoulos teaches political science at San Jose State University. He received his doctorate from the University of Missouri, Columbia, and has written extensively on international security and civil-military relations. Dr Danopoulos has also served as associate editor of the *Journal of Political and Military Sociology*.

Christopher Gregory was senior public affairs officer, Anglo-American Corporation of South Africa. Previously, he taught African politics in the Department of International Relations, University of the Witwatersrand, Johannesburg. His research and publications focus on southern African affairs. He currently resides in the United Kingdom.

Cobie Harris received his doctor of philosophy degree in political science from the University of California, Los Angeles. Currently, he teaches African politics at San Jose State

University. Dr Harris' research interests include constitutional development in postcolonial East Africa.

Herbert M. Howe has served as director of African studies at Georgetown University's School of Foreign Service since 1988. Professor Howe specializes on southern African's political and strategic affairs and has travelled extensively throughout the region. His present research focuses on civil-military relations in South Africa. Professor Howe received his doctorate and masters degrees from Harvard's Fletcher School of Law and Diplomacy.

Benjamin Kline holds a master's degree in modern European and African history from San Jose State University and a doctor of philosophy degree from University College, Cork, Ireland. Dr Kline has taught at the University of Alaska and California State University of Sierra Leone. His research interests include British imperial policy.

Ann Mosely Lesch teaches Middle Eastern politics at Villanova University. She has worked for the American Friends Service Committee in Jerusalem and the Ford Foundation in Cairo. Dr Lesch has written extensively on Middle Eastern affairs, especially on the Sudan, which she has studied with a grant from the US Institute of Peace.

Karl P. Magyar is professor of national security affairs at the US Air Force Air Command and Staff College. He has held academic and professional positions in the US and Africa. His research interests focus on strategic developments in the third world and US foreign policy. Dr Magyar holds a doctoral degree from The Johns Hopkins University.

Stewart Reiser is a faculty associate at the Center for Middle Eastern Studies and Committee on Degrees in Social Studies at Harvard University, Cambridge, Massachusetts. He is the author of *The Politics of Leverage: The National Religious Party of Israel and Its Influence on Foreign Policy (1984); The*

Israeli Arms Industry: Foreign Policy, Arms Transfers, and Military Doctrine of a Small State (1989); and various articles on the Arab-Israeli conflict and Middle Eastern politics.

Rebecca R. Ruelas is the Technical Processing Coordinator for the Government Publications Department at San Jose State University (SJSU). She is completing her masters in Political Science with a concentration in International Relations at SJSU where she also serves as a graduate research assistant.

Steffen W. Schmidt is professor of Political Science at Iowa State University. He is also director of research for the Master's Degree Program in Municipal Government at the Universidad Externado de Columbia. Dr Schmidt holds a doctoral degree from Columbia University. He has written more than 70 publications in political science, and he has a Latin America area interest.

Garth Shelton is a lecturer in the Department of International Relations at the University of the Witwatersrand, Johannesburg, South Africa. He teaches strategic studies and international relations of East Asia. He has conducted research in Germany and the United Kingdom and recently completed a doctoral dissertation on South Africa's nuclear weapons program.

M. A. Shahriar Shirkhani received his undergraduate education in Iran and his doctorate from the University of Idaho. His research interests include political movements in Middle Eastern societies and civil-military relations. Dr Shirkhani is teaching at the University of Tehran, Iran.

Charles L. Stansifer received his doctoral degree in Latin American History at Tulane University. He has taught at the University of Kansas and is currently chairman of the Department of History there. He was director of the Center for Latin American Studies at the University of Kansas from 1974 to 1989. During 1989–90 he was visiting professor of Latin American Studies at the Air War College. His research focuses on Central America, the Caribbean, and United States-Latin American relations.

Earl H. Tilford, Jr., is professor of military history at the US Air Force's Air Command and Staff College. He is the author of two books on the Vietnam War, the latest being *Setup: What the Air Force Did in Vietnam and Why*. Dr Tilford is at work on a book examining how the Vietnam War affected the mind of the American military. He teaches courses on the Vietnam War at the Air Command and Staff College and at Auburn University in Montgomery.

Frédérick Belle Torimiro is assistant professor of political science and coordinator of the International Studies Program at Ferrum College, Virginia. He has received grants for program development from the Association of American Colleges and from the Faculty Scholars Program of the University of Kentucky.

J. Richard Walsh is a foreign service officer with the US Department of State. Prior to joining the foreign service, he was professor of Asian studies at the Air War College. Dr Walsh has published on topics ranging from China's foreign policy to US foreign policy in East Asia.

Introduction
The Protraction and Prolongation of Wars

Karl P. Magyar

Studies of war focus decidedly on the conduct of wars. These studies examine doctrines, strategies, and tactics to learn what went wrong—or right—and for whom. They count casualties and assess physical damage. But the bulk of what is published concerns what may be physically observed or quantified.

They also examine causes for the outbreak of hostilities but in a much more speculative vein. Doing this presents a much more challenging task. It is tempting to argue simply that an inverse relationship exists between the gravity of the wars and the agreement on the explanations of their causes. Leaders on each side portray an obvious single enemy or offense, while each allied power participates for reasons often distantly removed from the major rationale for the conflict. With the causes being complex, and the reasons for the participants' involvements far from clear, the "fog of war" soon yields to the "mud of war"—as the rationale for the conflict becomes distorted and, most of the time, they get bogged down.

Lesser analytic attention has been given to the original perceptions of impending wars at the time of their planning. Invariably, the duration of the wars is projected to be short, "from a few weeks to a few months."[1] Some examples include the American Civil War, Europe's two world wars, and a potential war with the Soviets which, considering the nuclear prospect of it, was usually projected in yet shorter terms. Certainly America's war in Vietnam was not expected to last as long as it did. Nor is this situation unique to the United States. Napoléon never envisioned his Russian debacle in 1812, when he encountered a moderately capable match in Kutuzov's forces and the harsh environmental elements that should have been accounted for in the planning of what was foreseen as a short campaign.[2]

More recently, the Soviets got bogged down in Afghanistan, and, as in the case of the United States in Vietnam, a new Soviet administration inherited the conflict with different perceptions of cause, objective, and strategies. Iran and Iraq also fought a long war of attrition during the 1980s, employing full-conventional weapons and tactics in their battles. And early in 1990, Charles Taylor's rebel forces made surprisingly rapid advances against the highly vulnerable government of Samuel K. Doe in Liberia. But Taylor also encountered the mud of war as the prolonged conflict widened to include another opposition rebel group and then other external intervenors. *The Gambia,* a weekly newspaper, commented: "The Liberian nightmare has gone on for so long that the distinction between rebel and government soldiers is beginning to blur."[3]

There are short conflicts and wars. Examples include the Soviet Union's armed confrontation against Hungary in 1956 and the British war against Argentina over the Falklands in 1982. Other examples include the American invasions into Lebanon in 1958, the Dominican Republic in 1965, and Grenada in 1983. The US raid on Libya in 1986 and the ouster of Gen Manuel Noriega from power in Panama in 1989 are other examples. These interventions concerned clearly stated and limited objectives and were against comparatively weak opponents. The preponderant power easily won each of these battles.

However, when facing the uncertain and far more capable forces of Saddam Hussein across the borders of Kuwait (1990–91), the concept of "Vietnam" permeated the cynical appraisals of the war's likely course of events. In sharp contrast to those who held that the US had not learned the essential lessons of Vietnam, President Bush centered his war policy on avoiding a situation which would become progressively a war of attrition and would cause heavy military casualties. Due mainly to the lessons of Vietnam and a less menacing physical terrain, planners decided to invite a well-managed air campaign along with a cautious ground offensive—cautious in that the single objective of ousting Iraq's forces from Kuwait was to determine the length of the confrontation. This was not to be a prolonged war. Still, to prepare the US public for all contingencies, Gen A. Gray, Jr., commandant of the United States Marine Corps, cautiously estimated on the eve of the war that American forces "might have to stay in it for six months or longer."[4] Hussein had warned that his people would fight six or more years—and that his credibility was buttressed by his eight-year battle against Iran.[5] But Air Force Gen Andrew Dugan's controversial comments, made while inspecting troops in Saudi Arabia, projected a short, sharp resolution of the conflict with an attack on key central targets, should the war break out.[6] He was right.

The ground war lasted 100 hours and ended abruptly. Not surprisingly, many observers criticized this quick end, arguing that another four hours or one more day could have allowed for the attainment of a few more objectives. Before long, pressures emerged for the US and its allies to intervene on behalf of the Kurds in Iraq's north and on behalf of the Shias in the south for an assault on Baghdad to remove Saddam Hussein from power. Limited intervention on behalf of the Kurds and the Shias was in fact introduced, but here again, it was done most cautiously so as not to lose the ability to expand intervention on the allied powers' terms. To the analyst of prolonged wars, the debate between demands and resistance to escalation raised the altogether familiar components of conflicts which introduce new objectives, actors, tactics, and perceptions as these conflicts become bogged down.

Observers may be tempted to make a distinction between the length of conflicts fought among the great powers or the industrially developed states and those fought in the third world. Most European wars were hardly short, and interventions in the third world are not necessarily quick and decisive. Certainly, internal conflicts within third world countries are usually of long duration as are the wars between third world states. But these too count exceptions among them. By African standards, Nigeria's Biafran secessionist war may be viewed as a relatively short and concisely defined war, especially in light of its vast scale. Tanzania's ouster of Uganda's Idi Amin was a short, intense affair, even if Uganda's internal conflict continued. And numerous border skirmishes have flared up in Latin America and in Africa or between the Soviet Union and China. These conflicts may be part of much longer conflicts that fester but rarely erupt into full-shooting wars. But these comparatively rare cases of short, active hostilities are contrasted by the numerous lengthy conflicts which generally characterize the wars of the third world.

A distinction needs to be made between conflicts and wars. For our purposes, we define a conflict as an established attitude of contention between two or more groups, within one or among two or more societies. Hostilities will break out periodically but the disagreement is not quickly resolved. A conflict is more than mere spirited competition. Certainly there are nonviolent conflicts, but these will not concern us. More specifically, a conflict results in at least some casualties over an extended period of time.

Wars may be an integral part of such conflicts. In the wording of Air Force Manual 1-1, *Basic Aerospace Doctrine of the United States Air Force,* "War is a violent struggle between rival societies to attain competing political objectives."[7] A war is a more intensively fought engagement in which at least one side engages its full civilian and military resources. When part of a prolonged conflict, a war is not necessarily resolved—as in the case of Sudan, to cite one of many available examples. But when a war is fought as part of an extended conflict, it will either conclude without a clear victory, as in the case of the Iran and Iraq War, or with one side clearly victorious, an example of which is Britain's victory over Argentina.

There are different types of war; they may be classified as nuclear, full- or limited-conventional wars, or low-intensity conflicts or wars, or their illegitimate cousin, unconventional wars. These wars are all violent means for resolving conflicts instead of negotiated resolutions. The distinction between wars and conflicts often becomes obscured when analyzing the reasons for their prolongation. Both wars and conflicts in fact may get bogged down, especially when judged by the standards of the planners' initial expectations. An unresolved war may well become a conflict—if active hostilities break out again. That development, of course, ensures prolongation, which is far more damaging than if the war been fought to a clear, quick conclusion. Further subtle distinctions could occupy an entire volume, but the focus of these introductory comments is one of the reasons for prolonging hostilities—whether wars or conflicts. Hence, the distinction between the two can receive only limited treatment.

One may argue that a conflict by its innate nature is extended and that prolongation is to be expected. Wars, on the other hand, are the more interesting phenomena, since they appear in the planning stages to be much shorter than they usually turn out to be. Initially, the cost of prolonging a conflict seems cheaper than waging an all-out risky and aggressive war. But conflicts too are often prolonged beyond initial expectations; hence, they may be just as devastating. Conflicts may constitute the prolonged and less-intensive portion of unresolved wars. Facho Balaam, leader of Chad's Patriotic Front, expressed the entire phenomenon concisely when he referred to his country's 20-year-long conflict: "Frankly speaking, we think that this is a war without any result."[8] A prolonged war or conflict can degenerate into purposelessness when judged by the results—yet a war or conflict can take its toll in social devastation. This observation is not a recent one. Writing more than two thousand years ago in The *Art of War,* Sun Tzu noted: "In all history, there is no instance of a country having benefited from prolonged warfare. Only one who knows the disastrous effects of a long war can realize the supreme importance of rapidity in bringing it to a close."[9]

A further distinction needs to be made between the terms *protracted* and *prolonged.* Protracted may well be an appropriate term for conflict, while prolonged may be a more appropriate one for war. Most wars are usually planned for short duration, although hostilities exercised over a long time are often an integral strategy of insurgent forces when they confront a conventionally equipped superior enemy. This line of reasoning deliberately pursues a violent, protracted conflict in place of a standard, quick war—whose outcome would certainly favor the superior enemy. Time, terrain, and tactics are fused; therefore, there is no initial misassessment of the length of a conflict which extends beyond the usual short duration. A protracted conflict is planned and hence expected, but a prolonged war is not pursued as a matter of course.

Mao Tse Tung has popularized the concept of a protracted war, although this sophisticated military strategy had a rich history, especially in China. Mao recorded his thoughts on the subject in the 1930s while fighting the Japanese occupation of China and while his insurgent forces also opposed the Kuomintang government. He presented his analysis of this subject in *On The Protracted War*, a synopsis of speeches he made in 1938.[10]

With respect to fighting government forces in the civil war, Mao's strategy recognized two phases. The first of these phases employed guerrilla warfare; the second phase, regular warfare, was regular "only in the concentration of forces for a mobile war and a certain degree of centralization and planning in command and organization." He recognized that this did not rank with wars fought by foreign armies or even by the Kuomintang: "It was in a sense only guerrilla warfare at a higher level."[11]

Against the Japanese, he also formulated two similar phases of conflict. The first, the strategic defensive period, utilized guerrilla strategies. In the second, the strategic counteroffensive period, Mao relied on regular warfare, as by then, he claimed to have developed a more regular armed force capability which had acquired modern weapons. The second phase also showed evidence of centralization and a higher degree of organization. Mao believed there is little innate value in retaining a guerrilla strategy if the capability to wage conventional warfare is developed. Against the Japanese, he wrote, "regular warfare is the principal and guerrilla warfare the supplementary form."[12] Regular warfare is decisive, while guerrilla warfare is utilized while preparing for a counteroffensive.

Mao is the best-known exponent of the protracted war concept, but one may argue that he should have been credited with expertise on the subject of prolonged *conflict*. The Chinese civil war lapsed into a national war and reverted into a civil war, one occasionally interspersed with intense conventional warfare. Such caprice characterizes a classic prolonged conflict. Semantics aside, whether conflict or war, the object remains final victory. He warns that the theory of a quick victory is wrong: the "enemy is strong while we are weak."[13]

This observation calls for a "strategically protracted war." During the course of a protracted struggle, Mao counts on the enemy's position to weaken and his own to strengthen in three, somewhat obvious stages. Each stage requires a different form of fighting: mobile warfare, guerrilla warfare (which is a supplementary form), or positional warfare. Mao refers repeatedly to the factors that contribute to the success of a protracted war: the decline of the enemy's morale, tactics, finances, war-fighting ability, and optimism.[14]

Mao also considers important the support a nation receives from the international community. One may consider this an interesting point as it underscores the fundamental political nature of prolonged conflict, which, by engaging the superior enemy via guerrilla strategies, demonstrates at least a credible liberationist performance. In reality, as an integral tactic, international pressures help to resolve the conflict for the insurgents. This scenario was manifested more recently in the Southwest Africa People's Organization's (SWAPO) feeble efforts on behalf of Namibia's independence. At no time did SWAPO effectively threaten South Africa's defensive capabilities, but changing international configurations ultimately led to a diplomatic settlement with SWAPO emerging victorious in Namibia. Mao admits this point unequivocally when he observes that China "cannot win without utilizing the aid of international forces and the change within the enemy country."[15]

Mao divides the protracted struggle and terrain along Wei-ch'i lines: a classic board game played by Oriental peoples for thousands of years. The game is won through the slow, progressive accumulation of advantages as opposed to the capture of a single king in the cataclysmic finale of chess. Scott A. Boorman considers this Sinic tradition of waging war when mixed with Marxist-Leninist elements as "its use of fluid operational methods and yet its reliance upon relatively stable base areas; its emphasis on efficiency and yet its tolerance of protraction; and its delight in complexity in contrast to the simplicity of Western warfare."[16]

Mao stresses the spaces to be gained by guerrilla tactics. He elaborates on the political advantages of protraction in a passage reminiscent of America's frustration in Vietnam. He suggests that "Japan's military and financial power will be heavily consumed by China's guerrilla war, her home population will become more discontented . . . and her international position will become more isolated."[17] Superior weapons alone are inadequate he stresses and continues by arguing that "it is man and not material that counts." With reference to Mao and Vietnam, one may speculate that modern weapons actually limit the war-making options of a better-equipped force because the effective employment of modern weapons favors a quick, short war, but the innate composition of such forces handicaps them against a lesser-equipped force that still relies on men—whose effective utilization will be in a protracted situation. The 1991 Gulf War, in which both sides relied on modern high-tech weapons, amply illustrated this lesson.

Mao weaves an interpretation of his own strategic thoughts, fused with Clausewitz's teachings, which describe the contemporary third world revolutionary environment: "When politics has developed to a certain stage beyond which it cannot proceed by the usual means, war breaks out to sweep away the impediments in the way."[18] This statement suggests that the more normal condition prevailing in third world states is protracted conflict, something utilized for the development of effective and legitimate political institutions.

Adda Bozeman expresses the same concept: "Few modern theorists in the field of international relations or conflict resolution have bothered to explore the value content of conflict, war and violence." And, she continues, "Human dispositions towards stress, violence and death are by no means everywhere the same. . . . Nowhere outside North America and Northern Europe does one encounter the overriding desire to avoid armed conflict and to seek peaceful settlement of disputes that leading peace-minded scholars in our society assume to be generally present."[19] Conflicts, which concern the problems of national consolidation more than challenges to international strategic balances, will prevail in the developing areas. Expecting a much better-equipped external force to pave a path to victory through short, sharp strikes is a naive and possibly disastrous notion. Mao's vision of protracted conflicts envisions innate value in waging "innumerable but indecisive battles."[20] This is, of course, anathema to short-war doctrines.

Mao's legacy influenced subsequent generations, who came to look at protracted conflict as the predominating characteristic of the communist strategy for world domination. But where Mao had been concerned only with his particular dilemma against his own government and against the Japanese enemy, the new perspective elevated much of Mao's reasoning and strategy to global levels. Even such a militarily well-equipped power as the Soviet Union was expected to incorporate the advantages of the protraction rationale on the march to total victory. A quick, decisive, short-war showdown could defeat socialist gains, as the socialists of the world had not yet consolidated sufficiently to confront the superior capitalist enemy. Hence, insurgent fires would be kept burning around the world; peace would not be tolerated; the capitalist offensive would be neutralized; and eventually, victory would belong to the liberated peoples.

One of the more notable elucidations of this theme at the global level was the book, *Protracted Conflict,* by Robert Strausz-Hupe, William R. Kintner, James E. Dougherty, and Alvin J. Cottrell.[21] The publication of this volume was preceded by the testimony of Strausz-Hupe, Cottrell, and Dougherty at a hearing of the House of Representatives' Committee on Un-American Activities in 1958.[22] At the hearings, Strausz-Hupe postulated that the communist strategy had never been a strategy of limited war but a protracted conflict.[23] This strategy, he said, "prescribes the annihilation of the opponent by a long series of carefully calibrated operations, by feints and maneuvers, by psychological and economic warfare, and by diverse forms of violence." The weaponry employed ranged from seemingly innocuous political activities to the megaton bomb. The communist strategy of protracted conflict was an "organic scheme of conflict" aimed at the one goal of total victory. There was no difference between hot and cold war, nor between military and political means. The communists "hope to make small, steady gains and yet avoid the all-out conflict they do not want," according to Dougherty.[24] An important ingredient in this strategy was psychological warfare and its continuous use of threats of an all-out war. This strategy was dangerous brinksmanship, one in which the communists retain the offensive in true Maoist fashion.

Echoing Mao, Strausz-Hupe and his colleagues held that the choice of conflict mode was a matter of tactical expediency. Preferably, the communists would attain their goals without warfare. But a general war would be employed if they thought themselves capable of delivering a knockout blow. The conflict would be built up and revolutionary tensions increased. For this, the newly emerging third world states were opportune battlegrounds. Communists were counterrevolutionaries in that they captured revolutions made by others. Their participation would of course lead to the protraction of all conflicts and elevate history's inevitable class warfare within societies to a global level.

This was the stuff of classic cold war analysis. From the vantage point of the next three decades, it describes the Marxist image rather than the attained reality, as the fatal fissure in the Marxist world was developing at the time Strausz-Hupe and his colleagues were writing. But they must be credited with having identified communist strategies as protracted conflict and not as protracted war, as Mao had it.

In their volume on that subject, the authors further elaborate Mao's deliberate ambiguity between war and peace, which suggests that Western statesmen must appreciate a greater degree of analytic sophistication. A protracted conflict postpones the decisive battle until revolutionary forces are favored. The Russians and the Chinese "thrive upon conflict as the normal condition of the twentieth century."[25] The doctrine of protracted conflict includes the total objective; shifting battleground; and weapons and tactics which confuse the opponent, keeping him off balance and wearing down his resistance. The strategies for attaining these objectives should not be done by Europe's style of limited warfare. A conflict strategy gradually lapses into full war, if and when the opportunity arises. Global conflicts tend to be interlinked; regional conflicts are multifaceted but may well be an integral component of a specific conflict elsewhere. Strausz-Hupe and his colleagues noted that "the current struggle for the mastery of the globe has been waged for five decades." They also noted that "the festering sores on the international body politic cannot be healed by pious homilies on the blessings of peace."[26]

The implications for the Western strategic policy maker were obvious: We cannot counter this menace by being prepared only for war. Countering conflicts requires a new approach. But, their counsel did not make a sufficient impact before Americans escalated their involvement in Vietnam. Writing in 1959, they observed that the West had neither a doctrine of protracted conflict—nor a desire to produce one. To provide one would be to resort to the tactic of our enemy.[27] The communists had accepted the central formulation of Clausewitz regarding the interchangeability of military and political instruments. The communist doctrine of protracted conflict integrates war, politics, law, diplomacy, psychology, science, and economics in the conduct of foreign policy.[28] In short, the modern communist strategy of protracted conflict is a successful fusion of Clausewitz and Mao—perhaps the supreme synthesis of West and East.

Strausz-Hupe and his colleagues focused on the protracted global conflict as a centrally directed conspiracy by communism to take over the world. They did not, of course, appreciate modern third world liberation struggles which sought to exploit the opportunities that Moscow offered to them in their own conflicts for emancipation from colonial domination. Strausz-Hupe, et al., also did not appreciate the greater historical context of many conflicts, such as those in the Middle East, Africa, and East Asia, which preceded Marxism by several centuries. Many of these conflicts had been fought almost continuously.[29] Nor did these analysts investigate the nature of prolonged conflicts in the third world which were more than wars for independence. Many conflicts in Asia and in Africa concerned internal consolidation more than liberation from colonial rule. In the former case, conflicts of consolidation broke out simultaneously with the approach or attainment of independence. Many countries did not even have to fight wars for independence, yet they have been engulfed in conflicts ever since independence. Most of these countries had only national objectives and were not an integral component of global communist conspiracies.

More recent analyses of long, violent struggles continue to focus on protracted conflicts but not on prolonged wars. Edward E. Azar, Paul Jureidini, and Ronald McLaurin present a sociological perspective in their systematic study of conflicts. They define protracted conflicts as "hostile interactions which extend over long periods of time with sporadic outbreaks of open warfare fluctuating in frequency and intensity."[30] Conflict stakes are high; they involve whole societies, and they define national identity and solidarity. As they linger in time, they have no point of termination.

For the contemporary strategic policy maker, Azar and his colleagues maintain that these conflicts cannot be terminated by explicit decision; they will end by cooling off, transformation, or withering away. Protracted conflicts are characterized by duration, fluctuation in intensity of interaction, and spillover in all domains. They contain equilibrating forces which keep the interactions within established levels of conflict boundaries. Azar and his coauthors also recognize that strong forces operate to undermine attempts to bring about settlements: "It is the nature of the dynamics of protracted social conflict that the many benefits accruing from an institutionalized conflict are clearer or more immediate than those developing from peace." A current manifestation of this is evident in South Africa's rural, Zulu-based, long-term conflict against more radical opposition groups which are posturing to take over the leadership of South Africa under an African National Congress (ANC)-centered united front. Eventually, the institutionalization of the conflict becomes necessary to participants who find it in their interest not to resolve the dispute early.

Another more contemporary dimension of political protraction concerns the interrelation between conflicts and crises. Michael Brecher and Jonathan Wilkenfeld characterize conflicts by their lengthy duration, fluctuation in the intensity of interaction from violence to near accommodation and back to violence again; intense animosity among participants; and spillover into other issues. They were also characterized by the absence of discernible termination.[31] Crises within protracted conflicts emerge from violence, which itself is a response from one of the participants to a prior perceived threat. To understand protracted conflicts, they search for the triggers that spark the violence that leads to crises. Nonprotracted conflicts such as wars are probably easier to bring to a peaceful resolution than are protracted conflicts.

The work by Azar and his colleagues contrasts with the perspective offered by Strausz-Hupe and his team in one important respect. The former view protracted conflict as primarily a sociological phenomenon emerging from internal cleavages. Their "protracted social conflicts" focus on religious, cultural, and ethnic communal identities. Hence, terrorism and low-intensity warfare are common practices. Military or balance-of-power means cannot manage such conflicts.[32] Strausz-Hupe and his colleagues, however, stress the external, or global, dimension of the protracted conflict—of which local conflicts are likely to be an integral part.

These two perspectives need to be combined for analytic purposes. For example, where we may discount the grand conspiratorial nature of a centrally planned, global-level conflict today, we would err in glossing over the importance of the extensive international interactions which characterize so many prolonged conflicts. The Irish Republican Army has on occasion been tied to New York, Libya, and the Marxist world; Israel has been engaged in several conflicts in Africa to weaken the soft underbelly of Egypt; Saudi Arabia's financial largesse has reputedly financed wars from the Western Sahara to Mozambique and perhaps even in Nicaragua; and the legendary Carlos from South America operated from bases in north Africa and struck in Vienna as easily as at targets in the Middle East. Even the Japanese Red Army faction had a global mission. Though internationalized, these instances reveal no discernible centrally planned nature. Traditionally, protracted conflicts may have been almost purely parochial concerns, but today the internationalization of such conflicts may well form the central ingredient.

So far, this analysis has focused mostly on the identification and elaboration of *protracted* conflicts. Their nature has been developed by analysts, who have left us written legacies as well as substantial histories to ponder. A cursory review of history will no doubt verify that the weak pursued protracted conflicts. They would much rather wage full conventional wars, but in the absence of sufficient strength, a protracted struggle must serve as a supplementary strategy—in the formulation of Mao Tse Tung.

However, numerous struggles are in evidence which may properly be labeled as prolonged conflicts and *prolonged* wars. The former suggests conflicts of long but unplanned duration. In fact, they may follow the unsuccessful resolution of a war, but the key element is that prolonged conflicts should not be mistaken for the sophisticated strategic formulations of Mao and of other modern guerrilla organizations. Prolonged conflicts best characterize the numerous conflicts in Africa, most of which were not formulated according to a long-term plan. Military leaders like E. Mondlane in Mozambique and A. Cabral in Guinea-Bissau and Cape Verde implemented strategies of protracted conflict against the Portuguese colonial forces—hence, they are the exceptions. But one can't trace most of Africa's ongoing conflicts to strategically formulated protraction.

In the case of wars, one would be foolhardy purposely to wage them in protracted fashion, for that would then imply that they are in fact conflicts. However, wars may certainly be prolonged, as they tend to be the result of gross miscalculation by one or both sides. Should such a war continue as a series of related battles in which neither side clearly predominates, the war will continue until it results in mutual annihilation or it lapses into a prolonged conflict. Few of these, as Azar stresses, are conclusively terminated. Another alternative suggests that a prolonged war can terminate by mutual agreement, with no side being victorious; it may not be followed by a prolonged conflict. This was the case in the Iran-Iraq War of the 1980s, and it raises the prospect that a prolonged war is one fought continuously but with no clear victor emerging on the battlefield. The Cambodian war and the war between National Union for the Total Independence of Angola (UNITA) and governmental forces in Angola may be other examples.

Identifying reasons for a protracted conflict is relatively easy, as that strategy is one deliberately chosen by one side in a conflict. The weak rely on protraction as a calculated policy when they seek to exploit the advantages of time. But identifying the reasons for the prolongation of a conflict or war is much more difficult. A number of factors contribute to this difficulty, and no doubt each particular instance produces a unique combination of reasons. Yet, considering the numerous unresolved wars and conflicts in the world, agreement on the reasons for prolongation is notably absent; as a generic phenomenon, little academic insight has been offered. Histories and statistics of such wars abound, each attesting to unbelievable horror. Nevertheless, the phenomenon is repeated in virtually every region of the third world.

Although developed countries today manage to escape wars at least within their own boundaries, their frequent involvements in such conflicts attest that those conflicts are not only peripheral concerns. These prolonged wars and conflicts have the innate capability of considerable global involvement, disturbing regional balances, introducing new expansionists, impeding access to vital resources, and realigning political relations. In other words, such conflicts can't be ignored, and merely recording their histories hardly addresses the requirements for global stability. Understanding the reasons for prolongation of wars and conflicts may well rank as a major security concern on a par with the attempt to understand their causes in the first place.

In the emerging new world order of the postcontainment era, we may consider the termination of armed conflicts as a noble, but terribly naive, objective. In fact, most third world states fighting wars have pursued their own agendas apart from those of the cold war adversaries. However, the present political terms of the new world order allow for the isolation of these conflicts from global strategic concerns; therefore, understanding the prolongation phenomenon allows introduction of strategies for reducing the gravity of such wars by limiting their damage and by enhancing their prospects for an early peaceful resolution.

Notes

1. Maj Robert W. Hambridge, "World War I and the short War Assumption," *Military Review*, May 1989, 36.
2. Owen Connelly, *Blundering to Glory: Napoléon's Military Campaigns* (Wilmington, Del.: Scholarly Resources Inc., 1987), 173.
3. Reported in *Foreign Broadcast Information Service (FBIS), Africa*, 13 August 1990, 1.
4. *Atlanta Journal*, 27 September 1990.
5. *Atlanta Constitution*, 21 September 1990.

6. *Air Force Times*, 1 October 1990; *New York Times*, 18 September 1990.
7. Air Force Manual (AFM) 1-1, *Basic Aerospace Doctrine of the United States Air Force*, vol. 1, 1992, 1.
8. *FBIS*, 15 December 1988, 3.
9. James Clavell, ed., *Sun Tzu, The Art of War* (New York: Delacorte Press, 1983), 13.
10. Reproduced in Mao Tse-Tung, *Selected Works*, vol. 2, 1937–38 (New York: International Publishers, 1954).
11. Mao Tse-Tung, *Problems of War and Strategy* (Peking: Foreign Language Press, 1960), 18.
12. Ibid., 22–23.
13. Mao Tse-Tung, *On the Protracted War*, 179.
14. Ibid., 185.
15. Ibid., 187.
16. Scott A. Boorman, *The Protracted Game: A Wei-ch'i Interpretation of Maoist Revolutionary Strategy* (New York: Oxford University Press, 1969), 4.
17. Ibid., 190.
18. Ibid., 202.
19. Adda Bozeman, "War and the Clash of Ideas," *Orbis* 20, 1, (1976): 63–65.
20. Boorman, 242.
21. Robert Strausz-Hupe et al., *Protracted Conflict* (New York: Harper and Brothers, 1959).
22. 85th Congress, Second Session, 20 May 1958, Washington, D.C., United States Government Printing Office, 1958.
23. Ibid., 1.
24. Ibid., 2.
25. Robert Strausz-Hupe et al., 1–6.
26. Ibid.
27. Ibid., 28.
28. Ibid., 33.
29. Adda Bozeman presents a compelling case for this phenomenon. See Bozeman, 63–65.
30. "Protracted Social Conflict: Theory and Practice in the Middle East," *Journal of Palestine Studies*, no. 1 (1978): 50.
31. *Crisis, Conflict and Instability* (Oxford Pergamon Press, 1989), 188–89.
32. Edward E. Azar, *The Management of Social Conflict: Theory and Cases* (Aldershot, N.H.: Dartmouth Publications Co., Ltd., 1990), 2.

Iran-Iraq
Protracted Conflict, Prolonged War

M. A. Shahriar Shirkhani
and
Constantine P. Danopoulos

Centuries of historical rivalry, ethnic particularity, religious and ideological differences, and boundary disputes have been among the major forces contributing to a protracted conflict between Iran and its western neighbors—the Ottoman Empire from 1517 and Iraq since 1920. One or a combination of these factors have caused numerous military hostilities and many full-scale wars. The recent Iran-Iraq war (1980–88) was but the contemporary phase of the old conflict. Though bloody, none of the previous wars between Iran and its neighbors continued unabated for as long as eight years and were fought with such intensity as the Iran-Iraq war. While centuries of rivalry generated the potential and the psychological setting for this war, the Iranian revolution and the danger it posed to the interests of the two superpowers (US and the former USSR) and certain countries in the Middle East created a context which contributed to the length of the war. This chapter examines the factors that contributed to the prolongation of the 1980–88 Iran-Iraq war.

Historical Background

The Iranian plateau, the cradle of Persian civilization, was occupied in the seventh century by the Moslem Arabs. Their occupation set the stage for a long rivalry between Persians and Arabs. Mesopotamia (the land between the Tigris and Euphrates rivers) was annexed by the Ottoman Empire in the early part of the sixteenth century. The empire was dominated by political and economic elites drawn from the Sunni sect of Islam. At the same time Shah Ismail, founder of the Safavied Empire (1501–1722), adopted Shiism as the state religion in

Persia. The sectarian division between Sunni and Shiaa Moslems was used by both states for mass mobilization to support wars against each other, which took place in 1555, 1568, 1590, 1613, and 1618. Eventually, in 1636, a treaty for the first time laid down a vague border between the two empires.[1] This treaty secured relative peace for 200 years and remained the foundation for future treaties.

Hostilities resumed in 1722. Yet in spite of five wars and additional agreements, the 1636 treaty remained, for the most part, intact. In 1821 hostilities broke out again. The war ended with the First Treaty of Erzurum in 1823, which basically confirmed the previous treaties. The Second Treaty of Erzurum was signed in 1847 under the mediation of Great Britain and Russia. While it confirmed the 1823 treaty, the treaty of 1847 also made some adjustments to future disagreements. Iran ceded its claims on the Suleimaniya region (part of modern Iraq) to the Ottomans, and in return the Ottoman government formally recognized the unrestricted sovereignty of the Persian government on the left bank of the Shatt-al-Arab. Furthermore, Iranian vessels gained the right to navigate freely the Shatt.[2] The discovery of petroleum in Khuzistan in 1908 increased the economic value of Iranian ports on the Shatt for Tehran and its oil contractor, the United Kingdom. To secure the free navigation of oil tankers, Iran demanded a precise delineation of the borders on the Shatt. The Constantinople accords of 1913, and their modification in 1937, procured Iranian maritime access to the entire Shatt. These accords did not fully satisfy Iranian demands, as they mainly drew the boundary at the low-water mark line on the Iranian shore of the Shatt.

In spite of this, Shatt disputes remained dormant, as internal matters preoccupied both sides. Iran was facing an economic crisis, a political disorder caused by the Allied occupation of the country during World War II, and the forced abdication of the powerful monarch Reza Khan. His departure created a power vacuum which his young son was not ready to fill. Iraq, a new political entity carved out of the disintegrating Ottoman Empire, was in the first stages of state building. Following the British mandate, the dominant Sunni political elites in Baghdad were struggling to create a sense of nationhood and to establish a central authority in a country with ethnic and religious diversity. The Kurds concentrated in the north and the Shiaa in the south resented the authority of the minority Sunni Arabs of central Iraq. Lack of strong central control during the monarchy (1921–58), periodic political instability following the 1958 coup, and the establishment of a republic relegated the Shatt issue to a secondary concern in Iraq.

However, as the mid-1960s approached, the old disputes resurfaced again. Two factors contributed to their reemergence. First, increased support for Arab nationalism in Iraq was fueled by the nationalist movement in Egypt, led by the charismatic Gamal Abdul Nasser. The logic outcome of Arab nationalism was pan-Arabism, which helped to rekindle Baghdad's territorial expansion. The coming to power of the nationalist/socialist Baath party in 1968 strengthened Baghdad's claim over the entire Shatt. Hostilities with Tehran increased, and the probability of finding solutions based on mutual interests declined. Second, the modernization process in both Iran and Iraq improved the strategic importance of the Shatt and its ports on the east and west banks. Invoking international law, Iran demanded the establishment of a demarcation along the median or the thalweg line. Iraq objected and demanded sovereignty over the entire Shatt, basing its claims on previous treaties. Finally, in 1975, after Iran established itself as the hegemonic power in the Persian Gulf, Baghdad was forced to sign the Algiers accord in which it accepted the thalweg principle and settled for joint sovereignty of the Shatt.

Later events indicated the Algiers accord (1975) could not put an end to the long and deep-seated rivalry between Iran and Iraq. Then Iraqi Vice President Saddam Hussein, one of the followers of Nasser and a devoted Arab nationalist, felt personally humiliated when forced to give in to a determined adversary and to share sovereignty of the Shatt, which he considered to be an Arab waterway.[3] This psychological factor played an important role in Saddam Hussein's subsequent decision to attack Iran (1980) to terminate the Algiers accord, which he himself had signed.

Origins of the Recent Conflict

The Islamic Revolution and the turmoil it created provided a propitious opportunity for Iraq to gain the upper hand in its competition with Iran. The revolution's religious orientation imposed a threat to the ruling socialist/nationalist Baath party in Iraq. It questioned the legitimacy of both modern secular regimes as well as traditional monarchies in the region. The Islamic republic also potentially endangered the interests of the United States, Western European countries, and the Soviet Union by pressuring other Islamic nations to break off political and economic ties with these countries. Furthermore, the Iranian revolution alienated other countries by calling for tighter control over national resources, including oil. It also challenged OPEC (Organization of Petroleum Exporting Countries). The aim was to increase Tehran's revenues by driving up the price of oil.

The uncalculated consequences of these policies strained Iran's relations with its neighbors and isolated the Islamic republic from the world community. In this context, the prevailing view in the world community was to contain radical Islamic fundamentalism within the borders of Iran and to replace Tehran's theocratic regime with a secular and moderate government.

Alarmed by the threat Khomeini's fundamentalism represented to the Baghdad government, Iraqi policymakers felt they had the military muscle to bring down the Iranian regime. Besides the historical rivalry with Iran and Saddam's leadership ambitions, the Iraqi strongman feared his Shiaa and Kurdish population would be vulnerable to influence and manipulation by its Iranian brethren. Tehran was politically isolated, its military was in disarray, and there was considerable political disunity. Furthermore, the American hostage episode had deprived Iran of its major military supplier, the United States. All these factors combined to give Saddam Hussein a golden opportunity to strike. Even though Baghdad accused Iran of initiating the war, a recently completed United Nations investigation puts the blame squarely on Baghdad.

Like many other leaders, who decided to commit their nation's civilian and military resources to war, Saddam Hussein felt that, given Iran's problems, his forces would achieve a quick and decisive victory. Despite the Iraqi strongman's optimistic calculations, the Iran-Iraq war lasted eight years and earned the dubious distinction as "the longest conventional warfare of this century." More than one million people lost their lives and the direct and indirect cost of the conflict reached the "astronomical figure of $1,190 billion."[4]

Although the Islamic revolution in Iran and the threat it represented to Arab nationalism was clearly the fundamental and overriding factor, additional considerations also contributed to the prolongation of the war. These included: the beneficial and unintended consequences of the war in terms of regime consolidation in both Iran and Iraq, regional leadership considerations, and international involvement. The remainder of this essay analyzes in detail the role played by each of these factors in making the Iran-Iraq confrontation a long and bloody war.

The Iranian Revolution

The factor most responsible for the prolongation of the war was the Islamic orientation of the Iranian revolution, with its religious psychology, and the messianic zeal of its leaders and their followers, all of which created an environment hostile to a rational cost-benefit analysis of the war. The Iranian leadership's behavior and rhetoric were steeped in Islamic theology. Much like Christianity, which distinguishes between just and unjust war, Islam requires male Moslems to take part in the defense of Islamic soil. This holy war (Jihad fee Sabille Allah), often referred to as simply jihad, is the only legitimate type of war sanctioned by Islamic theology.[5] Ayatollah Khomeini, a faghih (jurisprudence) and the undisputed and charismatic leader of the revolution, pronounced Saddam as the enemy of Islam and identified the war as a jihad against Saddam's aggression. Khomeini announced that regardless of the final outcome of the war, Iran would be the real victor, as it was fighting for the sake of Islam. Moreover, Ayatollah

Khomeini refused to negotiate with Saddam Hussein, for he considered him an infidel and an agent enforcing the East-West conspiracy against Islam. Talking peace was regarded as blasphemous and counterrevolutionary.[6]

Saddam's actions reinforced Khomeini's beliefs. The execution of Ayatollah Mohammed Bagir Sadr, an eminent leader of Iraqi Shiaa, and his sister Fatemeh Benteh Hoda and the brutal treatment of the late Ayatollah Ozma Hakim's (grand ayatollah) family were seen by Khomeini and the ulema (scholars of Islamic law) to indicate the deep-rooted animosity the Baathist leadership harbored against Islam. At the same time, restrictions imposed on Shiaa Moslems in Iraq were seen in Tehran as an effort by Baghdad to suppress the establishment of an Islamic government in that country by the Shiaa majority. The zealous, devoted followers of Ayatollah Khomeini, known as Hizbollah, felt that a cease-fire would thwart their desire to see the establishment of an Islamic government in Iraq. Under the circumstances, cessation of hostilities was considered a retreat from religious duties.

Khomeini was in a strong position to carry out the fighting. His vehement refusal to negotiate with Saddam received popular support among Iranians, at least in the early years of the war. Iraq's initial successes, occupation of Iranian territory, the relatively low casualties, and destruction suffered by Tehran contributed to popular approval of Khomeini's position. With the exception of a few secular radical organizations—like Peykar, the Fedaiis, and the Organization of Communist Unity—many Marxist and other rival groups, who otherwise had ideological disagreements with the clergy, supported the Islamic republic's defense efforts. Iraq's full-scale war, atrocities committed against civilians, virulent anti-Iranian propaganda, and protection of Iran's territorial integrity sufficed to mobilize strong support behind Khomeini's jihad.

Moreover, revolutionary ferment and turmoil had transformed Iran into a highly politicized society "given to making sacrifices and adopting an ethic of social cooperation, so essential to waging a long war."[7] In this respect, the war became a prolonged contest between the national will of an Iranian community committed to preserve the nation's political unity and independence, and a well-equipped and supported Iraqi state apparatus threatened by the emergence and possible success of Islamic revivalism.

From 1982 onward, the outcome of military operations turned in Tehran's favor, further reinforcing Khomeini's willingness to continue the war. Iranian units overran Iraqi military bases, penetrated into Iraqi territory, and occupied the Majnoon islands, Fao peninsula, and strips in the northern section of the border. Tehran's successes occurred despite military support and intelligence information Iraq received from the Soviet Union and the United States. In the religiously charged atmosphere, these advances were interpreted as God's will. Iranian revolutionary leaders meanwhile increasingly entertained the belief that a final victory over Iraq was possible. Thus, it became even less justifiable to negotiate with Saddam, who they felt was attempting to enforce satan's will.

Under the circumstances, the numerous peace overtures pursued by a variety of third parties—including Pakistan, Turkey, Sweden, the Palestine Liberation Organization, and the Islamic Conference Organization—failed in their mission to persuade the warring parties to agree to a negotiated settlement.[8] The efforts ran aground because Iran questioned their intentions and because neither of the two warring sides was willing to agree on conditions set by the mediators. The Gulf Cooperation Council offered Iran $10–25 billion as war reparations in 1982 to entice Khomeini to agree to a cease-fire, but that offer did not appeal to the Ayatollah, either.[9] He did not view it as a genuine attempt to make peace, but rather as a ploy to save the Baath regime. Tehran argued that Saddam would use the time to reorganize his forces and would renew his attack against Iran when he saw fit.

The gradual victory Tehran hoped to achieve was not necessarily based on the Maoist notion of a protracted conflict, which would result in a victory sometime in the future. Iran's human and material commitment was designed for a full-scale conventional war. Accordingly, once Iranian troops overran Iraqi defenses and penetrated into enemy territory, Tehran stated that the "final" offensive would take place that year—an announcement repeated at the beginning of each of the following five to six years. This belief suggested that Tehran saw each annual campaign as the concluding round of the war. Iran's inability to achieve a decisive victory in part could be attributed to the fact that the theocratic leadership misjudged the determination of the international community to contain Islamic radicalism inside Iran. This issue will be discussed later in the chapter.

Despite the seemingly logical posture of the theocratic political elites, it is arguable that if the Iranian leaders could have predicted the outcome of the war, they would not have insisted on its prolongation. As the chances of a decisive victory became less and less remote, some members of the clergy voiced openly a distaste toward continuation of the war. For instance, Ayatollah Hossein-Ali Montazeri, Khomeini's designated successor, expressed his doubt about the benefits of prolonging the war. Khomeini reacted fiercely. He publicly denounced Montazeri and forced his resignation within a few days after Montazeri had aired his views.[10] Even though some members of the Iranian ruling circles may have shown some willingness to look at the war from a more realistic and cost-benefit point of view, the hard-liners, led by Khomeini, were firmly committed to their religious conviction. For them Saddam and his regime had to be eliminated regardless of the price.

The War and National and Regime Consolidation

Although unintended, the war provided a favorable opportunity for both regimes to consolidate their power. Baghdad presides over a country which is ethnically diverse and lacks a common political culture. Saddam exploited the war to send many Iraqis of Iranian origin back to Iran, and to relocate Kurds and non-Arab minorities from the northern mountainous areas to the predominantly southern Arab desert regions of the country. The resettlement policy was aimed to uproot the Kurds and to disperse them among the Arab majority to break down their solidarity and desire for regional autonomy. The settlements were located in harsh desert areas with scarce agricultural lands, far removed from urban centers. The aim was to put Kurds against the local Shiaa Moslems. Clashes between Sunni Kurds and the Shiaa Arab majority, Baghdad reasoned, in the short run would redirect frustrations from the political discrimination the ruling Sunni Arabs exercised against them. The strategy was designed to exploit the ethnic and religious diversity of Shiaa Arabs and Sunni Kurds in favor of the minority Sunni political elites in Baghdad. The Iraqi Baath party presumed Arab nationalism in time would overcome religious diversity and in the long run would create cultural homogeneity. The Iraqization of the Arab population, both Shiaa and Sunni, was one of the major goals toward state building the ruling Baath party pursued.

Islamic fundamentalism was represented in Iraq by the Daawa party. Ever since the revolution, Tehran had urged the oppressed Shiaa Moslems in Iraq, excluded from the nation's economic and political centers by the minority Sunni political elites, to rise up and seize control of their own destiny. Ayatollah Khomeini continuously called on the people of Iraq to topple Hussein's regime, and advised the Iraqi military not to obey the president's orders, calling Hussein and his supporters "the foes of Islam."[11] Furthermore, the Iraqi city of Najaf, the citadel of the Shiaa theology, as was the burial place of the first imam of the Shiaa, was considered sacred. The shrine of Hossein, third imam of Shiaa Muslems and sayyed-ol shohada (master of the martyrs), was also located in another Iraqi city, Karbala. Saddam feared that Shiaa fundamentalists in Iran would collaborate with the Daawa Shiaa militant party in Iraq to establish an autonomous Islamic government in these key cities. This fear convinced him that his socialist/nationalist regime was the prime target of Islamic Iran.

Saddam perceived the Ayatollah's aim of stirring up the oppressed, quiescent masses of the Shiaa Moslem majority of Iraq and feeding them with revolutionary values as a fundamental threat to his rule and his hope of creating a homogeneous secular political culture loyal to Arab nationalism. According to the *New York Times*, "Never in the 12-year history of the Baath regime in Iraq has the rule of the Saddam Husayn [Hussein] come under such a threat as it did since Khomeini came to power."[12] Therefore, it was natural for Saddam to resist his own demise by attacking Iran to cause the downfall of the Islamic regime. This was supported by an Iraqi Baathist in exile, who stated the single goal behind Hussein's attack on Iran was "to topple Khomeini."[13]

The war demonstrated that an external threat could overcome religious loyalties, as the Baath party successfully mobilized its Shiaa majority population in the war effort against the Islamic republic. Iraqi Shiaa did not switch sides and did not support their Iranian brethren in the war as expected. Interestingly enough, the recent Persian Gulf crisis had the opposite effect. The costly Iran-Iraq war deprived Saddam's ability to buy off his Shiaa subjects by doling out substantial monetary benefits as he had done before. Iraq's defeat by the coalition forces and the partial disintegration of its central government provided an opportunity for politicized Iraqi Shiaa to demand political recognition from Baghdad.

One of Tehran's conditions for cease-fire was the removal of Saddam from power. Such a demand could not realistically be achieved through peaceful means. Saddam had been the key leader in the Baath party and Iraqi politics since the 1960s. He has been ruling with remarkable authority and control, albeit ruthlessly, for over 12 years. Saddam's townsmen (from Takrit) and kinsmen had been instruments of control in the Baath party and military and the Iraqi military governmental structure. The authoritarian, sultanic, semifascist system of Baghdad heavily relies on fear and coercion, and ties the survival of the Iraqi state on the presence of Saddam Hussein. Under the circumstance, a political or military coup to topple Saddam was unrealistic simply because the Iraqi ruling elites realize his removal could bring peace but might also mean the end of their own political supremacy. Only a military victory by Iran could put an end to Saddam's rule. Khomeini's statements that Tehran intended not only to eliminate Saddam but the entire Baathist political structure as well did little to encourage Iraqi political and military elites to desert their leader, and provided good reason for them to rally behind Saddam's war effort.

The task of power consolidation was probably an unintended consequence of the Iraqi attack on Iran, but since it worked, it ended up contributing to the prolongation of the war. The Iranian clergy's drive to consolidate their rule also benefited from the war. At the beginning of the revolution, the popularity and charismatic appeal of Ayatollah Khomeini prevented even secular radical forces from attacking the clergy openly. However, by the end of the summer of 1979, less than six months after the Islamic revolutionary regime assumed power in Iran, it began losing popular support. A number of factors contributed to the disillusionment with the Mollahs, including a controversy surrounding the nature of the new political systems; autonomy demands in Kurdistan, Turkomn-Sahra, Baluchistan, Khusistan; and an economic crisis, which left two million people unemployed.[14] The new regime and its extremist Islamism alienated many of the social and political groups who found fundamentalism unbearable and those sectors who had benefited through economic trade with the West.

The takeover of the US Embassy raised the possibility of an American military attack and helped to mobilize people behind Ayatollah Khomeini. Documents seized from the embassy were skillfully used by such supporters of the clergy to discredit opponents of the Islamic regime as Velayat-e Faghin (the government of jurisprudence), attacking them as sycophants or lackeys of American imperialism. Liberals suffered the most. Prime Minister Medhi Bazergan resigned, Abbas amir Entezam, deputy prime minister, and Moghaddam Maraghei, one of the founders of the Moslem People's Republican party, both went underground. The hostage crisis helped the fundamentalists to strengthen their position against the left and the liberals.

Despite these developments, the official candidate of the Islamic Republican party (IRP) pulled in less than 6 percent of the vote in the presidential election of 26 January 1980. Bani-Sadr with Khomeini's tacit support was elected president of the Islamic republic.[15] He reorganized the army but failed to evict the Iraqi army out of Iran's territory. The well-organized IRP proved instrumental in mobilizing the masses against Iraq and created a new military organization composed of volunteers called Basij-e Mostazafin (the Mobilization of the Oppressed). The fundamentalists perceived Islam as the driving force behind the Iranian revolution and saw the war as a unique opportunity to imbue the masses with revolutionary values like shahadat (martyrdom), political independence, and xenophobia with regard to the West.

Their efforts proved successful. The masses mobilized behind the defense effort, and the regime's foreign policy of nonalignment, under the slogan: "no East, no West, Islamic Republic," gained support among the people. Ironically, Saddam's attack on Iran "changed the situation drastically on behalf of the fundamentalist mollahs."[16] Fighting the war received top priority at the expense of political reform, and helped to turn the balance of power overwhelmingly toward the Islamic Republican party. It became the vehicle of control and cohesiveness for Ayatollah Khomeini's supporters and established the religious fundamentalists as the dominant political group among the heterogeneous, divided forces of the revolution.

This served the IRP well in the first round of parliamentary elections, held on 14 March 1980. Its candidates won a majority of the seats. Using its strength in the Majlis, the IRP forced President Bani-Sadr to accept its candidate, Mohammad Ali Rejaii, as prime minister. The ensuing power struggle between Bani-Sadr and the IRP dominated domestic Iranian politics. Eventually, Bani-Sadr called for a referendum on the future course of the country, which put him in direct conflict with the Ayatollah Khomeini. The Ayatollah sided with the IRP and dismissed Bani-Sadr as commander in chief of the armed forces. IRP deputies in parliament "voted for the removal of Bani-Sadr from the presidency," which paved the way for his dismissal on 19 June 1981 by the Ayatollah Khomeini.[17] This strongly signaled that the fundamentalists had gained supremacy in postrevolutionary Iran.

Khomeini's charisma and sufficient mass-based support, along with the state's repressive capabilities inherited from the monarchy, were harnessed for power-consolidation purposes. These were further assisted by the numerous organizations that mushroomed after the revolution. They proved instrumental in policing citizen activity and potential antiregime plots from within the state security apparatus. Under these conditions, none of the regime's possible opponents could develop sufficient mass-based organizational strength to challenge the Islamic Republican party. Their inertia strengthened the hand of those who supported the continuation of the war. For instance, when Iraq accepted the appeal of the Islamic Conference for a cease-fire and return to the internationally recognized border, Tehran simply rejected the proposal. Although almost all radical, secular, and liberal organizations expressed antiwar positions, they remained weak, divided, and unable to press for an early end of the war.

By making the state bureaucracy the dumping ground for the unemployed, the Islamic regime compensated for the negative consequences of the war and at the same time pacified any potential for antiwar sentiment. In short, the Iran-Iraq war contributed to the durability of the system of the Valayateh Faghih in Iran, while fundamentalism contributed to the prolongation of the war.

Regional and Leadership Considerations

The secular and nationalist orientation of the regime in Baghdad and the personal aspirations of its leader were in direct conflict with the Islamic government in Tehran. Robert G. Neumann, a former US ambassador to Morocco, Afghanistan, and Saudi Arabia, noted that "quite apart from the long historical roots of the conflict, Iraq also aspired to regional leadership" that would put it in a direct conflict with Iran.[18] Simultaneously, the Islamic republic portrayed Ayatollah Khomeini as the leader of the Islamic brethren, which included the Arab world. Khomeini and other revolutionary elites perceived Islam as the motivating force behind political movements in the region and argued that only Islam could bring unity, glory, and real independence to Islamic countries. Traditional Moslems viewed nationalism as a negative force responsible for the fragmentation and division of Islamic society along ethnolinguistic lines. Baathist nationalism was condemned on those grounds, and Tehran advocated the overflow of the Iraqi regime.

In contrast, the Baath perceived nationalism as the driving force behind political movements in the region and strove to establish a united political entity, which would include all Arab states. Nationalism is viewed as the vehicle to revive the past glories of the Umayyad (A.D. 661–750) and Abbasid (A.D. 750–1258) dynasties and the caliphdoms, which at once ruled most of the Middle East. Saddam Hussein was a devoted follower of Gamal Abdul Nasser of Egypt, the undisputed leader of Arab nationalists in the 1950s and the 1960s. Secular nationalism advocated by Nasser was followed by Baath nationalism/socialism in Iraq and Syria and gave rise to authoritarian populist regimes in these countries. Anwar al Sadat, who succeeded after the death of Nasser in 1970, broke away from Nasser's Pan-Arabism and signed a separate peace treaty with Israel at Camp David in 1979. This isolated Egypt from the Arab world, undermined Cairo's historical role as leader of the secular Arab nationalist movement, and created a power vacuum in regional politics.

Saddam Hussein, considering himself the legitimate political heir of Nasser, attempted to fill the vacuum and continue the latter's mission. To establish himself as the leader of the Arab world, Saddam had to "orchestrate successfully an event or a series of events that would validate beyond all doubt his bid to be the Arab world's chief spokesman."[19] The government-controlled radio of Baghdad portrayed Jews and Iranians as the historical enemies of the Arabs. The military power of Israel and its geographic distance from Iraq made Iran the prime target. Saddam Hussein assumed that a quick military victory over Iran not only would regain Iraqi sovereignty over the Shatt al-Arab waterway but would establish him as the leader of the Arab world. Sovereignty over the entire Shatt was especially important to Saddam, as he himself had signed the Algiers accord in 1975.

In the same vein, Hussein presumed a speedy military victory could bring him control of the three islands in the Persian Gulf that Iran had occupied one day before the British withdrew from them in 1971. Occupation of these islands would give Iraq control over the Strait of Hormuz and supremacy in the Persian Gulf. Finally, Hussein assumed that a military victory would put Iraq in a position to demand autonomy for the Arab minorities living in the Khuzistan Province of Iran and thus portray himself as a defender of Arab rights. He even went as far as to change the name of the Khuzistan Province to Arabistan.[20] Saddam and his colleagues hoped that a "quick victory on the battlefield, coupled with increased support for the anti-Khomeini forces inside Iran, would further weaken the regime in Tehran, and thus force the Iranian government to accept Iraqi demands."[21]

The fall of Mohammad Reza Shah and the turmoil that followed eclipsed Iran as the dominant power in the region. Saddam believed that the time had come to realize "Iraqi aspirations to become the neighborhood's new primary power and protector" and to fill the vacuum created by Iran's weakness.[22] By attacking Iran, Hussein let it be known that a peaceful coexistence with an Islamic Iran was less desirable than the risk of an all-out military conflict. Hussein's drive for leadership of the Arab world and for regional power, coupled with the advantages gained from a victory over al-Ajam (non-Arabs), were powerful incentives.

Political events, Saddam's ambitions, and his psychological frame of mind made compromise and consensus with Iran difficult. None of these, however, meant that Baghdad was interested in a prolonged war. Although Hussein's personal ambitions were the underlying cause of the war, he called for a cease-fire followed by a negotiated settlement when it became apparent than an all-out victory was not likely. A cease-fire satisfied Hussein as it could be sold as a victory to the Iraqi public. He could claim he had contained Persian expansionism and had stopped Tehran's aggressive designs. Saudi Arabia and other moderate oil-rich Arab states were also apprehensive. They feared fundamentalism and Khomeini's strident rhetoric. Gulf state governments counteracted by calling the revolution in Iran a Shiaa revolution, hoping this sectarian categorization would restrict its mass appeal.

Islamic fundamentalism throughout the Middle East had been a minority movement with marginal social and political impact until 1979, when the popular uprising in Iran gave it a boost. For the past two centuries, reformist Islamic governments and secular nationalism were the dominant ideological traditions in the region. However, conflicts among the Arab states and their inability to secure a homeland for the Palestinians had dealt a severe blow to Arab nationalism/socialism and the dream of Arab unity. A new generation of

Arabs, disappointed with Arab nationalism, saw in Islam the only vehicle to achieve Arab aspirations.

Pan-Arabism was being replaced by Pan-Islamism. The Islamic Republic of Iran hoped to capitalize on this sentiment by pushing the idea of a united Islamic state.

The Iranian clerics believed "that their revolution had stemmed from the universalist idea of Islam. They saw the change in Iran as the first step towards the recreation of the Domain of Islam of the seventh century."[23] Tehran wanted to radically change the political structure of the region along the lines of an Islamic model. Khomeini and his colleagues planned to export the revolution and to achieve fundamental changes through a mass uprising. Ayatollah Khomeini used his religious authority and issued a fatwa (a decree or religious declaration), making it a religious duty for Shiaa Moslems to refrain from sectarian rivalry. He encouraged his followers to work toward unity with Sunni Moslems. In a similar vein, Tehran designated a Moslem unity week (Hafteh-e Vahdat), issued stamps, and presented cultural programs to convince Sunni Moslems that the Iranian revolution was not exclusively a Shiaa affair. Khomeini believed that popular revulsion against the pro-Israeli stance of the United States and the failure of the moderate Arab states to secure unity would "become powerful enough to sweep across the region beyond the Sunni-Shiaa divide."[24]

The Islamic revolution initiated fundamental and profound changes in the Middle East region. It replaced the secular hereditary and Western-oriented Pahlani regime with an Islamic republic. The latter aimed to reverse the penetration of Western values into Iran and other Islamic cultures. Revolutionary leaders viewed the Islamic values of the revolution in sharp contrast to both communist and Western values. Khomeini and his colleagues regarded secularism as evil and as such antithetical to the ordinances of Islam. The United States, the Soviet Union, the United Kingdom, and France were pronounced Shaytan-e Bozorg (Great Satan). Accordingly, Iranian leaders felt that there was a joint East-West conspiracy to undermine and destroy the Islamic revolution. Saddam Hussein was portrayed as a puppet agent carrying out foreign satanic orders when he launched his attack against the Islamic government.

These beliefs helped the revolutionary leaders to develop and propagate a set of values and attitudes, which defined the political behavior of the elites and their followers and in turn contributed to the prolongation of the Iran-Iraq war. The core of the orientation of the revolutionary leaders and their devoted followers could best be identified as Hizbollahi (members of the Party of God). The Hizbollahis were inspired by the Third Imam and the Prophet's grandson, who refused to surrender to Omayyad Caliph Yazid (683 A.D.), the symbol of Zolm (tyranny and injustice), and stated "death is better than life under an oppressor." The imam and his disciples were massacred. Since that time shahadat (martyrdom) has become an important aspect of the political culture of Shiaaism. Khomeini exemplified his belief when he rejected Saddam's cease-fire offer stating, "we cannot compromise with 'Hussein' a perpetrator of corruption, . . . we [are] bound by our religion to resist as much as we [can]."[25] These utterances prompted Tehran to cast the Iran-Iraq war as a battle between Islam and "Satan." Death in the defense of Islam was honored as shahadat. To the Hezbollahi, the highest honor of martyrdom, the key to Heaven, could not be compensated by worldly, material well-being. Shahadat remained the motivating force for millions of Hizbollahi, who made up the backbone of support for the Islamic regime during the war.

Associating Saddam Hussein with the West and East served a number of other purposes. First, it provided a legitimate reason for Islamic fundamentalists to continue the war as a war between Islam and "The Great Satan." Second, it served as a mobilizing force and created stronger ideological resistance among ordinary Moslems against Western infidels. Third, it minimized and rationalized the inability of the military and the Revolutionary Guard to win the war against a smaller country, Iraq, in a short period of time. Beneath all that, Tehran assigned itself the role of the regional leader striving to free the region of foreign domination and to unify the "House of Islam" against the infidel.

Tehran's strident rhetoric and its efforts to export its revolutionary values to the neighboring countries frightened many moderate Islamic regimes. Ayatollah Khomeini's declaration that the liberation of Karbala and Najaf (Shiaa holy cities in Iraq) would pave the way for the liberation of Jerusalem did little to allay negative impressions regarding Tehran's intentions. The Islamic republic claimed that it, alone, represented "Islam-e Mohammandi" (the true righteous Islam), while other Islamic regimes were actually the enemies of Islam and agents of the West. These regimes represented "Islam-e Americaee" (false alarm). Stephen R. Grummon wrote in 1982 that "the Khomeini regime in Iran adheres to a particular brand of theology that denies the legitimacy of most of the current governments in the Islamic world."[26]

Less radical Iranian leaders like Ali Akbar Hashemi Rafsanjani, then speaker of parliament, played down these declarations as presented only for public consumption and intended to generate public support, did not reassure governments in neighboring countries. According to Christoper C. Joyner,

> Save for Syria, Libya, and (what was then) South Yeman, all the Arab governments supported Iraq in its war effort, politically and rhetorically if not materially or militarily. In large part this pro-Iraqi attitude stemmed from anxieties that Iran's Islamic fundamentalism might spread throughout the region, infecting neighboring Gulf states with domestic unrest.[27]

Saddam sought and obtained support from fellow Arab states, including oil-wealthy Persian Gulf states and Saudi Arabia. Though fearful of Khomeini's fundamentalism, most Arab regimes also viewed Baghdad's brand of Pan-Arabism and Saddam's personal ambitions with considerable concern. Arab governments, minus Syria and Libya, rushed to Saddam's side, providing him with money and diplomatic support to prevent an Iranian victory. But support for the Iraqi dictator was never enough to score a decisive defeat against Khomeini's forces. Arab leaders did not want either side to win and seemed to have reasoned that prolonging the war would weaken the warring sides and would force Khomeini and Saddam to shelve their respective expansionist and hegemonic designs. Ironically, stalemate and bloodshed appeared more desirable to a clear victory by either side, as far as neighboring Arab regimes were concerned. Tehran's provocative policy denied Iran access to the financial and diplomatic resources necessary to break the stalemate.

International Factors

If external intervention had not altered the expected payoff matrix for the dominating parties, the Iran-Iraq war would have continued until one side decisively won or both sides collapsed. The two superpowers (US and USSR), France, and to a lesser extent the United Kingdom played a major role in preventing the war from proceeding along that path. It is reasonable to argue that the war would have ended sooner if Iraq had not received generous financial, military, and diplomatic support after 1982; Iran had not been subjected to an economic blockade and diplomatic isolation; and arms suppliers had given Iran free access to military hardware so that Tehran could score a military victory. The existence of the external factors caused the stalemate and contributed to the prolongation of the war.

The West perceived Islamic fundamentalism as a reaction to modernity. European and American policymakers saw Islamic revivalism as inherently opposed to science and reason. Politically, the theocratic regime in Iran was perceived as an ideologically totalitarian system, bent on indoctrination, antithetic to pluralism, and contemptuous of rational discourse and electoral politics. The Iranian revolution undermined America's once powerful influence in Iranian domestic developments. The takeover of the US Embassy in Tehran by Khomeini's supporters and the hostage ordeal that followed altogether suspended relations between the two countries. Iran became increasingly isolated and anti-American. Uncertain and shaken, Washington feared that Tehran may overthrow pro-Western regimes in the area, which "contained more than half of the world's known oil reserves." Such an outcome would have amounted to "an unprecedented catastrophe," and Washington was not about to let it happen. By 1983 the Reagan administration began making overtures toward Saddam at a time when Baghdad was facing an "acute crisis" due to significant losses in the battlefront. Washington had few options and considered "an Iraqi defeat . . . as a blow to U.S. interests." And, despite protestations to the contrary, the Reagan administration set in motion "plans to shore up Iraq morally and materially."[28]

Washington's posture appeared to be driven by considerations similar to those of the Arab states. Saddam received intelligence information and enough supplies, which helped him gain enough ground to force Khomeini to accept a cease-fire in August 1988, but not score an outright victory. Washington's stand seemed ambivalent: the Reagan administration wanted Khomeini defeated but did not wish to see Saddam emerge as the clear victor. Moreover, the fate of Western (including American) hostages by the Hizbollah in Lebanon prompted the Reagan administration to give arms to Tehran secretly, hoping that the Khomeini regime would use its leverage to obtain their release. Information of the deal leaked out, becoming an embarrassment to the Reagan administration and did little to end the war.

West European governments reacted in a similar manner, throwing their support behind Iraq. Having invested heavily in the Iraqi economy, Paris took the lead. By 1983 "France emerged as a vitally important military-cum-financial prop for Iraq's long term war efforts." While the Mitterrand government backed Saddam to the "hilt," French and other European defense contractors clandestinely provided Tehran with "huge" amounts of explosives and other war material.[29] Needless to say, this behavior made it possible for the Khomeini regime to satisfy its war-making needs and added substantially to the prolongation of the war.

The anti-American and, to a lesser extent, anti-European climate in Iran appeared at first to benefit the Soviet Union, which had long sought to have a say in Iranian political developments. However, the Islamic revolution broke "the chain of anti-Soviet forces [in the area] surrounding the U.S.S.R."[30] Moscow's efforts to play the role of a neutral arbiter met with little success. The Kremlin maintained its pro-Iraq position throughout the war.

A number of factors contributed to this situation. First, Islamic revivalism had spilled over into the southern republics of the Soviet Union, creating the possibility of ethnic conflicts.

The Kremlin feared that Iranian leaders would stimulate the consciousness of the 40 million Moslem minorities living in the Soviet Union. Second, Iran took an uncompromising position with respect to the Soviet intervention in Afghanistan. Tehran supported the Mujahedin against the Soviet-backed Kabul government. Third, Moscow had signed the Friendship and Cooperation Treaty with Baghdad in 1972, which committed the Kremlin to protect Iraq's security. The supply of military equipment to Iraq by the Soviet Union played an important role in Baghdad's ability to prolong the war. Finally, Arab socialism, the ideology of the Iraqi ruling Baath party, had more in common with Soviet official ideology than it did with Islamic fundamentalism. In due course, "the Kremlin lost any hope of furthering its ties with Iran [and] came to perceive the possible fall of the Baathist regime as an unmitigated strategic loss offering nothing in compensation."[31]

Summary and Conclusion

A combination of domestic, regional, and international factors was the rise of Islamic fundamentalism in Iran and the threat it presented to regimes in the Middle East as well as in the world community. Islamic fundamentalism coupled with Arab nationalism created the conflict of interest that led to intolerance and eventually to war. Then the uncompromising positions of their leaders prolonged the war. While Saddam's personal ambitions were the main reason the war broke out, the fundamentalist line of reasoning adopted by the revolutionary leaders of Iran and their refusal to agree to a negotiated settlement caused the war to go on for eight years. However, the war had unintended but positive effects for each of the two regimes. It provided the opportunity for Saddam and especially for the Iranian clerics to further penetrate their respective societies and to consolidate their power.

The involvement of the regional and international actors also added to the prolongation of the war. Although most Middle Eastern countries, European states, and the two superpowers lined up behind Iraq, their support was enough to create a stalemate and eventually to tip the scales slightly in favor of

Baghdad but not adequate for the Iraqi dictator to score an outright victory. Clandestinely or otherwise, Iran, with a population three times the size of Iraq's, received enough war material from North Korean, Chinese, French, German, and Austrian corporations as well as Western goods to make up for its isolation and to hold its own against a better-supplied and -equipped enemy. It should also be noted that neither superpower had an overriding interest in stopping the fighting. At no time did the Iran-Iraq war increase the likelihood of a direct US-Soviet confrontation. In fact, both superpowers benefited from the war. Iran and Iraq were dependent on the international military market and had oil revenues to finance the war. Both superpowers were among the major suppliers of sophisticated military hardware to third world countries.

The war ended when a much-weakened Iran felt the US was about to become directly involved in the conflict on the side of Baghdad. The downing of an Iranian airbus on 3 July 1988 convinced the reluctant Khomeini that further prolongation of the war would lead to Iran's defeat and would destroy the revolution. He accepted United Nations Resolution 598 unconditionally, and a cease-fire went into effect shortly thereafter. This century's longest conventional war came to an end without resolving any of the issues that brought it about.

Notes

1. Barry Rubin, "Iran's Revolution and Persian Gulf Instability," in Shirin Tahir-Kheli and Shaheen Ayubi, *The Iran-Iraq War, New Weapons, Old Conflicts* (New York: Praeger Publisher, 1983), 13.
2. Quoted in C. J. Edmons, *Kurds, Turks, and Arabs* (London: Oxford University Press, 1957), 132.
3. M. S. El Azhary, ed., *The Iran-Iraq War, An Historical, Economic and Political Analysis* (New York: St. Martin's Press, 1984). The approach of the book, especially chapter three, is beneficial to readers who are interested in an Arab nationalist's point of view toward the Shatt-al-Arab dispute.
4. Dilip Hiro, *The Longest War, The Iran-Iraq Military Conflict* (New York: Routledge, Clapman, and Hall, Inc., 1991), 1.
5. Mohammad-Reza Ziaee-Bigdelli, *Islam and International Rights* (Islam va Hoghoghe Beinolmelal), in Persian, second edit. (Tehran: Inteshar, 1987, 1366 H. S.).
6. Rouhollah Khomeini, *Islam and the Revolution*, trans. Hamid Algar (Berkeley, Calif.: Mizan, 1981); R. Khomeini, *Selected Messages of Imam Khomeini Concerning Iraq and the War Iraq Imposed Upon Iran* (Tehran: Ministry of Islamic Guidance, 1981).

7. Hiro, 225.
8. Christopher C. Joyner, ed., *The Persian Gulf War: Lessons for Strategy* (New York: Greenwood Press, 1990), 9.
9. Fred Lawson, "Using Positive Sanctions to End International Conflict: Iran and the Arab Gulf Countries," *Journal of Peace Research* 20, no. 4, 311–28.
10. Abol Hassan Bani-Sadr, *My Turn to Speak, Iran the Revolution, and Secret Deals with the U.S.* (New York: Brassey's Inc., 1989), 219.
11. *New York Times*, 10 April 1980.
12. Ibid., 26 October 1980.
13. Ibid.
14. Mohammad Amjad, *Iran, From Royal Dictatorship to Theocracy* (New York: Greenwood Press, 1989), 140.
15. Ibid., 142.
16. Ibid., 144.
17. Ibid., 145.
18. Stephen R. Grummon, *The Iran-Iraq War, Islam Embattled* (New York: Praeger Scientific Studies, 1982), vi.
19. Ibid., 11.
20. Nicola Firzli, Nassim Khoury, and Elias Dib, eds., *The Iran-Iraq Conflict* (Paris: Editions du Monde Arabe, 1981), 48–54.
21. El Azhary, 2.
22. Rubin, 126.
23. Hiro, 255.
24. *World Press Review*, November 1991.
25. Khomeini, *Selected Messages*, 68.
26. Grummon, vi.
27. Joyner, 9.
28. Hiro, 119.
29. Ibid., 123–28.
30. Ibid., 265.
31. Ibid., 122.

The Longevity of the Lebanese Civil War

As'ad AbuKhalil

The only observation one can make with certainty about Lebanon is that it is too early to declare the end of the Lebanese civil war, although the Lebanese government has succeeded in ending armed combat in most parts of the country. The Syrian military intervention in October 1990 against the army of Gen Michel 'Awn, who was defiantly resisting the central Lebanese government, helped to spread the authority of the Lebanese government over areas that have not seen government troops since at least 1975. The war that has raged since 1975 has claimed the lives of more than 144,000 people and has injured nearly 200,000 between 1975 and 1990.[1] And, as much as the Lebanese people want to believe that this bloody chapter of their history is over, there is evidence that many of the conflicts that have manifested themselves in violent eruptions throughout Lebanese history have not been decisively resolved. The accords reached at Ta'if in Saudi Arabia in 1989 only devised a formula for internal reforms. These reforms cannot guarantee the end of hostile sentiments and frictions among the various Lebanese confessional communities.

This paper examines the underlying causes of conflict and the reasons for the prolongation of the Lebanese civil war. No blow-by-blow account is provided here, as numerous books and articles have adequately documented the chronology of the conflict. Instead, the emphasis is on the factors that made an early resolution of the Lebanese war impossible and which allowed for the perpetuation of the Lebanese war beyond the intentions of some of the protagonists themselves.

The Emergence of the Lebanese Political Idea

In establishing the nature of the Lebanese conflict, one must outline the general features of the Lebanese civil war. It is clear that Lebanese society has been riddled with conflicts and tensions since before the creation of the Lebanese state in 1920 under French mandate auspices.[2] The nineteenth-century history of Lebanon chronicles massacres, communal bloodshed, foreign intervention, and displacement of population. The two major protagonists at the time, the Druzes and the Maronites, dominated the area of Mount Lebanon, which was designated as the Lebanon area.[3]

The major problem in Lebanese history and politics is that the idea of Lebanon is relatively new, an idea born early in this century primarily of Maronite lobbying efforts directed towards the Western colonial powers and French regional interests. There was no consensus among the various groups inhabiting the area that is today Lebanon about the identity of the new state or about the formula for power sharing in government. There was substantial opposition among the Lebanese Muslim faction to the creation of a state called Lebanon because it would lead to the fragmentation of the Arab world. Furthermore, the creation of a Greater Syria was deemed desirable at the time, particularly with the efforts of Prince (later king) Faysal of Iraq to create an independent Syrian princedom based in Damascus.[4]

France created the Lebanese state in 1920 by adding to the historic area of Mount Lebanon the four other provinces: Beirut and its surroundings, the Biqa', the South, and the North. This new entity became known as Greater Lebanon, and the annexation of the new provinces was intended to provide the new entity with economic viability. The system was consolidated in 1926 with the promulgation of the Lebanese constitution, which established the juridical legitimization of the sectarian system according to Article 95 of the constitution which stated (before its amendment in 1990): "As a provisional measure and for the sake of justice and concord, the sects shall be represented justly in public posts and in the formation of the cabinet without harming the interest of the state." This vague stipulation became the cornerstone of the Lebanese political system and was used to justify the distribution of political power on a purely sectarian basis.[5]

French mandate authorities clearly wanted a system that would impose Maronite political supremacy on the political system since the Maronite community was regarded by them as the "protected community" and since that community always viewed France as "the tender mother." Since the new state of Greater Lebanon had a substantial Muslim majority, the French authorities needed to justify the Maronite political privileges by establishing demographic evidence. A 1932 census (often described—and rightly so—as "highly dubious")[6] revealed that the Lebanese population was split almost evenly between Christians and Muslims and that Christians enjoyed a slight edge, with the Maronites constituting the single largest sect. The census remains the only official source for the sectarian distribution of the Lebanese population. The Christian establishment in Lebanon has consistently refused over the years to conduct another census; the high birth rate of the Muslims (particularly the Shi'ites) would have quickly guaranteed them a comfortable majority. A 1986 estimate by the United States Central Intelligence Agency of the confessional distribution of the population showed 27 percent Sunnis, 41 percent Shi'ites, 7 percent Druzes, 16 percent Maronites, 5 percent Greek Orthodox, and 3 percent Greek Catholics.[7]

The census results were used—and continue to be used—to justify a system of government that affixes sectarian tags to all official posts and offices with a clear advantage to the Maronite community. This system was officially set on foot in 1943 when the National Pact was reached. The pact, an unwritten agreement, came into being in the summer of 1943 because of numerous meetings between Bishara Al-Khuri (a Maronite and the first president after Lebanese independence in 1943) and Riyad As-Sulh (a Sunni and the first prime minister after independence). The Christians' fear of Arab/Muslim domination of Lebanon and the Muslims' fears of Western hegemony lay at the heart of the negotiations. In return for the Christian promise to forego foreign (Western) protection and to accept Lebanon's "Arab face," the Muslim side agreed to recognize the independence and legitimacy of the Lebanese state in its 1920 boundaries and to renounce any Arab (or Syrian) national aspirations that could compromise the viability of the Lebanese entity.[8]

The pact also formalized the confessional distribution of high-level posts in the government on the basis of the 1932 census and a six-to-five ratio favoring Christians over Muslims. The most notable formula of the pact was the stipulation that the presidency should be reserved for the Maronites, the speakership of Parliament for the Shi'ites, and the prime ministership for the Sunnis. This formula, as well as the idea of sectarian distribution of power, remains at the heart of the Lebanese problem.

Origins and Evolution of the Lebanese War

While observers regard the Lebanese political system as the most democratic in the entire Middle East, they also note that the contemporary history of Lebanon—even before the outbreak of the civil war in 1975—has been marred by tensions and conflicts in society and in the body politic. The system functioned according to the sectarian structure of the National Pact and of the much publicized Article 95 of the Lebanese constitution. The degree of satisfaction with the political system and with the economic life of the country was an integral function of one's sectarian affiliation, as there was an imbalance in socioeconomic justice based on the sectarian formula for the distribution of political power and economic benefits. The most serious crisis occurred in 1958, when then-president Kamil Sham'un intended to amend the constitution to allow himself another term of office.[9] Sham'un's political ambitions triggered an internal rebellion—with its external, regional, and international dimensions—and eventually led Sham'un to request the military intervention of US Marines, thereby violating a major principle of the National Pact.[10]

The increase in the percentage of Muslims in the Lebanese population and the stagnation of the Lebanese political system, coupled with the opposition by the Maronite political establishment to any meaningful political or economic reforms to accommodate the rising demands of the Muslims, augmented the stresses on the political system. The Palestinian presence in Lebanon and the rise of their armed groups also increased pressure on the system and presented angry Lebanese with a model of armed resistance. The resentment of many Lebanese, particularly the Shi'ite faction, whose region in the south became a battleground for Palestinians and Israelis, increasingly manifested itself in strikes, armed attacks, and the proliferation of political parties and organizations.

The rising opposition in Lebanon articulated a variety of issues: improving the military abilities of the Lebanese army, resisting Israeli military actions in the south, and vitiating the need for radical socioeconomic reforms in the Lebanese system. Lebanese President Sulayman Franjiyyah (1970–76) initially attributed the social turmoil in the country to a Palestinian-Soviet conspiracy; he subsequently began an unprecedented arming of Maronite right-wing militias in Lebanon to help suppress Palestinian military forces in Lebanon. Franjiyyah and his right-wing allies failed in 1973 when the Palestinians received wide popular support from many Lebanese who did not want the Lebanese army to repeat what the Jordanian army did to the Palestinian Liberation Organization (PLO) forces in Jordan in 1970–71. The failure of the Lebanese army campaign strengthened the resolve of the Lebanese president and Pierre Gemayyel, the leader of the main right-wing militia, the Phlanges party, which was modeled after Spanish fascist youth organizations.

The continued deterioration of the situation in South Lebanon, the worsening economic crisis plaguing the entire country, the continued conflict between the president and the prime minister, or in other words between the Maronite and the Sunni political elites, the frustration and mounting radicalization of the Palestinians and Shi'ites, and the intervention of Arab states (particularly Syria and Iraq) and Israel in internal affairs of Lebanon led to the eruption of the civil war in the spring of 1975. The first phase of the Lebanese war lasted from 1975 until the fall of 1976, when Syrian military forces were deployed in most places in Lebanon under the umbrella of the League of Arab States and with the support of the US and France. The first phase of the civil war was fought between Maronite militias supported by some brigades of the Lebanese army, on the one hand, and Palestinian forces supported by left-wing and Islamic militias, on the other. Syria intervened just when the Palestinian/leftist coalition was about to prevail in Lebanon and was close to defeating the right-wing forces. The Syrian intervention saved the Maronite establishment from an unsalvageable defeat, if not outright extinction.[11]

After a few years of friendship between the Syrian regime and the right-wing militias, Israel became an important player in the south, where the Israeli Defense Forces set up their own military centers and began arming and financing a surrogate militia. Israel also began cultivating close relationships with the right-wing militias in East Beirut. Lebanon then became subject to Syrian and Israeli political interests, although the local militias were able at times to continue to influence events according to their own interests.

The next phase of the Lebanese civil war shifted the network of alliances as the right-wing militias (then under the leadership of the youthful and charismatic Bashir Gemayyel, who headed the Lebanese forces) severed their ties with Syria and consolidated their relationship with Israel. Syria in turn moved to improve its ties with the PLO and the left-wing/ Muslim coalition. Numerous rounds of fighting between those enemies continued between 1978 and 1982 in various parts of Lebanon. In 1978 Israel invaded Lebanon to drive the PLO from Lebanon, but was forced by the administration of then-president Jimmy Carter to withdraw and to accept the deployment of UN forces (known as UNIFIL) to separate the Israeli and the Palestinian forces. Israel insisted, however, on maintaining its troops in the southern strip of south Lebanon.[12] The continued hostilities created the context in which the Israeli invasion of 1982 took place.

Israel invaded Lebanon in 1982 in the wake of the attempted assassination of its ambassador in London by the anti-PLO Abu Nidal organization. The invasion was intended to expel the PLO from Lebanon and to install a pro-Israeli government in Lebanon under the leadership of Bashir Gemayyel. Militarily, Israel swept through south Lebanon and imposed a siege on predominantly Muslim West Beirut to put pressure on the PLO to evacuate and to influence the results of the presidential elections. Eventually, Bashir Gemayyel was elected president, and the PLO agreed to leave Lebanon under international guarantees. Nevertheless, Israel's chief ally Gemayyel was assassinated before he assumed his constitutional responsibilities, and his armed men entered the Sabra and

Shatila Palestinian refugee camps and engaged in brutal vendettas against Palestinian civilians, an event that shocked the world at the time and forced then-president Ronald Reagan to redeploy US forces to Lebanon. Bashir was succeeded by this brother, Amin, another major figure of the Phalanges party.[13]

In the administration of Amin Gemayyel (1982–88) economic and political conditions became worse than they were in 1975–76. Gemayyel relied on the military powers of the US Marines to crush any domestic opposition, and he dragged the US into participation in the civil war, which resulted in the bombing of the Marine barracks in 1983.[14] But the withdrawal of the Marines in 1984 helped the Lebanese opposition, supported by Syria (which had been strategizing for the defeat of the Israeli plan for Lebanon), to take control of West Beirut and other areas formerly dominated by Gemayyel's forces. From 1984 to 1988 Lebanon witnessed yet another violent chapter of the civil war, with car bombs and indiscriminate shelling being the favorite weapons of the militias, none of whom refrained from brutal and savage acts.[15] In 1988 Gemayyel's term ended and he refused to allow the Muslim prime minister to take charge while Lebanon prepared for a presidential election. Instead, he appointed Gen Michel 'Awn, the Maronite commander in chief of the Lebanese army, the prime minister in charge of the affairs of the country. However, a Sunni prime minister remained in power in West Beirut, and the country was officially split into two contending governments, each claiming full constitutional legitimacy. Arab states intervened with US blessing to sponsor a national accord to establish a central government. Lebanese parliamentarians were invited to Ta'if in Saudi Arabia to work out a formula for political reforms and presidential elections to end the civil war. General 'Awn ignored these developments and launched his war of "national liberation" against the Syrians and their allies.

The accords of Ta'if were received positively in Lebanon and the rest of the world, although General 'Awn rejected them and continued his war against his enemies, considering himself the savior of Lebanon.[16] After Rene Muawwad was elected president and then shortly thereafter assassinated, the parliament elected Ilyas Hrawi president in the fall of 1989. 'Awn was finally ousted in October 1990 by Syrian troops while the world was preoccupied with developments in the Persian Gulf.

Underlying Causes of the Prolongation of War

The complexity of the Lebanese civil war rendered all mediation attempts ineffective during its course. As of 1992 it is still too early to determine whether the stability that was brought about by the accords of Ta'if will last and whether the civil war has ended. The situation in the south continues to be violent; Israel still maintains its military presence there; and its surrogate militia (the south Lebanon army) still operates under Israeli aegis. Furthermore, various Shi'ite militias still roam south Lebanon and insist on remaining armed until Israel leaves the country.

Multiplicity of Dimensions

A main reason for the prolongation of violence in Lebanon, a conflict which produced the civil war and which could produce more bloodshed and strife in the future, lies in the many dimensions of the conflict. Reference to the conflict focuses on centuries-old tensions between the various confessional groups that have sought refuge in Lebanon. The war, of course, refers to the savage civil strife that erupted in 1975 and continued in various forms and shapes until 1990 when General 'Awn suffered defeat and expulsion from Lebanon.

The many causes for the civil war resulted from the connection between the various internal and external factors that have perpetuated the war.[17] The source of tension remains the failure of peaceful solutions to address simultaneously the external and internal factors behind the war. While the external factors in the Lebanese war always have been exaggerated by the Lebanese to absolve themselves from any responsibility for the war, outside parties and powers have played crucial roles in abetting (but rarely in solving) the conflict.[18] Even in the nineteenth century, European powers and the Ottoman Empire had given themselves the right to intervene in the most minute affairs of the country, and several powers assigned to themselves the role of protectors of some sects.

The Lebanese participants, on the other hand, must accept responsibility for sometimes allowing outside powers to intervene in Lebanon "on their side." Lebanese factions have thereby legitimized foreign intervention in their affairs to bolster their local standing or to strengthen their military postures during the war.[19] The link between the internal and the external dimensions of the war has obstructed reconciliation attempts, as outside powers have insisted on ostensible "neutrality" to distance themselves from problems in Lebanon they helped to create or to maintain. Thus, both Syria and Israel, the two regional powers most responsible for sponsoring clients in the country to further their own agendas, have pursued policies in Lebanon while maintaining their innocence by promising to refrain from intervening in the local affairs of Lebanon.[20]

The multiplicity of causes for the war has obfuscated the Lebanese reality; it has confused observers who want one simplistic reason for the breakdown of the political system. It has also confounded observers who want to promote an economic explanation for the civil war. For example, providing a class analysis of the Lebanese civil war that still explains the intensity of hatred and conflict between Maronite and Shi'ite workers is problematic. Ignoring the sectarian consciousness of the Lebanese people and its impact on political mobilization obscures the origins of political movements in Lebanon.

The intensity of conflict and the prolongation of war in Lebanon stem not only from the multiplicity of cleavages in Lebanese society but also from the overlapping of the various cleavages. There is evidence that socioeconomic cleavages, for example, tend to overlap with sectarian cleavages: most Shi'ites tend to occupy the lower stratum of Lebanese income groups.[21] The overlap, however, is not complete; there are some wealthy Shi'ites and some poor Maronites. Peaceful reforms in Lebanon have tended to focus on one of the two important dimensions by either looking at the issue of sectarian injustices or by treating the problem as one of income distribution. The inability to analyze the connection between sectarian frictions and economic injustices has rendered most attempts at reform obsolete.

On the question of foreign intervention, outside powers interested in solving the Lebanese problem have tended to attribute the ills of Lebanese society to external factors; Israel believed Lebanon's problems were produced by the Palestinians' presence in Lebanon, while Syria focused on the role of Israel in Lebanon without looking at the reasons that led some right-wing forces to seek an alliance with Israel. When opposition was mounting against the pro-US government of Amin Gemayyel, the US government insisted that opposition to his government was merely Syria's way of furthering its own narrow political interests in the region. In reality, Syria did use the opposition for selfish purposes, while the Muslim/leftist opposition used Syria in turn for the opposition's agenda. In Lebanon a convergence of interests always cements the relationship between a local player and an outside actor.

Lebanon as an Arena of Foreign Conflicts

Notwithstanding the inherent problems in Lebanese society and economy, regional and international powers have historically used the open and free environment of Lebanon as an arena for settling scores and for foreign intrigues. Many Lebanese tend to resort to conspiratorial scenarios to explain the protracted nature of the war. While the conspiracy theory underestimates the impact of the internal dynamics of conflict in Lebanon, the evidence suggests that outside powers have—before and during the war—exploited the Lebanese scene to promote various agendas and policies.[22]

While the phenomenon of foreign sponsorship of Lebanese clients did not start in the present century, the intensity of conflict and hatred among the various Lebanese sectarian groups and their desire to impose their will against the wishes of members of the "other sect" has invited intervention in the past few decades. The strategic location of Lebanon and its uniquely open socioeconomic environment have made it a tempting place for foreign intelligence services. Beirut became a place to direct foreign intrigues at the Middle East: coup d'états in the entire Arab world were plotted in Lebanon, and international intelligence services operated in the Middle East out of Lebanon. Moreover, the spread of socialist regimes in the Arab East and the influx of capital into Beirut, which had banking secrecy, made Lebanon a perfect place for Arab dissidents. Its free press reflected the opinions and views of its foreign financial patrons rather than the views of local actors; a Lebanese president once welcomed members of the press syndicate by saying, "Welcome to your second country, Lebanon," to remind the press of its allegiance to non-Lebanese patrons.

The most damaging result of the exploitation of Lebanese politics was the intensification of the Israeli-Palestinian conflict on Lebanese soil. Consequently, the painful longevity of the Arab-Israeli conflict was mirrored in the prolongation of the Lebanese conflict. It was less costly for both the PLO and Israel to fight their bitter wars and conflicts in Lebanon rather than somewhere else. This is not to say that the Lebanese conflict is primarily a by-product of the Arab-Israeli conflict, but it is clear that this intractable problem has prolonged the conflict. Yet, the evacuation of PLO forces from Lebanon in 1982 proved that those who insisted that the PLO presence in Lebanon was responsible for the eruption of the civil war were wrong. Some of the most brutal and savage battles of the Lebanese civil war were fought between Lebanese groups (primarily Druzes and Maronites) without the direct involvement of PLO forces.

But there is another repercussion of the Arab-Israeli conflict; Arab states engage in fighting one another either through their propaganda outlets in Beirut or through their client militias operating in Lebanon. The Syrian-Iraqi conflict, for example, has always been manifested in armed confrontations between pro-Iraqi and pro-Syrian Lebanese and Palestinian forces in Lebanon. Furthermore, the Egyptian and Syrian governments have utilized the Lebanese arena for their own purposes since as early as the 1950s. Lebanon became the stage from which Arab government asserted—often in bizarre fashions—their alleged dedication to the solution of the Palestinian cause. Lebanon provided Arab governments with elements for foreign policy that they lacked in their own countries. The Lebanese press could, for example, engage in debates and discussions that the Syrian and Iraqi press could never print by virtue of the strict control of the press in those two countries.

Furthermore, the Lebanese arena also allowed non-Arab players to engage in regional bidding for power. The Iranian revolution was quickly reflected in Lebanon not only because Iran valued the strategically important location of Lebanon but because the Shi'ite population in Lebanon was receptive to the ideas of Khumayni and was willing to be used by Iran for its own regional ambitions. The Iran-Iraq was fought on Lebanese territory in the late 1970s before the real war erupted on the Iranian-Iraqi front. Similarly, Armenian groups fought Turkey on Lebanese soil, while Kurdish groups have found Lebanon the only nonrestricted environment as far as their own political and guerilla activities are concerned. All these dimensions have compounded the Lebanese problem and made a solution more remote even if the internal (purely Lebanese) dimension were adequately addressed.

Absence of the "Neutral" Mediator

A peaceful resolution to any prolonged conflict requires the assistance of a neutral observer—someone trusted by all sides. Peace proposals for Lebanon and the numerous reconciliation missions to Lebanon by foreign mediators failed because they represented interests linked to Lebanese players; their agendas were often partially or fully reflective of the interests of their local allies and clients. Of the non-Arab players, American, French, and Vatican authorities were the most frequent mediators. But those parties had close ties to various Lebanese militias and politicians, and their intentions were distrusted by the other side. The US, for example, was closely allied to the Lebanese right-wing coalition in the 1975–76 phase of the war and became closely associated with the government of Amin Gemayyel (and with the militia of his brother Bashir before his assassination); all of whom had a lot of enemies inside and outside Lebanon. As the US was regionally allied to Israel, the US faced a huge credibility gap in the eyes of many Lebanese and in the eyes of those Arab states involved in Lebanese affairs.

Similarly, the Vatican, which took on several mediation roles in the course of the Lebanese civil war, while highly trusted by the Maronite establishment, is regarded as an adversary by the Muslim establishment due to the long tradition of religio-political ties between the Holy See and the Maronite church in Lebanon. Many Muslim politicians suspected that the Vatican was interested in undermining the Arab/Muslim influence in Lebanon to bolster the position of the Maronite establishment and the Maronite church in Lebanese life. France, which also dispatched several mediators to Lebanon during the course of the war, was not popular among the Muslim population because of the traditional French-Maronite alliance and because the French socialist government openly supported the Maronite establishment. Moreover, the French expressed sympathy with the highly controversial Gen Michel 'Awn and granted him asylum (along with perks usually reserved for heads of states and monarchs) after he was expelled from Lebanon.

The two most important players in the Lebanese arena are, of course, Syria and Israel. Syria has always portrayed a neutral role in Lebanon, a role that seeks to end bloodshed and restore stability and order to the country. But the Syrian role in Lebanon is one of shifting alliances and feuds; Syria aligned itself with the left-wing/Palestinian coalition in 1975, then switched to the Maronite/right-wing side in 1976, and was later embroiled in bloody confrontations with the right-wing forces between 1978 and 1982. One could argue that the nature of the Lebanese conflict and the highly charged emotional issues that split the Lebanese along sectarian, geographic, and socioeconomic lines make it impossible for any one party to be acceptable by all sides with the same enthusiasm. It is not possible to identify with the positions of all sides.

As for Israel, it never attempted to play the role of the mediator simply because Israel has never claimed neutrality about Lebanon and has consistently argued that "Palestinian terrorism" was responsible for the breakdown of the Lebanese political system. Israel's unpopularity among most Lebanese also would negate any Israeli mediator's role. When PLO forces were evacuated from Lebanon and Israeli occupation of parts of Lebanon intensified Lebanese popular antipathy towards the government of Israel, Israel then insisted that the rise of a South Lebanese resistance movement against the Israeli presence was simply the product of a "Shi'ite terrorist" movement and thus a legitimate target of Israeli strikes and incursions.

One could argue that only such an international body as the United Nations could play a neutral role acceptable to all sides in Lebanon. However, the Syrian government was adamant in its rejection of a UN role in Lebanon because it fears a diminution of its influence there. Syria has always persuaded and urged its Lebanese allies to reject a UN role (aside from the limited role in South Lebanon in the wake of the 1978 Israeli invasion of Lebanon) from fear of losing its leverage in the country's domestic affairs. Syria even discouraged the League of Arab States from taking any serious mediation role in Lebanon because Syria sought to monopolize Arab influence in Lebanon.

Rise of the War Elite

One of the most damaging factors as far as Lebanese civilians are concerned is the rise of the war elite in the wake of the 1975–76 phase of the war. While the historical system of *zuàma'* (political bosses) was maintained in some of the communities (primarily Maronite and Druze communities), the war has led to the emergence of a new breed of leaders who spoke for the war activists or for the street thugs, as most Lebanese came to refer to the salaried combatants of all sides.

The Lebanese civil war resulted in a loss of popularity for many members of the old families who have monopolized political representation in Lebanon for almost two centuries. Those families were associated with the rampant corruption of the government and Parliament and were blamed for the inefficiency of the Lebanese administration. The decline, and in some communities the total demise, of the role of the traditional elite led to the rise of a new generation of leaders who sought to replace the old guard. The new elite was produced by the various militias that were born in the various communities. The new leaders tended to be youthful and experienced on the battlefield. They were trusted by their fighters and were close to the average people on the streets. The new elite represented the militancy of the civil war, a militancy responsible for the brutality and savagery that characterized the war.

As the war progressed, the new elite developed a sophisticated bureaucracy within the structure of the militias to cope with the rising needs of the thousands of fighters and families of "martyrs" for whom the militia leadership was responsible. Furthermore, the militia leadership had to inherit state responsibilities in various regions of the country. In many cases, the militias deliberately disrupted the governmental process to cripple the state bureaucracy so that the militia could fill the vacuum. The militias—those of the right and the left—developed their own economic networks that relied on revenues from narcotics and arms smuggling. The revenues from the underground economy and from the looting of state funds provided the militias with large fortunes. The continuation of the war and the continued paralysis of the Lebanese state guaranteed the militias their precious sources of income. One can assert without exaggeration that militia leaders were the most stubborn opponents of any reconciliation, for an end to the war would have divorced the new war elites from their basis of financial and material support and from resources that enabled them to impose their political will. In other words, the war became necessary both politically and financially for militia leaders. An end to the war would have revived the civilian (nonmilitary) leadership that had been competing for political leadership with the war elite since 1975.

Syria's Belief in the Efficacy of "No Victors, No Vanquished"

The best elaboration of Syrian policy towards Lebanon has been a famous speech made by Syrian President Hafidh Al-Asad in June 1976. Speaking at Damascus University, Al-Asad promulgated publicly his belief in the notion of "No Victors, No Vanquished." This principle has guided politics in Lebanon since at least 1958, when the Lebanese political elite decided the resolution of the brief civil war of that year would be based on the rejection of domination of one group by another. The politicians argued that this principle would maintain social harmony accord; any victory by one party would tear the nation apart.

The Syrian government has pursued its objectives in Lebanon under this slogan, but for different reasons. While the Maronite elite was fearful in 1958 of a Muslim takeover of the country, the Syrian intervention in Lebanon was genuinely motivated by Syria's rejection of domination by any of the sectarian groups or militias. The Syrian military intervention in 1976 crushed the PLO/Lebanese leftist alliance, which was about to control all Lebanese territory. The Syrian government opposed the victory of the Muslim/leftist coalition because Lebanon under Muslim/PLO control could drag Syria into an unwanted confrontation with Israel. Moreover, a radical regime in Lebanon could pose an ideological challenge to the Syrian regime, the legitimacy of which is predicated on the notion that Syria represents the most pro-Palestinian, pro-Arab nationalist regime in the entire Arab world. The victory of the Lebanese left in Lebanon could have undermined Syria's propaganda claims. Syria was also concerned that the defeat of the Maronites in Lebanon could aggravate the already critical problem of minorities in the region, which could, in Asad's mind, lead to futher fragmentation in the Arab East and in Syria, in particular. A member of a minority sect himself, Asad has been especially sensitive to the minority question in the region.

Without addressing the politico-ethical implications of a leftist victory in 1976, the Syrian intervention in that year undoubtedly prolonged the war and prevented a decisive, albeit a violent, resolution of the civil war. One could argue that Syrian intervention (as an external factor) was one of the most important reasons behind the prolongation of the Lebanese civil war. In other words, Syria did not allow the contradictions in Lebanese society and policy to clash to produce a radical change in the social and political order. Had Syria remained unengaged, the civil war could have ended as early as 1976, although the price of a decisive end might have been exorbitant. The price could have been especially high with respect to the well-being of Christians in Lebanon and even with respect to the Christians of the Middle East. Following its military intervention in 1976, Syria pursued a policy that allowed Lebanese factions and militias to combat one another without allowing any side to achieve total victory. While the end of the war in 1976 would have been costly in human life, prolongation of the war beyond the first phase increased the suffering of all Lebanese. The individual Lebanese will have to decide whether an early end of the war in 1976 would have been more desirable than the current situation in which the motto of "No Victors, No Vanquished" has not been entirely abandoned.

Sectarian Agitation and Mobilization

One of the weapons that has been effectively used by Lebanese politicians—whether they belong to the traditional elite or to the war elite—is sectarian agitation and mobilization. In time of crises, Lebanese politicians have often resorted to sectarian agitation of the masses to bolster their own standing within their respective communities. Unfortunately for Lebanon, the Lebanese people have always proved susceptible to sectarian mobilization. The argument that blames members of the other sect for the problems facing different confessional communities tends to appeal to the masses; people in all cultures cling to easy answers even if the answers are rooted in prejudice and bigotry. The demonization of the other sect has helped traditional leaders in presenting themselves as champions of the interests of the community. The narrow electoral districting in Lebanese elections has promoted sectarian agitation and mobilization, and the Lebanese political system itself was the product of a sectarian arrangement. The level of sectarian agitation and mobilization during the civil war was the highest it had been since the sectarian wars of the nineteenth century. The agendas and outlooks of the various sectarian leaders and parties were different, which accounted for the fragmentation in the Muslim and Christian communities. The Shi'ite agenda was different from the Sunni agenda, and there were divergent interests between the Maronite and the Greek Orthodox communities. Unlike previous situations in Lebanese history, no monolithic Muslim and Christian blocs existed. Ambitious sectarian leaders found it convenient to agitate the masses in a narrow, sectarian fashion to advance their political careers. Claims of narrow sectarian concerns are almost always rewarded among the Lebanese communities, and much of the war elite is comprised of individuals who rose to power because of hateful, sectarian agitation.

It is, of course, economic and political frustrations and resentments that allow for the exploitation of the religious factor in Lebanese politics. Lebanese political leaders and post-1975 war leaders have found engaging in narrow sectarian argumentation too tempting, because the Lebanese have always felt that the other sect harbors hostile intentions. The multiplicity of political identities and the "fragmented political culture," to use the language of Michael Suleiman, have promoted political representation according to sectarian affiliation. Members of a sect assume that genuine representation of their interests requires the election of leaders who champion the interests of the sect. Because socio-economic standards vary in Lebanon among the sects and since there are regional imbalances in economic development, the lines between class and sectarian oppression have become blurred. This blurring leads people to attribute their dissatisfaction to the tyranny or misguidedness of the other sect.

While the tendency of the war elite to engage in sectarian agitation and mobilization to perpetuate their dominance has obstructed the resolution of the Lebanese conflict, the relationship between the length of the conflict and the intensification of sectarian agitation is dialectical. The Lebanese people themselves have become more willing to receive and adopt sectarian arguments after their long years of civil strife. The war elite cultivated what was already a fertile ground. This situation then led to the demonization of the enemy, which makes compromise unacceptable.

The time factor, however, proved to be crucial in this case; evidence suggests that by the late 1980s, with the intensification of the armed conflict and material destruction, many Lebanese reached a point of exhaustion. There came a point during the Lebanese civil war when most Lebanese were simply fed up with the situation. They became increasingly impatient with the enthusiasm that members of the war elite exhibited towards the continuation of the war. By 1990 many Lebanese became physically and psychologically fatigued, and political considerations (and sectarian considerations as well) became irrelevant. In other words, sectarian agitation worked up to a point in the course of the civil war, but the people proved that there was a saturation point as far as mobilization behind the slogans of hate and demonization.

Fear of the Return to the Status Quo

Another factor that helped to prolong and intensify the war in Lebanon was the fear among many political/militia leaders and among Lebanese that emanated from the nonresolution of the Lebanese conflict. Some peace plans were dismissed because they were not regarded as comprehensive enough to address the roots of the Lebanese problem. There were fears that some of the peace plans, if implemented, would not resolve the Lebanese war. Past historical experiences were too unacceptable for many Lebanese; the tendency of the traditional political leaders to accept the tribal-style entente (sulh), based simply on embraces between the individuals themselves, made many Lebanese (particularly the young fighters in the militias) suspicious of the motivations of the peace proposals. The youth of Lebanon considered the price of the war itself too much in terms of human life and physical destruction to justify a return to the status quo ante. Had there been a peace proposal to be accepted—some Lebanese argued—it had to have the depth and the scope to deal with the Lebanese problem from all its aspects. An incomplete or partial peace plan would result in a cease-fire and not offer finality to the war according to many demanding Lebanese skeptics.

The experience of the 1958 conflict, which ended in the formula, "No Victor, No Vanquished," was too fresh in the memory of the Lebanese. Many Lebanese argued that had the 1958 mini-civil war led to a rearrangement of the power-sharing formula and to a real rectification in the socioeconomic imbalances in the country, the war of 1975 could have been avoided. Many Lebanese feel obliged not to accept a solution that would serve as a prolonged cease-fire rather than a final, definitive solution—if there is such a thing—to the protracted conflict.

Militia leaders in Lebanon have said to the Lebanese people that solutions that would not address the deep, underlying causes of the Lebanese war were not worth considering. They argued persuasively that unless someone addressed the root of the Lebanese problem, the war was destined to recur. Militia leaders, of course, had their own reasons for blocking peace proposals, as these proposals would undermine their positions of prominence. But the Lebanese people themselves were unwilling, at least in the early phase of the war, to accept a mere prolonged cease-fire for fear of a renewal of the civil war.

The Ta'if accord which—officially at least—ended the civil war was accepted by most Lebanese, since it came at a time when the people were fatigued by the protracted conflict. Some observers believe the accord represents a disguised return to the status quo. Therefore, one must not proclaim an end to the Lebanese civil war; the Syrian-sponsored accord may merely provide a respite for the war-torn country and its people.

Arab Official Antipathy to the Lebanese Model of Democracy

While Arab governments have consistently paid lip service to the necessity of restoring peace and tranquility to Lebanon, they have without exceptions long resented the Lebanese system of political pluralism and the press freedoms that Lebanon enjoyed. Arab governments have used the civil war in Lebanon to argue against democracy; they claim that the civil war is itself the direct result of the Lebanese democratic experience. The Arab regimes found that the political system in prewar Lebanon threatened the despotism that had been imposed on the Arab citizen. Lebanon presented the Arab citizen everywhere in the region with an alternative model that, despite its weaknesses and imperfections, was admired by Arab intellectuals and ordinary people alike.

The second element in the Lebanese prewar system which presented a danger to Arab officialdom was the press and publishing freedoms in Lebanon. Before the Lebanese civil war, Arab publications in Lebanon served as mirrors of Arab public opinions at large; they also served as voices against many of the regimes. Additionally, Lebanon before 1975 was a haven for Arab dissidents expelled from their countries because of opinions they expressed or activities in which they participated. It was also a place where Arab capital was concentrated due to Lebanon's excellent banking system and its unique (within the region) banking secrecy. Arab dissidents in Lebanon constituted a nuisance to the regimes that did not—and still do not—tolerate criticism. Thus, the end of the Lebanese civil war, which discredited the idea of democracy according to the thinking of Arab despotic regimes, was not a priority for the regimes. The end of the civil war posed a challenge to many Arab governments, a challenge that none of the Arab governments were willing to tolerate at a time when domestic opposition was rising because of the emergence of the Islamic factor in popular politics.

The Politics of the War Generation

The continuation of the Lebanese civil war brought about a generation of Lebanese, many of them who have served in the various militias, who have known only the politics of violence. The present generation of Lebanese have been insulated from the pluralistic experience of prewar Lebanon, where Lebanese from various sects at least met one another, even if they did not get along with one another. At the cessation of the fighting, many in this generation had never even met a member of another sect; this group was more susceptible to sectarian agitation and mobilization than other groups in Lebanon. Demonization of the other sect was easier among youths who had never interacted with members of other sects due to the separateness of the Lebanese communal existence in the wake of the war. And these youths had a say in issues of war and peace, as they constituted the fighting backbone of the militias.

The group of militia fighters also was salaried, thanks to the largess of the patrons of the various militias. And many of the fighters were unskilled and uneducated; a situation of normalcy would have meant a sharp decline in their living standards, particularly with the extensive use of narcotics among many of the armed youths. Lebanese leaders were fully aware of this factor when the Ta'if accords were being negotiated and foresaw that thousands of Lebanese would need rehabilitation desperately. The Lebanese government decided to absorb the armed youths in its armed forces. The process already has met some success, although it is shortsighted to assume that a few months of training within the Lebanese army would erase the traces of militia agitation of hate and discord as well as financial security.

Conclusion

In analyzing the cause behind the prolongation of the Lebanese war, one must make the early distinction between the Lebanese war and the Lebanese conflict. The Lebanese conflict began as early as the nineteenth century, when the two major groups occupying the area of Mount Lebanon (the Druzes and the Maronites) fought over the domination of the small piece of land. The creation of Lebanon in the twentieth century expanded the Lebanese conflict to include the numerous sectarian communities that have been separated by geographical lines and by fear and suspicion. The Lebanese conflict—or more accurately conflicts—was not resolved with the creation of the new state, nor even with the Lebanese republic after independence in 1943. Rather, the Lebanese political system seemed to incorporate these outstanding social conflicts into the Lebanese polity, thereby perpetuating the tradition of hate and hostility.

The Lebanese war refers to the strife that began in 1975. This strife was produced by the conflicts among the Lebanese and by the external stresses on the Lebanese system, which exacerbated the already existing tensions and hostilities. I have emphasized that the Lebanese war resulted from internal dynamics and socioeconomic schisms within the Lebanese society; the idea of the war as an external conspiracy—an idea that is still popular among the Lebanese politicians who wish to absolve themselves from any responsibility for the war—has served to obfuscate the Lebanese reality.

The Ta'if accord has certainly helped to end the blood and destruction in Lebanon, although there is still a part of the country (the south) where combat between the Israeli army and its surrogate militia and the Lebanese groups who wish to rid Lebanon of Israeli occupation still takes place. The presence of the Syrian army, however, has been legitimized by the comprehensive Lebanese-Syrian treaty that was signed in the wake of the Ta'if accord. The Syrian presence, however, is still preferred by the Lebanese to the oppressive, thuggish rule of the militias, who terrorized the population. The resentment of the Lebanese people against the various militias served to give the Syrian presence the popular legitimacy that it needed, although a substantial section of the Christian community in Lebanon still opposes on principle the presence of the Syrians in Lebanon. The support of the Lebanese presidency for Syrian military presence makes any withdrawal of Syrian troops from Lebanon unlikely. Syria also uses Lebanon as a valuable bargaining chip in regional negotiations.

Finally, we can assume the Lebanese conflicts have not been resolved. As for the Lebanese war, there is a possibility that the prolonged cease-fire the Lebanese people have been enjoying could be transformed into a real end to the war. But this possibility requires radical internal political and economic reforms in Lebanon and a curtailment of regional (primarily Syrian and Israeli) interference in Lebanon's internal affairs. The administration of President Hrawi does not seem, however, willing or even able to lead Lebanon out of its long conflict and to put a final end to the war.

Notes

1. These figures represent the most reliable, official estimates of the Lebanese government. The figures were published in the Beirut daily *An-Nahar* and reported in the *New York Times,* 10 March 1992. The figures exclude some 6,630 people who were killed and about 8,000 wounded in conflicts involving Palestinians. Also, the report stated that some 17,415 people remain missing.

2. For the history of Lebanon, see Meir Zamir, *The Formation of Lebanon* (London: 1965); M. Daher, *L'Histoire Sociopolitique de la Republique Libanaise sous Mandat Francais: 1926–1943,* 2 vols. (Paris: 1980); and Adel

Ismail, ed., *Documents Diplomatiques et Consulairs Relatifs a L'histoire du Liban* (Beirut: 1979).

3. For the nineteenth-century history, see Samir Khalaf, *Persistence and Change in 19th Century Lebanon: A Sociological Essay* (Beirut: A. U. B., 1979); and Antoine Khair, *Le Moutacarrifat du Mon-Liban* (Beirut: 1973).

4. See Q. Qarqut, *Tatawwur Al-Harakah Al-Wataniyyah Fi Suriya: 1920–1939* (The Evolution of the National Movement in Syria: 1920–1939) (Beirut: 1975).

5. For the constitutional development of Lebanon, see Edmond Rabbath, *La Formation Historique du Liban Politique et Constitutionel* (Beirut: 1973).

6. See Tabitha Petran, *The Struggle over Lebanon* (New York: Monthly Review Press, 1987), 31.

7. This author received the data through the Federal Research Division of the Library of Congress. It was printed in As'ad AbuKhalil, "The Society and Its Environment," Federal Research Division of the Library of Congress, *Lebanon: A Country Study* (Washington, D.C.: Government Printing Office, 1989), 48.

8. The classic study of the National Pact is Basim Al-Jisr, *Mithag 1943* (The Pact of 1943) (Beirut: Dar An-Nahar, 1978).

9. For Sham'un's views, see Camill Chamoun, *Crise au Moyen Orient* (Paris: Gallimard, 1963); for the views of his critics, see Isma'il Musa Al-Yusif, *Thwrat Al-Ahrar Fi Lubnan* (The Revolution of the Free in Lebanon) (Beirut: Manshurat Az-Zayn, 1958).

10. For a discussion of the 1958 crisis, see Michael C. Hudson, *The Precarious Republic: Modernization in Lebanon* (New York: Random House, 1968).

11. For the 1975–1976 phase of the war, see Kamal Salibi, *Crossroads to Civil War: Lebanon 1958–1976* (New York: Caravan, 1976); and Marius Deeb, *The Lebanese Civil War* (New York: Praeger, 1980).

12. For events in the south and the developments within the Shi'ite community, see Augustus Richard Norton, *Amal and the Shi'a* (Austin: University of Texas Press, 1987).

13. For the 1982 invasion, see Ze'ev Schiff and Ehud Ya'ari, *Israel's Lebanon War* (London: Allen and Unwin, 1985); and Rashid Khalidi, *Under Siege* (New York: Columbia University Press, 1986).

14. For an account of the Marines' role, see Michael Petit, *Peacekeepers at War: A Marine's Account of the Beirut Catastrophe* (London: Faber and Faber, 1986).

15. For this phase of the war, see Robert Fisk, *Pity the Nation* (New York: Atheneum, 1990).

16. For an analysis of the reforms, see Augustus Richard Norton, "Lebanon after Ta'if," *The Middle East Journal 45, no. 3 (Summer 1991).*

17. For the external dimension of the war, see P. Haley and L. Snider, eds., *Lebanon in Crisis: Participants and Issues* (Syracuse: Syracuse University Press, 1979).

18. For this point, see As'ad AbuKhalil, "Arab Intervention in the Lebanese Civil War: Lebanese Perceptions and Reality," *The Beirut Review* 1, no. 2 (Fall 1991).

19. See, for example, B. J. Odeh, *Lebanon: The Dynamics of Conflict* (London: Zed Press, 1985).

20. For a treatment of sectarian consciousness, see As'ad AbuKhalil, "Druze, Sunni, and Shi'ite Political Leadership in Present-Day Lebanon," *Arab Studies Quarterly*, Fall 1985.

21. As'ad AbuKhalil, *The Politics of Sectarian Ethnicity: Segmentation in Lebanese Society* (PhD dissertation, Government Department, Georgetown University, 1988).

22. For a well-informed, journalistic account of the involvement of foreign intelligence services in Lebanese affairs, see Said K. Aburish, *Beirut Spy: International Intrique in the Middle East* (London: Bloomsbury, 1989).

The Arab-Israeli Wars
A Conflict of Strategic Attrition

Stewart Reiser

Few modern regional conflicts have endured the tenacity of the Arab-Israeli conflict. To understand its duration and durability, one must examine the conflict on two levels. One level focuses on how the antagonists defined the issues. The second level of analysis examines why this particular conflict still has not been resolved during its cycle of warfare. Historically, war does not resolve many conflicts. Apparently, both parties to this particular dispute assume war can settle their differences; therefore, why, to date, has it not? This study examines the conflict to determine whether its duration has been caused by design, as part of a policy of strategic attrition, or as a product of blunder and circumstances beyond the control of policymakers on each side.

Concerning the first level, one can argue this is an evolving and dynamic conflict. In this context, it should be seen and defined as a conflict because the many wars between the Arabs and Israelis failed to resolve their underlying concerns, and these wars were regularly separated by periods best characterized as phases of "no war, no peace," which later broke down into active conflict. The conflict began with the Arab side anticipating a rapid conventional victory. As neither side achieved this during the first two wars (1948–49 and 1956), both sides began to engage in conflict that gradually developed into one that was characterized by strategies of protraction. These strategies reflected each side's perception of their respective natural advantages as well as their abilities to extract resources from the great powers.

Thus, time and circumstances have changed the reality of the Arab-Israeli conflict. It has evolved through several distinct phases. One reason for this evolutionary nature is that the conflict is influenced by and, in turn, influences other political dimensions in the Middle East region. These dimensions include the internal character of the major states in the region, the inter-Arab political dynamic, and, finally, the interests and involvements of the great powers in the region. A major change on one political level, or dimension, usually causes changes in the other three.

Pursuing this approach, one can see that there have been three phases to the conflict. The first phase began in 1917 with the issuance of the Balfour Declaration and ended in 1948, the year the first Arab-Israeli war began or when the internationalization of the conflict occurred. The three intervening decades were characterized by an intercommunal conflict between the Jewish and Arab communities in Palestine under the British mandate. Each side during this stage received aid from external support groups, the Zionist organizations, on the one hand, and the Arab states still encumbered by the colonial occupation of Great Britain and France, on the other. Each side also experienced periodical shifts by Great Britain toward and against their particular interest. The conflict during this period focused on the meaning of the Balfour Declaration and the League of Nations mandate within the context of the changing circumstances in Palestine, as well as the altering British interests in both the region and in Europe.

The second phase spanned from the war of 1948–49 until the Six-Day War of 1967. The conflict became interstate in nature during this period. It was also defined by the rise of militant Pan-Arabism and the objectives of Egypt's charismatic leader, Gamal Abdul Nasser, which included the integral unity of the Arab world and the elimination of Israel. This stage was also characterized by the beginning of the cold war and was accompanied by a spiraling arms race. Thus, there was a change in both the issues at stake within the conflict and the conditions that surrounded it. This phase ended with the Israeli capture of Arab national territory (beyond Palestine) in June 1967. The consequences of this war created the conditions for the third, or current, phase.

The third phase of the Arab-Israeli conflict has lasted from 1967 to the present. The results of the 1967 war has brought about three major changes in the overall conflict, changes that have neutralized one another in terms of conflict resolution. First, there was an alteration in the balance of incentives for the major national actors within the conflict. Prior to the Israeli occupation of (non-Palestinian) Arab territory, the Arab side could maintain the status quo of "no war, no peace" in the belief that time, as well as justice, was on its side and wait for more propitious conditions to renew active warfare. There were no conditions or factors within or surrounding the conflict that could force the Arab side to negotiate with and recognize the state of Israel.

The seizure of Egyptian, Jordanian, and Syrian territory in 1967 altered this equation. The new status quo benefitted Israel as long as the United States sustained it with diplomatic, military, and economic support. From this point onward, the three Arab confrontation states bordering Israel had to reconcile their "national" interests with their broader "Arab" interests, the recovery of Palestine. Each state—Egypt, Jordan, and Syria—operated under a different calculus of incentives in their choice of maintaining the status quo, proceeding along a diplomatic path, or using coercive means. Each state based its strategic decisions on domestic, regional, and international factors.

This seizure highlights the second complicating feature of the post-1967 period. This feature was the reemergence, following two decades of dormancy, of the Palestinian issue as a national issue. As the Palestine Liberation Organization (PLO) gained recognition as the "sole legitimate representative" of the Palestinian people, Israel and Jordan could not just "trade land for peace," as Israel and Egypt had, given Jordan's domestic political divisions and the relationship between the Palestinians in the East Bank Kingdom and those in the Israeli-occupied territories. This was exasperated by the PLO's refusal to recognize Israel's right to exist, within *any* boundaries, until December 1988.

Third and finally, the occupation of the West Bank and Gaza stimulated the growth of both the revisionists and religious Zionists within Israel. These smaller groups had lived under the shadow of labor (or social) Zionism from 1920 onward but had never given up their aspiration of an expanded Greater Eretz Yisrael (Whole Land of Israel). Thus, the same Israeli expansion and occupation of territories that reawakened the Palestinian national movement simultaneously catalyzed the significant growth of Israeli parties who had historical and religious attachments to the same land.

These three local and regional factors have characterized the conflict since 1967. Each factor has contributed to the protraction of the conflict. Attendant to each has been the collapse of the cold war and the emergent American dominance of the region, particularly after the defeat of Iraq in 1991 and the earlier start of the intifada (uprising) in the territories in December 1987. While the latter variable was believed, particularly during its first 12 to 18 months, to have been costly and disruptive enough to alter Israel's balance of incentives for maintaining its presence in the territories, this scenario has proven not to be the case. Even the Palestinian leadership now perceives that it has extracted maximal political gain from the uprising. It is currently looking to foreign (Arab and American) sources to alter Israel's stance.

On the international power level, the general Soviet collapse has already contributed to one conflict, the Iraqi invasion and occupation of Kuwait, which in turn brought about the American diplomatic and military riposte that enhanced United States political and military hegemony in the entire Middle East. The United States, alone, has the capacity to alter the balance of incentives for the local actors. Time will determine whether it has the will, whether it decides that the push toward a settlement is worth the political capital that it would warrant, and finally whether it perceives that a shift from the status quo is in the interest of the United States.

While the Arab-Israeli conflict at its earliest stages appeared to have shared certain characteristics with other prolonged conflicts, it soon reflected many features of a protracted conflict. This is because during different phases the involved major Arab actors reached the conclusion that the status quo of "no war, no peace" gave its side the time during which it could improve the conditions required for a next round of war. As for the Israelis, they too favored a protracted strategy, since it seemed their best policy for both survival and territorial expansion. This calculus changed for Egypt and Israel following the 1973 war largely because of the particular type and level of American involvement as well as the consequences of that round of combat. The conflict may now be at a similar historical crossroad. At this point, it seems necessary to summarize the series of choices taken in earlier stages.

The Intercommunal Phase, 1917–48

This period was characterized by a three-cornered struggle between the growing Jewish and Arab communities within Palestine and the British mandatory power over the meaning of the Balfour Declaration. The central and defining mission of the declaration and the preamble and articles of the League of Nation's mandate focused on "the historical connections of the Jewish people" with Palestine and "the grounds for reconstituting their national home in that country."[1]

While this mandate appears straightforward and supportive of a pro-Zionist interpretation, articles no. 1 and no. 6 also state that "nothing should be done which might prejudice the civil and religious rights of existing non-Jewish communities in Palestine" and that promotion of Jewish immigration and land settlement was not to prejudice "the rights and position of other sections of the population." Thus, sufficient ambivalence was created to give scope to a wide interpretation by both sides to the conflict as well as to Great Britain.

Britain's response to the heightened violence by each community immediately before and after WWII was military repression of the Arabs (and later the Jews) on the one hand, and a series of white papers that reversed the original content of the original Balfour Declaration, on the other. The conflicting responses of the Arab and Jewish communities to Great Britain's own strategic reversal set the tone for the Palestinian and Israeli approaches for the following decades.

When Zionists were offered concessions but didn't care for their specific content, they accepted such principles of the concession as territorial partition and statehood for each community but rejected any concessions on boundaries. The Arab side rejected both the principle and the content of the concessions, making it difficult for the international community to accept their absolutist and maximal objectives. Both before and following WWII, the Palestinian leadership refused to view the nature of the conflict as changing and insisted that justice and their interests could only be served by the creation of a single exclusively Arab Palestine. Not only had they rejected the recommendations of partition in 1937 and 1947, which would have resulted in a miniscule and contained Jewish state in one corner of the vast Arab Muslim world that was coming into its own, they also refused the Anglo-American Commission's recommendation of a binational federal state in 1946. They also rejected Great Britain's (mostly pro-Arab) white paper of 1939, which would have given them an independent Palestine and would have frozen the Jewish population into a permanent minority status within an Arab state whose leadership could control immigration policy as well as unite with neighboring Transjordan.

The question is, why did the Arab states and the Arab Higher Committee take such a hard noncompromising line? There are both practical-political as well as historical-cultural reasons for this approach during the first phase of the Arab-Israeli conflict. These reasons may apply for the two phases that follow.

On the practical level, the governments of the Arab states were divided among themselves on the issue of Israel, and they could unite, to the degree that they did on the surface, solely on the most extreme program, which was the prevention of the creation of a Jewish polity in a historic Arab-Muslim territory. This feature of the conflict prolonged it, since the lack of real Arab unity, based on these and other political differences, has contributed to the difficulty of containing, no less defeating, the Israelis.

A reflection of the political culture that also has prolonged the conflict has been the fact that the more moderate Arab leaders were in physical and political danger if they publicly embraced a position that did not absolutely reject the legitimacy of a Jewish state of any proportion within Palestine. Following the 1948–49 war, King Abdullah of Jordan, Prime Minister Riyad al-Suhl of Lebanon, and Prime Minister Nuqrashi Pasha of Egypt were assassinated for not adhering to the most extreme position concerning Palestinian rights. (One should add Anwar Sadat and Bashir Gamayel in later years.)

Perhaps a more profound reason for the uncompromising position of the majority of the Arab and Palestinian leaders was their moral belief that "rights" could not be compromised, that they were absolute, and that the pursuit of these rights warranted a totalistic approach which conflicted with other cultural notions which held a more relativistic approach to justice.

The first "Arab-Israeli war" actually began in November 1947 as a Palestinian-Jewish "civil" conflict and became international when the Arab armies crossed into Israel in March 1948. The Arab intervention that was designed to reverse the outcome of the United Nations decision and the earlier Palestinian Arab-Jewish struggle failed. The Arab defeat leads to three questions. First, why did the Arab states fail in their objectives during this war? Second, why did the Israeli military victory fail to settle the conflict? Third, was the continuation of the conflict based on design or happenstance?

The Internationalization of the Conflict, 1948–67

Two factors highlight the defeat of the Arab states in their first war against Israel. Each factor contributed to the conflict's prolongation as well as to the choice of protraction as a strategy by both sides. One factor is that there had been an almost total absence of contact between the emerging Jewish society and the Arab leaders in Palestine. As a result the Arab side of the conflict, with the exception of King Abdullah of Transjordan, knew little of the organizational strengths or limitations of their adversary.[2]

The Israeli strategic ignorance of their enemy was almost, but not quite, equal. On the Israeli side, only David Ben-Gurion, who assumed the positions of prime minister and minister of defense, perceived the need to prepare the Yishuv for an all-out conventional war against the regular armies of the Arab states. This meant a full mobilization of Jewish manpower (women were conscripted as well as men) and the procurement of heavy weapons for land, sea, and air warfare.[3]

The second important factor is that the Arab states were united only in their negative purpose of destroying Israel (and Transjordan didn't even share that goal) but had no unity of purpose after the anticipated victory. Transjordan and Iraq (at that time also headed by a branch of the Hashemite family) desired to annex as much of Palestine as possible for themselves, even at the cost of splitting the territorial difference with the Israelis. Syria, Egypt, and Saudi Arabia, on the other hand, wanted to establish a puppet Palestine, but they primarily wanted to obstruct the Hashemite objectives. The Palestinian leadership, demoralized and divided, was highly suspicious of the Arab states but leaned toward Syria and Egypt to prevent Hashemite annexation of Palestine. These conflicting interests led to uncoordinated military planning, mistrust, and a certain degree of deception between the armed forces of the Arab states. This position of dependence and minimal leverage on the part of the Palestinian national movement prevails even today.

The second issue is why did this battlefield defeat not lead to a negotiated peace between the two sides. With the exception of Iraq, the Arab states separately signed armistice agreements with Israel. Neither the armistice agreements nor the talks that followed led to peace agreements between the antagonists. Historically, peace treaties usually follow armistices because the defeated party fears continued punishment if it doesn't accede to all or most of the demands of the victor. While it is likely that some Arab governments believed for several months following the defeat that they might have to grant concessions, including that of recognition, to the victorious Israelis. It is also likely that the Arabs realized that despite their defeat they were in a novel historical situation. The Arab states realized that Israel, despite its military superiority, was politically incapable of renewing hostilities due to the public and private warnings from the United Nations, Great Britain, and the United States. At this point in its history, Israel could afford neither the international condemnation nor the possible and threatened (British) intervention that might have followed a renewed offensive.

The Arab leaders concluded that they were not confronted with the usual choices of continued (and punishing) war or peace. Rather, they could opt for a status quo of "no war, no peace" that served several purposes. They didn't have to take the highly unpopular act of signing peace treaties with Israel and recognizing its legitimacy; also, the future of the conflict was left open. The following quote from the former secretary general of the Arab League, Azzam Pasha, summarizes this strategy of protraction in a 1960 interview.

> We have a secret weapon which we can use better than guns . . . and this is time. As long as we do not make peace with the Zionists, the war is not over; and as long as the war is not over there is neither victor nor vanquished. As soon as we recognize the existence of the state of Israel, we admit by this act that we are vanquished.[4]

This nonacceptance of the defeat does not mean that the Arab states planned for an immediate renewal of the fighting. Rather, the Arab strategy of protraction began when their leaders realized that they had time on their side and that they could wait until circumstances were more favorable for the pursuit of their goal. The Arab side had assumed a lightning conventional victory as it entered the 1948 war. Having failed in their "preventive" effort, they moved into a "restorative" phase. Thus, once the first effort to prevent Israel from coming into being failed, the Arab side shifted to the goal of restoring the "pre-Israel" situation.

Regional Changes, 1949–67

Between 1949 and 1967 three crucial changes occurred that altered the Arab perspective. The first change focused on the issues at stake between the Arabs and Israel. The second change surfaced in the conditions surrounding the conflict. The third change appeared in the festering of the conflict.

In terms of the first change, the introduction of Pan-Arabism and integral Arab unity into the region as a central and defining political force for the Arab people escalated the residual issues of Palestinian refugees and boundary disputes, which remained the official position of the League of Arab States until the late 1950s, into a clash of national destinies between Zionism and Arab nationalism. Once Egyptian President Gamal Abdel Nasser called for integral unity with the Arab east, the Mashriq, and once this call found a receptive ear among the Arab populations, Israel's boundaries and the Palestinian refugee problem were no longer the point. Israel became a physical as well as a political obstruction to Arab unity.

From the Arab perspective, there was little or no incentive toward peace and a limited capacity to wage war between 1949 and 1956. From the Israeli perspective, there was little leverage by which to induce the Arabs to accept peace and no political freedom to go to war and force the issue. In terms of the absence of war during this period, one can say the Arab side was limited in its instruments of force, and Israel for its part did not have the freedom to use its military superiority until Great Britain and France not only permitted Israel to strike at Egypt but actually assisted for their own reasons in October 1956.

From the mid-1950s onward, the conditions surrounding the conflict changed dramatically. What altered the dynamic between Arab and Israeli most, as well as the inter-Arab dimension, was the introduction of the cold war into the Middle East. In short order, Great Britain, supported by the United States, attempted in 1954–55 to organize a Muslim anticommunist containment alliance, centered in Iraq and called the Baghdad Pact.

Nasser, fearing a reinforcement rather than a decline of Western, particularly British, power in the region, succeeded in mobilizing the Arab masses in Jordan, Syria, and Saudi Arabia against the installation of Western bases in their countries. He succeeded despite sympathy for the concept by several of the leaders (except Syria). In turn, the Soviet Union rewarded Nasser for his obstruction of Western bases in the region with major arms supplies. This shattering of the Western military and diplomatic monopoly in the region led to the unsuccessful invasion of the Suez region and Sinai by Great Britain, France, and Israel.

The United States and Soviet Union each applied pressure on the three invaders for similar, albeit competitive, reasons; each wanted to expand its influence in the Arab world at the other's expense, as well as that of Great Britain, France, and Israel. Thus, détente between the two great powers did not follow, and one major result of this hectic cycle of activity was a marked belittling of diplomatic power surrounding the core Arab-Israeli dispute. In short, the Western monopoly, as well as unity of purpose in the region, was shattered, and *both* major actors in the cold war had extended their rivalry from Europe and northeast Asia into the Middle East and southwest Asia.

The second consequence of this series of events was the beginning of an accelerated and comprehensive regional arms race that has not abated to the present time.[5] Israel's decision to attack Egypt in 1956 was based on several criteria. A preemptive strike against Egypt was chosen before its army had time to assimilate the Soviet arms and exercise its own first strike.

Third, the Soviet rearming of Egypt as well as the groundswell of Arab popular support for Nasser following the Western withdrawal renewed the possibility of war for Egypt. Whereas the economic boycott of Israel (that included blockades in the Suez Canal and through the Gulf of 'Aqaba) and the support of Palestinian border raids had characterized Arab attrition tactics during the early 1950s; these tactics were never perceived as part of a grand strategy to eliminate the state. Rather, they served the purpose of maintaining the status quo.

Arab Strategy from 1957 through 1967

Egypt's grand strategy had room within it for parallel sets of maximal and minimal goals. The maximal goal was to liquidate the state of Israel.[6] The minimal goal sought to alter Israel's boundaries by seizing the south Negev, enabling Egypt to have territorial contiguity with Jordan, the oil-rich state of Iraq, and the Arabian Peninsula.

From a military point of view, the means required to reach this more limited goal were similar to those needed for the maximal; in either case, the IDF would have to receive a crushing blow. However, Nasser acknowledged there were some circumstances wherein the minimal goal could be reached without a total victory over Israel. These circumstances would include an early Egyptian ground gain followed quickly by the intervention of the great powers. Under these circumstances a negotiating situation might occur with Egypt trading a nonbelligerency pact, short of both a peace treaty and the recognition of the state of Israel, in exchange for the territorial gains made in the early stages of the war. However, Nasser appeared wary of this as an overall strategy and opted instead for what Gen Yehoshafat Harkabi has called a "war a' outrance," a war to the bitter end.[7]

Nasser considered the liquidation of Israel indivisible, requiring one major stroke; not an incremental strategy as part of an overall protracted process. The choice between liquidation "by event" versus "by stage" dominated Arab debate until the 1967 war. Those advocating a protracted approach expressed it in two forms during the 1960s. The former president of Tunisia, Habib Bourguiba, expressed one form when he proposed in 1965 that Israel accept and implement the first United Nations resolution (of 1947) for partitioning of and withdrawal to those boundaries called for in the UN debate. This was in exchange for an undefined "form of peace." Bourguiba insisted that if Israel accepted the resolution, it would be dramatically weakened for a later stage, a stage that Bourguiba alluded to but did not detail. If Israel rejected the overture, the Arab manifestation of peace then would constitute an Arab diplomatic victory in the international community. However, Nasser rejected the entire enterprise since he believed that an Israeli acceptance could actually lead to a political settlement, one that the international community would find in its own interest. In short, the first stage in the process of weakening Israel for a future round might actually terminate the conflict and allow for the existence of Israel.[8]

The second type of phased process that Nasser opposed was the "incremental violent process," an openly protracted struggle that was proposed by Syria at frequent intervals between 1959 and 1966. Syria proposed starting limited military actions that would incrementally weaken Israel. Nasser, however, feared Israel's escalatory ability and believed that although the Arab states could sustain and absorb battlefield defeats as well as material losses better than Israel due to the demographic asymmetry between the two societies there was a chance of devastating internal political effects on the Arab societies.[9] The consequences of the June 1967 war justified these fears.

Thus, Nasser believed that the war against Israel should be protracted because the Jewish state would be gradually weakened (morally, financially, and psychologically) by economic embargo and guerrilla warfare while Arab unity and internal development would strengthen the Arab side. At the right time, and only the right time, a conventional military strike would win the war.

Nasser eventually conceived of four interacting conditions as essential for his long-term grand strategy. First, the Egyptian home front had to be politically consolidated; there could be no widespread social fissures as existed in Syria, Iraq, and the Sudan. Second, the Arab side would have to acquire superior military force. Third, there had to be a united Arab front. This meant a revolutionary change in the social structure of the Arab world since "the unity of the objectives" required homogeneity of regimes. In this estimate, the modernization of the Arab world was a prerequisite to the showdown with Israel. Fourth, the United States had to be neutralized. Conditions had to prevail in which the United States would not intervene to save Israel from military defeat. Nasser also wanted to use the Soviet Union to nullify an intervening American power.

Israel's Strategy, 1949–67

From 1949 to 1967 Israel also underwent a debate regarding its own strategy as to how to preserve and consolidate the status quo. Prior to 1967 Israel had nothing to offer the Arab states in exchange for peace and viewed the surrender of any of its 1949 territories (as proposed by the Bourguiba plan) as the first increment of an Arab strategy meant to weaken Israel "a piece at a time" (known as the "salami tactic"). Therefore, deterrence, through a variety of means, characterized the Israeli strategy for these two decades. The security debate within Israel was over these means.

Until 1955, Israel attempted to attain security by joining a collective alliance. In vain, it tried to attain membership in NATO as well as a mutual defense pact with the United States. From 1956 onward, it altered its approach to one characterized by self-reliance but with a close relationship with one great power that would lend it diplomatic support in the United Nations and afford it military and economic aid. France played that role for a little over a decade, beginning shortly before the 1956 war, and the United States followed suit after the 1967 war. However, the most important and interesting aspect of the Israeli conception of deterrence was the internal debate over a nuclear versus a conventional strategy.[10] The debate, as far as one can tell from limited public sources, was formed by two divergent approaches to deterrence—compellence and conflict resolution. In their purest form, one resolution is represented by Ben-Gurion's strategy and the other by general, and later labor politician, Yigal Allon.

Shlomo Aronson's voluminous research led him to conclude that Ben-Gurion opted for an "opaque" nuclear policy; he chose to proceed with the development of the weapons and delivery systems but to use the concept of a "bomb in being" in lieu of developing a public war-fighting doctrine.[11] In this context, opacity meant that Israel would not openly threaten the Arab side with nuclear strikes since this would both humiliate the enemy *and* hasten its own quest for a nuclear arsenal which he believed inevitable in the long run. Instead, Israel would allow the enemy sufficient information regarding its nuclear capacity to influence the Arab will to fight.

Second, Ben-Gurion believed that Israel's possession of nuclear weapons would be necessary, although not sufficient, to make the Arabs *eventually,* albeit grudgingly, accept Israel's existence in the region. Israel's diplomatic positions, including maintaining the principle of the partition of Palestine, in conjunction with the opaque nuclear strategy, would eventually bring about a change in the Arab position regarding the acceptance of Israel. Thus, Ben-Gurion's long-range strategy pulled closely together the two concepts of "deterrence" and "conflict resolution."

Ben-Gurion's position represents one of the two major schools of thought within Israel's strategic elite, who saw the conflict as one requiring prudently managed protraction. Yigal Allon articulated the contending school of thought. He consistently advocated a strategy characterized by *conventional* deterrence and compellence. He believed that if Israel started the regional nuclear race, it would ultimately lose it. Basing his position on his historical-cultural assessment of the Arab world, Allon believed that the nuclear "balance of terror" achieved between the East and West was not replicable between the Arabs and Israelis if and when the Arab side gained access to nuclear weapons.

Allon for his part contended that once Israel obtained a bomb, Egypt would do all it needed to provide one for itself and that Nasser would quickly use it since Israel's small size prohibited a second-strike capability. Allon also believed that a "balance of terror" was impossible in the Middle East for more profound cultural reasons. Part of this second calculation was based on his belief that the Arab value system would be willing to suffer the enormous casualties accompanied by nuclear warfare—so long as Israel was destroyed—and therefore would not be deterred by an Israeli retaliatory capacity. In either case, Israel had to prepare itself for a lengthy and manageable conflict that should avoid nuclear arms as a strategic component and advocated Israel "maintaining a last resort" option if the Arabs did go nuclear. However, he came to base his overall strategic doctrine on the principles of Israel's striving for conventional superiority in military terms, improved territorial holdings for security purposes, and the ability to trade extra territory for peace.

Allon succeeded in prevailing over Ben-Gurion and laid the basis for Israel's conventional strategic doctrine in the early 1950s. Five major factors provided the rationale for his particular conceptions of security, war doctrine, and the planned structure of the IDF. Central to this study was Allon's own concept of the "war of attrition." Essentially, Allon contended that Israel could not create a doctrine that itself was based on the belief in a "final decision" over the Arab side. This final decision would never be actualized due to the material asymmetry between the two sides.

Allon's strategy of protraction was based on his belief that in the long run deterrence would eventually lead to resignation on the part of the Arabs and that resignation would ultimately lead to acceptance and peace. The deterrence would be made up of astute political maneuverings and *an unknown but manageable number* of Israeli battlefield victories over an *unspecified but reasonable* period of time. Allon believed this application of deterrence-compellence would eventually alter the Arab strategic calculus. Finally, if Israel captured and occupied Arab territory beyond the 1949 armistice lines, they would provide the final lever that would extract a contractual peace from the surrounding Arab states.[12]

Finally in 1962 Ben-Gurion agreed that the Israeli nuclear program and strategy did face some of the limitations elaborated on by Allon and agreed to an enhanced conventional effort. (It was Moshe Dayan who asked for a reduction in Israel's conventional capabilities and who actually claimed that Israel should give up its doctrine of opacity [attributed to his mentor, Ben-Gurion] and opt for an open nuclear strategy. Dayan's view was based on his own contention, shared by Ben-Gurion, that Israel could not economically sustain a conventional arms race with the surrounding Arab world, given the asymmetrical land, economic potential, and demographic imbalances. Dayan's view was rejected.) The strategic debate ended in compromise—"an undeclared bomb with an enhanced conventional effort."[13]

The War of 1967

Given the overall strategies of the two contending sides, the 1967 war was inevitable. While Nasser may have wanted to await a more propitious time (in terms of Arab strength and unity of ranks), he was captive to too many smaller actors in the Arab system. By retaliating against these actors, Syria and al-Fatah, Israel stayed true to its doctrine and made it difficult for Nasser not to become prematurely overengaged.

Both al-Fatah and the PLO (before their merger in February 1969) appeared to want to instigate a conventional war out of their fear that the Arab frontline states would abandon the national cause once Israel "went nuclear."[14] Syria, as already noted, also favored an escalated conventional attrition of Israel.

President Nasser of Egypt scorned precipitous action against Israel. His initial actions placed the two sides on their collision course despite the lack of the conditions that he believed Egypt required for initiating a "war of destiny" with Israel. Nasser calculated that by closing the Gulf of 'Aqaba (one of the *casus belli* in 1956) and concentrating Egyptian troops in the Sinai he could gain a *diplomatic* victory over Israel by contending that Egypt's forward deployment (and subsequent withdrawal) stopped Israel from launching a major offensive against Egypt's sole Arab ally, Syria. (While there was no immediate evidence that Israel was planning such an offensive, observers noted a marked escalation in both clashes and rhetoric between Israel and Syria since 1966.)

Nasser had alienated the conservative Arab world by his costly intervention in the Yemen civil war from 1962 onward and indeed needed any type of visible gain. Or, Nasser may have come to believe that a limited war, with accompanying limited Pan-Arab gains, was possible because of his misperception that he had more Soviet support than really existed.

Confronted by this challenge, Eshkol expanded his left-center cabinet to include Dayan (to replace him as minister of defense) and Menachem Begin, both maximal territorial expansionists (to add to the "minimalist" expansionism of Allon). This new "wall-to-wall" coalition (minus the Israeli Communist party) conducted the preemptive war of June 1967. Dayan believed that Israel could win by conventional means and that a defensive posture (that may have been recommended by his former mentor, Ben-Gurion) was inappropriate for the challenge as the IDF would have lost its deterrent value, despite the opaque threat of a nuclear strike. The result was the destruction of the three frontline Arab armies and the occupation of the Golan Heights, West Bank, and Arab Jerusalem, as well as Sinai and the Gaza district. This occupation and the forces that it catalyzed altered the Arab-Israeli conflict in several fundamental ways and brought it into a phase that has lasted until the present time.

The Year 1967 to the Present

The same questions arise from the consequences of the 1967 war as from the 1948 war: why didn't the outcome of the war end the conflict? and why did the "no war, no peace" stalemate set in again creating the conditions for the 1973 war?

Regarding the first question, the Soviet Union immediately rebuilt the Egyptian and Syrian armies and rescued their political leaderships from a feeling of total helplessness. This same material and diplomatic support by an external power enabled the Arab side to continue its protracted struggle despite the severe losses of the 1967 war. Furthermore, even during the course of the war, Israel did not seize as much territory as it could have. Thus, with the advantage of hindsight, an observer could say that Israel didn't attain sufficient leverage to impose the start of negotiations.

In addition, by war's end the Soviet Union was threatening to intervene—a replay of the veiled threats of Great Britain in 1949. The convergence of these factors saved the Arab side from the feeling of powerlessness and the need to settle immediately under these highly unfavorable conditions.

Thus, the questions for the Arab side could be reduced to the following: (1) Was the stalemate of lost national territories possible to live with? (2) Were the prospects for a protracted war good? and (3) What were the potential costs and returns of a peace strategy?

The New Arab Calculus

Given the profound material and economic losses suffered during the war, Egypt had to choose between forms of coercion against Israel and real peace. The former included a war of attrition along the Suez Canal, the support of the Palestinian guerrilla groups to irritate Israel, and a limited conventional war (as in 1973) to catalyze a peace process. However, Egypt could not tolerate an indefinite occupation of the Sinai accompanied by the closure of the Suez Canal as well as the prolonged evacuation of its industrial cities along the canal's West Bank.

Syria, on the other hand, not only had a stronger ideological commitment to "Arabism" and the Palestinian cause but could sustain the (much less significant) material loss of the Golan Heights to a far greater extent than Egypt could endure its own losses. Thus, Syria sustained the status quo of "no war, no peace" (broken by Israel during the 1982 invasion of Lebanon and not by itself) until the collapse of the Soviet Union as its financial and diplomatic support system in the late 1980s. The loss of the Soviet support system in the late 1980s, followed by the United States's destruction of Iraq's offensive capability, altered Syria's own balance of incentives and drove it toward the negotiating table in 1991.

Jordan, for its part, occupied a somewhat middle position between its two stronger frontline confrontation states. While Jordan's material and economic losses surpassed even those of Egypt, the country's incentive toward retrieving its territories through a diplomatic approach to Israel was more than neutralized by the demographic split on the East Bank between the original East Bankers and Palestinians. This split occurred when the return of Palestinian nationalism was a central issue of the overall conflict.

Therefore, three major changes occurred in the conflict after 1967. The first, Israel's capture of Arab state territories, opened the possibility of a trade of "land for peace." However, the other two changes have complicated this "rational-state" model of analysis. After two decades of dormancy, the Palestinian issue, as *national* issues, rather than a refugee issue, reemerged central to the conflict. As the PLO gained legitimacy as the "sole legitimate representative" of the Palestinian people, it found that Israel and Jordan could not trade "land for peace," as Israel and Egypt had done. The PLO exasperated this condition when it refused to recognize Israel's right to exist, within *any* boundaries until December 1988, following the first year of the intifada.

Third, the occupation of the West Bank and Gaza district stimulated the growth of both revisionist and religious Zionism within Israel—smaller political and social groups that had lived under the shadow of the dominant Labor Zionism since the 1920s—but movements that had never given up their aspiration for a whole land of Israel. Therefore, international, regional, and domestic forces pulled the states central to the conflict in contradictory directions, both toward conflict resolution and continued conflict, but away from the status quo.

The Arab States

From Egypt's perspective, the prospects of the status quo prevailing in the post-1967 period were poor. These prospects had built-in ingredients that would lead it toward either war or peace. Both the war of attrition conducted along the cease-fire line of the Suez Canal and the Egyptian support of Palestinian *fedayeen* raids from Jordan were asymmetrical in the nature of their action and the Israeli reaction.

In the war of attrition, which spanned from 1968 until 1970, Egypt used its large standing army and long-range, heavy, Soviet-made artillery to barrage the Israeli side of the canal to inflict an intolerable level of Israel Defense Force (IDF) casualties and force Israel to soften its own demands. Israel, on the other hand, relied, then as now, on a small standing army, since a mobilization of its reserves (i.e., nearly its entire civilian male population) is possible only in time of full crisis since its economy can not bear the strain of too frequent interruption. Thus, Israel, which had a much less substantial artillery corps than Egypt, responded to the Egyptian artillery barrages by using its air force as mobile artillery. Soon the Israelis were bombing deeper into the Egyptian heartland until the Soviet Union intervened with both pilots and weapons systems to save the regime of Gamal Nasser.[15] In sum, the military responses were different from the original actions in scale and elevated the "limited" war of attrition to a new and far more dangerous level. What had begun as a form of protracted strategy for Egypt threatened to erupt into a full, regional, and even international conflict by 1970.

The same escalatory cycle of action and reaction occurred with the guerrilla raids from Jordan by the Palestinian organizations. Israel began striking and punishing Jordanian as well as Palestinian assets across the river to force King Hussein's army to police the Palestinian raiders at the incursion's original source of location. When the United States came forth with a peace initiative by Secretary of State Rogers to cut short the escalating war of attrition between Israel and Egypt, as well as to initiate a peace process, the Palestinian guerrilla organizations in Amman Jordan attempted to block the negotiations by skyjacking several international aircraft and landing them in Jordan. King Hussein decided that the time was right to crush the Palestinian "state within a state" that had formed since 1967 and the result was the 1970 Jordanian civil war.

These chains of events actually confirmed Nasser's earlier apprehension of using incrementally calibrated coercion against

Israel as part of a protracted strategy. He had resisted Syria's call for "incremental violence" against Israel during the 1960s, prior to the June 1967 war, on the grounds that the Arabs would have no control over Israel's escalated level of retaliation. In addition, he had feared the domestic political consequences of a long, drawn-out, costly struggle with Israel. The weakening of his own regime and the Jordanian civil war indicated that his fears were justified and that another negative consequence of an extended conflict, characterized by both protracted and prolonged features, is the spread of the conflict across the region and the enhancement of domestic sociopolitical strains.

The talks were stillborn for several reasons. The paramount reason was the death in 1970 of President Nasser during his attempt to negotiate a halt to the Jordanian civil war. However, despite the fact that the combatants felt relief from the respite of the limited wars along Israel's southern and eastern borders, the renewed status quo was, again, politically and economically intolerable for Egypt, and to a lesser extent, Jordan and Syria. Of particular importance was the Egyptian fear that the longer the new cease-fire held, the more the world would accustom itself to shipping its merchandise around the Suez Canal rather than through it. Of a more profound nature, Arab governments feared that the international community might become accustomed to Israel's consolidation of its 1967 territorial acquisitions as it did to those added following the 1948–49 war.

Of the three confrontation states, Syria could bear the economic costs of the stalemate best of all and therefore didn't accept the 22 November 1967 United Nations Resolution 242 until the midst of the 1973 war. However, Syria needed assistance from other Arab states in its fight against Israel, as the Golan range was too narrow a front. Help finally came in 1973 when Nasser's successor, Anwar Sadat, reached a point of almost final desperation in Egypt.

The Israeli Perspective

The territorial consequences of the 1967 victory altered Israel's own strategic balance of incentives. To begin with, the territorial alterations gave Israel the "strategic depth" it required and therefore provided a doctrinal victory of the "conventionalists" over those advocating a more active nuclear deterrence policy. While this turn of events didn't halt Israel's development of its nuclear capability, it gave Allon's doctrinal conventionalists the upper hand in keeping "the bomb" as the weapon of last resort.[16]

The Allon faction within the Labor alignment called for a partition of the territories with Israel retaining those parts required for security in return for the heavily Arab populated sectors to contiguous Arab states. Allon sought this in return for peace and recognition of Israel. However, no Arab governments were ready to negotiate with Israel in principle, nor were any governments ready to divide the territories that they had just lost. At the same time, Moshe Dayan ascended within the rival Ben-Gurion wing of the Labor movement. Supported by Shimon Peres, Dayan favored a permanent Israeli presence in the West Bank accompanied by an autonomy scheme for the Palestinian inhabitants but with neither full self-determination for the Palestinians nor a return to Jordanian rule.

Just as the Labor party was split over the future of the territories, increasing numbers of Israelis shifted toward such parties as the Herut (now the Likud bloc) that had held onto their Greater Israel creed since 1949. Israel now held the lands that these parties had consistently claimed as fulfilling the Jewish historical heritage. Thus, just as the leadership of the central Arab state, particularly Egypt, began moving (sincerely or not) toward a form of conflict resolution based on a formula of trading land for peace as well as undefined "justice for the Palestinians," it noticed that increasing numbers of Israelis, for historical, strategic, and religious reasons, became fixed on the retention of the "trump cards" that would allow for such a trade. The result was the spread of Jewish settlements within and outside the areas Allon said Israel required for strategic defense and deterrence.

The 1973 War

The 1973 war was fought, from the Egyptian and perhaps even the Syrian perspective, for political objectives rather than the traditional Arab war aim of eliminating Israel. The aftermath of this war once again altered the strategic calculus for each of the regional participants. The peace process between Israel and Egypt reflected several complicated factors. Israel made full territorial concessions in the Sinai because of mutually acceptable security arrangements within the returned territories. In addition, while such important civilian settlements, as Yamit had been established in the Sinai, the territories captured from Egypt in 1967 had little historical or religious significance for Israel. In exchange for the Sinai, Israel obtained what the now-ruling Likud bloc considered a separate peace with Egypt, allowing Israel greater latitude to its north and east.

From the Egyptian perspective, Sadat had delivered the necessary psychological victory to his public at the outset of the 1973 war, enabling him to make the concessions essential to retrieve national territory and start the arduous task of restoring the economy. However, this "trade" would not have been possible without the Israeli nuclear shadow that convinced the Egyptian leadership that a more complete victory over Israel was impossible. This shadow drove a wedge between Sadat and the more militant Arab leaders. At the same time, and even with Israel's nuclear arsenal, the process probably would have still run aground without the unprecedented level of United States diplomatic involvement throughout the talks.

Egypt's withdrawal from the conflict altered the balance of incentives for the remaining Arab states. However, the combination of the American effort, the pressures emanating from Arab economic hardship, and the Israeli nuclear deterrence (opaque as it was), as well as the altered regional balance of power, was still not sufficient to further the peace process. These factors that seemed to reinforce the peace incentives were countered by a formidable array of forces that continued to obstruct progress. First, there was the far different Israeli view of the remaining occupied territories for which Israel had more intense ideological and historical ties. Second, these remaining territories had more strategic importance than the returned Sinai. Third, the formidable domestic and inter-Arab political forces, particularly within Jordan and Syria, opposed recognition of Israel.

The continued Soviet assistance to Syria and the PLO enabled them to continue the status quo, if not to wage war. While the status quo was no more uncomfortable for them than before the Egyptian-Israeli peace, it appeared unlikely that a war option was feasible. Confronted with this reality, Syria attempted to create an eastern-front coalition that had both offensive and defensive capabilities and one that would compensate for the loss of the western (Egyptian) front.

The United States became increasingly central to any hypothetical peace scenario as *both* Israel's and Egypt's economic dependence on Washington grew following the 1973 war. Israel became increasingly tied to the American economic and military lifelines, despite increased efforts to minimize its arms dependence on the United States through the expensive development of its own indigenous weapons industry.[17] However, well into the 1980s the Israelis prevented the US from using this economic dependency to push it toward an unfavored settlement by relying on the pro-Israeli lobby in the Congress and the powerful anti-Soviet instincts and policies of the Reagan administration.

Because of these instincts, the Likud bloc believed the United States had given it the license to enter Lebanon in 1982 to eliminate the PLO and push the Syrian army out of the small buffer state.

The 1982 Israeli Invasion of Lebanon

Israel sought to redraw the political map of the Arab states to the east and north of Israel during the war of Lebanon. By crushing the PLO militarily in Lebanon and by forcing it to remove its political headquarters from Beirut, Israeli Minister of Defense Ariel Sharon, supported by Prime Minister Begin, wanted to fully and permanently suppress any spirit of Palestinian nationalism in the occupied territories and eventually transfer the Palestinian national problem into the East Bank Hashemite Kingdom of Jordan. Within the Likud bloc, Ariel Sharon most openly and ardently advocated the

"Jordan is Palestine" thesis, desiring the establishment of a Palestinian government and state with its capital in Amman, a stroke that envisions ending, once and for all, the international and Arab pressure for Palestinian statehood within the occupied territories for the Palestinian people.[18]

However, the initial success the IDF experienced against both the Syrian and PLO forces during the war was more than offset by the setbacks the IDF faced during the following occupation. Having misread the degree of support the Christian right could and would offer Israel in the remaking of a "new order" in Lebanon and having overlooked the revolutionary changes that the poorest and most populous community in Lebanon, the Shi'ites, had undergone, the IDF was incrementally forced southward by acts of violent resistance and terrorism until it was left with the security zone that it currently occupies with the proxy support of the Christian-dominated South Lebanese Army (SLA).

This swirl of events set in motion several forces within the Palestine national movement and Syria that remained under the surface for much of the remainder of the 1980s but, when reinforced by other regional and international currents, led to a new strategic calculus for both the PLO and Syria.

As for the Palestinians, the 1977 visit by Egyptian president Sadat to Jerusalem and the peace treaty that was produced with Israel two years later split the Palestinian leadership ranks. Some observers saw the end of armed struggle and openly advocated coexistence with Israel and an independent Palestinian West Bank state. Other observers refused to accept the notion that the Palestinians did not have a military option against Israel. These internal PLO conflicts surfaced following the Israeli invasion of Lebanon.

Negotiations in Amman between Arafat and King Hussein regarding future relations between the two peoples and banks of the Jordan River following a hypothetical Israeli withdrawal led to a military clash between Palestinian rejectionists, supported by Syria, and the forces still supportive of Arafat. The PLO and its leader were expelled a second time from Lebanon, this time by Arab forces from Tripoli.

Two lessons emerged for the PLO leadership from the Lebanese quagmire. First, when it was under assault by the

IDF *within* an Arab state, no Arab army came to its defense; the Syrian armed forces merely defended itself when attacked by Israel. The follow-up Syrian assault on the residual PLO in Lebanon reinforced this sense of isolation when, again, no Arab state attempted to block Syria's actions. Second, the PLO learned that an occupied people, in this case the Shi'ites of Lebanon, could create sufficient penalty and pain for an Israeli occupation *if* the occupied party itself was willing to pay an even greater price by forcing the Israelis to recalculate the costs and benefits of retaining the occupation. This was one of several important inputs into the intifada that began in December 1987.

The Intifada

There is a multifarious variety of immediate as well as more underlying developmental causes to the outbreak of the popular uprising (known as the intifada) in the occupied territories.[19] As for the deeper underlying causes, the following considerations should be taken into account: The nationalistic impulse of a people under occupation for over two decades that includes the built-up frustration and despair that accompanied the occupation; the development of a new generation of local Palestinian leadership in the territories, a leadership, however, that was structurally organized through many new voluntary associations, such as unions; the deteriorating living conditions, particularly in Gaza, caused by the population explosion and the failure of structural economic development to keep pace with the demographic increases; and the decline of the IDF's deterrent profile based somewhat on the lessons of the Lebanon experience as well as more localized episodes that included the General Security Services (Shabak).

The more immediate causes were also domestic and regional in nature and included: the April 1987 London Agreement between then Foreign Minister Shimon Peres and King Hussein that reinforced Israeli and Jordanian control over the Palestinian political future (the implementation of the London Agreement was torpedoed by both Prime Minister Yitzhak

Shamir and Yasir Arafat); the influx of released Palestinian guerrillas from Israeli (and South Lebanese) prisons into the occupied territories, as part of a prisoner exchange; and last, the November 1987 Arab summit conference in Amman that placed Arab support for Iraq against Iran as the highest "national" priority and gave little attention to the worsening Palestinian plight in the territories.

There was, and remains, divided opinion within Israel as how to respond best to the intifada. Political hardliners advocated a full-military response and a rapid repression. The IDF general staff, supported by the then-minister of defense Yitzhak Rabin, preferred a containment policy that utilized many tactics but which would keep casualties low, on both sides, and would simultaneously attempt to limit the erosion of international good will that was sure to accompany the containment of the uprising. Therefore, the IDF, border police, and Shabak relied on methods that ranged from live ammunition to rubber bullets, beatings, the use of Palestinian informants (resulting in mass arrests), the closure of universities and schools, and economic and financial embargoes.

On the other side, the Palestinians formed the United National Command (UNC) which attempted to organize as well as mobilize a mass base in the occupied territories. The goal of the internal Palestinian structure was to increase areas of Palestinian authority and wrest control of functions as well as territory from the IDF and the civil administration. At first, the UNC demonstrated a good deal of decision-making independence from the PLO-Tunis.

The UNC stoned IDF personnel as well as Jewish-settler civilians who traveled through the territories; printed and distributed pamphlets throughout the population to create unified action as well as to demonstrate to the Israeli authorities that UNC had control over the Palestinian population rather than the occupying forces; attempted to set up an embryonic autonomous Palestinian economy in the territories; and used civil disobedience through the nonpayment of taxes to the civil administration.

Conflict on each side has periodically escalated: the Palestinians with the use of firearms and firebombs and the Israelis with the demolition of homes and the expulsion of suspected intifada leaders. These actions have caused a stalemate. The Israelis have learned to live with the costs and discomforts of the intifada in terms of a low level of casualties, a significant percentage of the annual defense budget, a cleavage within the security establishment in regard to the continued value of the occupied territories, and a pronounced decline in public morale. The Palestinians have clearly registered certain gains. From their perspective, they have, in a certain manner, recreated the "green line" (the 1967 boundary) because travel by Israelis into the territories after the start of the uprising is limited to the settler population and the army. In addition, Palestinians, at great human and material cost, have succeeded in placing their cause near the top of the international agenda; they also have raised the cost of the Israeli occupation.

Of greatest significance on the diplomatic level has been the development of a parallel Palestinian leadership within the territories during the intifada that forced Yasir Arafat to finally accept a two-state solution in public at a UN General Assembly meeting in Geneva in December 1988. This development in turn began a series of official negotiations between the PLO and the United States that were discontinued when the PLO failed to officially and publically denounce Palestinian acts of terrorism against Israel that followed.

The Consequences of the 1991 Persian Gulf War

While the intifada met with certain important successes during its early stages, it ultimately has not been capable of altering the strategic or economic calculus of Israeli policymakers. Failure to alter this balance of incentives has thrown the Palestinians, once again, into a dependency relationship with the Arab "confrontation" states and the United States. However, the allied destruction of most of the offensive capabilities of the Iraqi armed forces in turn has reinforced the shifting strategic orientation of the other key Arab confrontation state, Syria. The results of the Western military reaction not only eliminated Iraq as an "expeditionary" force member of an eastern front against Israel but also further placed the Soviet Union on the diplomatic sidelines in the region.

Furthermore, the rapidly deteriorating economic (as well as political) situation within the USSR increased the Soviet reliance on the West for its economic reconstruction, and this economic crisis, in turn, reinforced the Soviet-political forces who favored enhancing the flow of Soviet-Jewish emigration to Israel. Any of these fast-moving events and forces had the potential to alter the policy frame of references of the major actors in the region. Since they occurred in such a compressed period of time, they created a unique situation for breaking the status quo of the Arab-Israeli conflict.

The Arab *state* perspective found the war option far less realizable in the present and the foreseeable future than in recent memory. However, the Arab states, with the exception of Jordan, can live with the status quo, if a peace process fails to produce results, until the point that Israel assimilates or annexes the occupied territories. At the point of Israeli annexation, the possibility of future peace not only would be eliminated, but the Arab world would be impelled to plan and prepare for a war similar to that of 1973.

The Palestinians, on the other hand, not only have no war option but can only see the extension of the status quo, under the current circumstances of the rapidly accelerating settlement policy within the territories, as disastrous to their future hopes for self-determination of any sort west of the Jordan River. Yet, Yasir Arafat's support of the Iraqi invasion of Kuwait reinforced the mistrust of the Israeli "man in the street" regarding the true intentions of the Palestinians despite the PLO's official recognition of Israel's "right to exist" in December 1988.

The United States and the Future Balance of Incentives

The Arab-Israeli conflict became prolonged because of the particular confluence of domestic, regional, and international forces. This confluence has both shielded the Arab side from having to make domestically unpopular political decisions aimed at terminating the conflict as well as reinforced its ability to pursue a protracted strategy. For its part, Israel may have wanted to force a conclusive decision at any point during the decades-long conflict but realized that the demographic and geographic asymmetries between the antagonists prohibited such a conclusion. This led to a protracted defensive strategy on the part of Israel; however, following the successful offensive of 1967, a significant growth began of the social and political forces aspiring to retain the captured territories.

The cultural and psychological views each side had of the other also influenced the duration of the conflict. For a long time, the Arab side saw the Israelis as a "Jewish version" of the crusades. By believing that Israel was an artificial entity, foreign to the history and culture of the region, Arab political leaders and intellectuals believed that time was on their side, as the "artificial entity" would weaken from its own internal contradictions and from Arab tactics. This assumption reinforced the view that a protracted strategy was conducive to the "withering away" of the artificial entity.

As noted, just as the Israeli strategy appeared to be bearing fruit and the ideological fervor in parts of the Arab world seemed to be losing some of its strength, changes occurred within Israel's domestic political system that reinforced ideological claims to the very territories required for a future bargain with the Arab side. Thus, one may conclude that one consequence of this prolonged conflict, best characterized by protracted strategies for each antagonist, is the formation of domestic forces on either or both sides who in turn alter their policymakers' agendas in dealing with the other side.

External forces helped to prolong the conflict by giving each side the means to exercise their protracted strategies. At the present time, despite its hegemonical position in the region, the United States cannot sufficiently alter the balance of incentives for the local actors to bring about a change of heart. However, the United States can manage the status quo so that the local actors find it disadvantageous not to enter and stay in a peace process that takes on a life of its own. The process, if prudently guided, could alter the beliefs, fears, and ambitions of the respective populations and could lead to a comprehensive regional settlement that could allow for Israel's real security needs and the political rights of the Palestinian people.

Notes

1. J. C. Hurewitz, *The Struggle for Palestine* (New York: Schocken Books, 1976), 18.
2. Nadav Safran, *From War to War: The Arab-Israeli Confrontation, 1948–1967* (New York: Pegasus Press, 1969), 28–29.
3. Yoav Gelber, "Ben-Gurion and the Establishment of the IDF," *The Jerusalem Quarterly* 50 (Spring 1989): 62–63.
4. Safran, 39.
5. J. C. Hurewitz, *Middle East Politics: The Military Dimension* (New York: Praeger Pubishers, 1969).
6. Professor Yehoshafat Harkabi, former chief of IDF intelligence, *Arab Attitudes Toward Israel* (New York: Hart Publishers Co., 1971).
7. Gen Yehoshafat Harkabi, *Arab Strategies and Israel's Response* (New York: Macmillan Publ. Co., 1977), 9.
8. Ibid., 10.
9. Ibid. See also Harkabi's assessment of Nasser's preconditions for a lighting victory over Israel.
10. Shlomo Aronson, *The Politics and Strategy of Nuclear Weapons in the Middle East: Opacity, Theory, and Reality, 1960–1990—An Israeli Perspective* (New York: SUNY Press, forthcoming); Leonard Spector, *The Undeclared Bomb* (Cambridge, Mass.: Ballinger, 1988); Louis Rene Beres, ed., *Security or Armageddon: Israel's Nuclear Strategy* (Lexington, Mass.: Heath, 1986); and Robert Harkavy, "After the Gulf War: The Future of Israeli Nuclear Strategy," *The Washington Quarterly* 14, no. 3 (Summer 1991): 161–79.
11. Aronson, 467.
12. Allon's strategic concepts are developed in many professional forums. However, *A Curtain of Sand* (Tel Aviv: Hakibutz Hameuchad, 1959, Hebrew) remains the blueprint of his thoughts.
13. Aronson, 143–44.
14. For an organizational history of the Palestinian movements, see William B. Quandt, Fuad Jabber, and Ann Mosely Lesch, *The Politics of Palestinian Nationalism* (Berkeley: University of California Press, 1973).
15. Nadav Safran, *Israel, the Embattled Ally* (Cambridge, Mass.: Harvard University Press, 1978), 437.
16. Aronson, 221.
17. Stewart Reiser, *The Israeli Arms Industry: Foreign Policy, Arms Transfers, and Military Doctrine of a Small State* (New York: Holmes & Meier, 1989).
18. Ze'ev Schiff and Ehud Ya'ari, *Israel's Lebanon War* (New York: Simon and Schuster, 1984); and Itamar Rabinovich, *The War for Lebanon, 1970–1983* (Ithaca: Cornell University Press, 1984).

19. Books and articles abound regarding the uprising by Arab, American, and Israeli scholars. The Palestinian perspective is best articulated by a diverse and extensive collection of articles in the *Journal of Palestinian Studies: A Quarterly on Palestinian Affairs and the Arab-Israeli Conflict.* Ze'ev Schiff and Ehud Ya'ari have written *Intifada: The Palestinian Uprising—Israel's Third Front* (New York: Simon and Schuster, 1989); Don Peretz has contributed *Intifada: The Palestinian Uprising* (Boulder, Colo.: Westview Press, 1990); and Israeli Brig Gen (Res) Aryeh Shalev, *The Intifada: Causes and Effects* (Israel: Jaffee Center for Strategic Studies, Tel Aviv University, 1991).

Prolonged Conflict in the Sudan

Ann Mosely Lesch

Protagonists in internal conflicts that lead to high levels of violence tend to view their situation in zero-sum terms. Defense of national identity, the survival of the community, or the future of the economic class appear at stake in the struggle. Thus, although the intensity of the violence can fluctuate, the stakes are so high that a negotiated resolution is difficult. Powerful forces undermine attempts to resolve the conflict, in part, because the benefits from conflict are more immediate than the benefits from peace and, in part, because each group believes that only a clear victory for its own side can be the acceptable outcome.

When conflicts start, protagonists often expect them to end quickly. Governments may view a rebellion as an illegitimate action by outlaws that can be thwarted by police action or as a manifestation of grievances that can be contained by coopting rebels into the political system. Groups that challenge the regime may consider the system already so decayed that it can be easily toppled, may believe that a simple change in the top leadership will accomplish their goals, or may anticipate that the government will soon accede to their demands.

In some situations, however, rebels prepare a protracted strategy to achieve their aims. They anticipate a lengthy period of guerrilla warfare and other unconventional means of pressure through which public attitudes will be transformed and the morale of government officials and troops undermined. That strategy, when coupled with intensive diplomatic contacts, may lead to the anticipated transformation of relationships and foster alliances that cut across the violent divide. However, the strategy may also increase polarization and harden opposition in certain sectors of the public. Tactics on both sides may alienate rather than attract support and diminish the prospects for mutual accommodation.

Moreover, the country may bog down in a seemingly endless struggle, in which casualties, social dislocation, and economic hardships burden everyone. The struggle may become prolonged far beyond either side's desire to continue, with polarization so profound that neither side can end the contest without admitting defeat. No resolution would then be possible short of the total defeat of one side by the other.

The internal war that began in Sudan in 1983 fits the patterns of protracted conflict. The military officers who launched the guerrilla struggle sought to transform the country's political structure. They recognized that an extended time period would be required and systematically built up their fighting capacity. Their efforts brought them to the brink of success in 1989 when a broad range of sociopolitical groups pressed the government to negotiate a fundamental resolution of the issues that had caused the civil war. But the seizure of power by hardline politico-military forces that summer preempted negotiations and exacerbated polarization. The new government insisted on a total victory and wanted to impose its own ideological vision on the society, a vision diametrically opposed to the rebels and to the social forces that supported negotiations. Since then, the struggle has been prolonged in ways that destroy the already weak economy, undercut its sovereignty, and damage the body politic.

Overview

The Sudan is the largest country in Africa, covering 1 million square miles.[1] Its 23 million residents, scattered across that wide expanse, derive from a complex mix of ethnic groups. Arabic culture and ethnicity predominate, even though 60 percent of the population belongs to such African peoples as Nubian, Fur, Nuba, Dinka, Shilluk, and numerous others in the far south. Moreover, more than 70 percent of the citizens are Muslim by religion, including many non-Arab peoples. Traditional African religions predominate in the south, and perhaps 6 percent are Christian, of whom most derive from churches established during the Anglo-Egyptian condominium (1898–1956).

During the British-dominated colonial era, the western districts of Dar Fur and Nuba Mountains and the entire southern one-third of the country were kept isolated from the central Nile Valley. At independence they were incorporated into a unified political system, in which Arab politicians, with their well-articulated political structures and leading role in the negotiations for independence, assumed dominant positions. Political decisions were made in Khartoum and economic development was concentrated in the Nile Valley. Peripheral areas—which tended to be non-Arab and/or non-Muslim—felt marginalized.

Residents of the south rejected that status most forcefully. They waged guerrilla warfare for 17 years, from mid-1955 until 1972. Having been denied their demand to decentralize authority onto the regions, many southerners pressed for separation and the formation of their own state. They were also angered by government measures to Arabize the educational and administrative systems in the south and to restrict Christian churches. In the mid-1960s, officials and political groups in Khartoum began to respond to southern demands, when proposals for regional self-rule were discussed in a round-table conference and other political fora.

No conceptual breakthrough occurred until 1969, when the new military ruler, Col Ja'far Muhammad Numairi, declared that the government "recognizes the historic and cultural differences between the North and South and firmly believes that the unity of our country must be built on these objective realities. [Therefore, the south should have] regional autonomy within a united Sudan."[2] That approach was embodied in the Addis Ababa accord of February 1972 that ended the civil war. The entire south would comprise one region, with its own assembly and elected executive. The region had an independent budget and tax sources to control internal security and local administration in the social, cultural, and education fields. English, rather than Arabic, was recognized as the principal language in the south. (The Sudanese constitution of 1973 explicitly accorded respect to Christianity and traditional beliefs as well as Islam.) Moreover, the Addis Ababa accord specified that the guerrilla forces, known as the Anya Nya, would be gradually absorbed into the army and would serve in the south. Southerners thereby relinquished their demand for independence in return for gaining substantial self-rule and protection from pressure from the center.

Numairi never allowed the system to function as intended. He frequently interfered in the operation of the regional government to prevent independent power bases from emerging. By the 1980s, as Numairi strengthened his alliance with radical Islamic political forces, he actively undermined southern autonomy. That effort culminated in his unilateral redivision of the south into three provinces in June 1983. Redivision undermined the already limited self-rule and freed Numairi to institute a version of an Islamic criminal code in September 1983, a crucial step towards establishing an Islamic state in which he would be the *imam* (religiously sanctioned ruler), and non-Muslims would have second-class status. The "September laws" were widely opposed by Muslims in the north as well as by citizens in the south and contributed to increasingly overt public opposition that culminated in a popular uprising in Khartoum that overthrew Numairi on 6 April 1985. Public discontent was also galvanized by economic crises triggered by drought, failed agro-industrial projects, high-level corruption, and renewed war in the south.

Fighting erupted there for the second time after the forcible suppression of mutinies in Bor and Pibor on 16 May 1983. The soldiers, who came from Anya Nya units absorbed into the army after the Addis Ababa accord, had resisted illegal orders to be transferred north.[3] Rather than negotiating a resolution of the standoff, Numairi repressed the mutineers. After a day-long battle, the commanders and soldiers evacuated Bor and Pibor and regrouped in Ethiopia, where they coalesced with soldiers who had fled to the bush after earlier mutinies. Overall command was assumed by Col John Garang de Mabior, an officer from the absorbed forces who deserted his post in Khartoum to join the rebels. Garang welded the disparate troops into the Sudan People's Liberation Army (SPLA) with its political wing, the Sudan People's Liberation Movement (SPLM).

The Bor mutiny triggered an uprising that caused unprecedented political turmoil, social disruption, and economic collapse. Unlike Anya Nya, the SPLM did not want the south to secede but sought to restructure the bases of political power in

Khartoum. Its opposition to the September laws and Numairi's autocratic rule won support from the northern dissidents who led the popular uprising that overthrew Numairi in April 1985. But the SPLM felt the uprising was incomplete. Garang criticized the transitional government set up in April 1985, which combined a Transitional Military Council (TMC) with a civilian cabinet.[4] Even though the TMC pledged to return power to an elected government within one year and even though activists from the professional and trade union movements were influential in the cabinet, the SPLM mistrusted the transitional government. The power of Numairi's generals was intact: the SPLA could not forget that the chairman of the TMC had urged Numairi to suppress the Bor mutiny in 1983. And the September laws were not rescinded.

The SPLM also criticized the parliamentary elections held in April 1986, since the war prevented most southerners from voting and since no constitutional transformation had yet occurred. The elected prime minister, al-Sadiq al-Mahdi, also failed to reach an understanding with the SPLA. Head of the Umma party and great-grandson of the religiopolitical leader who had ousted the Turco-Egyptian rulers a century earlier, Mahdi articulated a vision of a liberal Islamic government that would respect the rights of religious minorities within a relatively centralized Muslim state. That approach was suspect to the SPLM as well as to regional and secular political groups. They argued that a political system had to be constructed that would reflect the multireligious and multiethnic realities in Sudan. The situation polarized further in May 1988, when the National Islamic Front (NIF) joined the cabinet on a platform committed to instituting a comprehensive Islamic legal system within two months. NIF, a pillar of the Numairi regime, had rejected negotiations with the SPLM that would restructure the political system.

Meanwhile, the conservative Democratic Unionist party (DUP) feared that its support among Muslim religious orders was being undermined by the NIF and that the Umma-NIF alliance would relegate the DUP to a minor role. DUP leaders were also concerned that NIF's absolutist approach would tear apart the country, and they believed that a pragmatic accommodation was required in the multiethnic Sudan. The DUP therefore negotiated a path-breaking accord with the SPLM in November 1988 that promised to freeze Islamic laws until a national constitutional conference could make fundamental decisions concerning the legal system and the nature of the state. Mahdi and NIF rejected the DUP-SPLM accord, which forced the DUP to pull out abruptly from the government in late December 1988.

By then the SPLA controlled 90 percent of the countryside in the south. A dozen army garrisons surrendered to its forces that winter. By February 1989 the commanders of the armed forces were fed up with Mahdi's alliance with NIF and his unwillingness to negotiate. They believed that a negotiated resolution of the conflict was preferable to an endless, draining, and unwinnable war. The defense minister resigned abruptly and nearly 300 senior officers issued an ultimatum to Mahdi in which they demanded that he negotiate peace if he could not arm them adequately. They pressed him to implement the steps needed to conclude an agreement and convene a constitutional conference. Under acute military pressure, Mahdi removed NIF from the cabinet and formed a broad-based government that began to adopt the measures specified in the DUP-SPLM accord. When the foreign minister met with SPLM leaders on 10 June 1989, they agreed to finalize arrangements on 4 July and to convene the constitutional conference on 18 September.

NIF rejected the terms of the officers' ultimatum, which required shelving Islamic law until the constitutional conference could decide on the fundamental bases for rule. NIF activists, recognizing that they could not institutionalize their views by democratic means, conspired with hard-line army officers to overthrow the regime. The coup d'etat on 30 June 1989 cast aside parliamentary institutions and banned all political parties and unions. The new leader, Brig Gen Umar Hassan Ahmad al-Bashir, tore up the DUP-SPLM accord and accelerated military operations against the SPLA. In 1991 NIF consolidated its hold by proclaiming Sudan an Islamic republic and organized it on a nominally federal basis. The central government retained overwhelming financial and executive powers, but states with non-Muslim majorities could exempt themselves from certain provisions of the Islamic criminal law. Popular committees were formed on the Libyan model to mobilize and control the public.

Despite the government's sweeping arrests of political activists and intellectuals, banned political and union forces created a national democratic alliance (NDA) in October 1989 that called for the restoration of democracy by a campaign of civil disobedience against the regime. The NDA charter was formally endorsed by the SPLM in March 1990 and, in September 1990, by the high command of the armed forces that had been ousted after the coup. The officers even urged army garrisons to stop fighting the SPLA and join forces against the government. Thus, the political forces that sought to restore democracy aligned with the leaders of the violent rebellion in their common aim of destroying the NIF-led military government.

The SPLM/SPLA had grown from a small band of mutineers in 1983 to a broad-based movement that controlled nearly all the south and allied with all the political groups opposing the regime. If that alliance were to hold together and overthrow the government, the SPLM might realize its far-reaching aspirations. If, however, the government warded off those challenges, the protracted conflict would continue to wreak havoc on the society and the economy. With government and opposition pursuing their struggle in zero-sum terms, no compromise appeared possible.

The Sudan People's Liberation Movement

The aims of the SPLM crystallized soon after its establishment in 1983. Garang articulated comprehensive goals: the creation of "a united Sudan under a socialist system that affords democracy and human rights to all nationalities and guarantees freedom to all religions, beliefs, and outlooks. A united and socialist Sudan can be achieved only through protracted revolutionary armed struggle. Peaceful struggle has always been met with ruthless suppression and callous killing of our beloved people."[5]

Goals and Structures

Garang emphasized that his aim was not merely to destroy "Numairi's one-man system of dictatorship" but also to overthrow "any other minority clique regime in Khartoum" that might attempt to replace Numairi. He stressed that the SPLM was not a southern movement that focused on regional issues. Rather, as a national Sudanese movement, the SPLM happened to emerge in the south where exploitation was especially intense: "The marginal cost of rebellion in the south became very small, zero or negative; that is, in the south, it pays to rebel." As such, the SPLM was the "vanguard movement for the liberation of the whole Sudanese people."

Garang declared that the "New Sudan" would be democratic and guarantee equality, freedom, and economic and social justice and respect human rights. The monopoly of power by any one group must end, whether that monopoly is held by "political parties, families' dynasties, religious sects or [the] army."[6] Consequently, Garang criticized the Transitional Military Council as a "gang of generals," Mahdi's Umma party and DUP as invidious exemplars of family dynasties linked to religious sects, and NIF as an ideological sectarian movement.

The SPLM also rejected tribalism and racial distinctions as bases of rule: "The emergence of regional political groups [is] a natural revolt against the appalling conditions in which the masses live in those areas," but those conditions cannot be overcome by viewing each region and group in isolation. Rather, "the root causes of Sudan's chronic social and political instability are essentially national. As such, they should be tackled nationally."[7] Once power was restructured in Khartoum, each region could achieve genuine autonomy. Then the central government would not monopolize power, and the economies of the less-developed peripheries would benefit. Since the SPLM rejected the limited approach embodied in the Addis Ababa accord, Garang criticized government proposals to negotiate solely concerning the south.

The SPLM's aims were all-embracing and highly political. Yet the movement was organized along military lines, since force was its primary means to pressure and overthrow the government. Diplomacy was initially viewed as secondary, since SPLM leaders believed that no negotiations would succeed without control over territory and considerable military leverage. Although a joint SPLM/SPLA high command governed the movement, the primary responsibilities of its senior members were to command particular battlefronts. Decision making was complicated and slow, since messages had to be sent to far-flung officers, responses collated, and further discussion carried out before agreement could be reached on major policies and diplomatic issues. Meetings were logistically difficult to arrange and relatively infrequent. Garang wielded special power. As the premier commander articulated the goals of the movement with authority, he provided direction of the overall military campaigns and served as the leading diplomatic envoy. Senior officers played prominent public roles in meetings with political groups and in negotiating significant agreements.[8]

Until May 1991 the SPLM had its political headquarters in Addis Ababa and maintained liaison offices in Nairobi and London. The Sudan Relief and Rehabilitation Association (SRRA), which provides humanitarian aid in the SPLA-controlled areas, also has offices overseas. In practice, SRRA operations are controlled by SPLA officers in the field, even though the SRRA is legally independent. Similarly, the SPLM did not develop an autonomous administrative structure in the territories under its control. SPLA commanders encouraged civil administrators, health personnel, and teachers to return to their posts, once the area had been secured by the SPLA. But no SPLM government was set up, even though a substantial number of former high-level administrators in the southern regional government joined the movement. Considerable tension between civilian cadres and leaders with solely military backgrounds therefore emerged.

Garang's concern for political coherence within the movement merged with his belief that maintaining the unity of political and military cadres was essential for long-term success. SPLM leaders remembered that the Anya Nya rebellion suffered from military fragmentation and the conflicting ambitions of rival politicians; the movement could negotiate effectively with the central government only after Col Joseph Lagu forcibly united the factions. The SPLM/SPLA faced competition initially from the reemerging separatist Anya Nya movement, called Anya Nya II, whose leaders had deserted the armed forces shortly before the Bor mutiny. They expected him—senior in rank and age to Garang and the 1,200 men from Bor and Pibor—to come under the authority of Anya Nya II. Instead, in Garang's words, the SPLA waged a "bitter struggle" from June to November 1983 before the "correct direction prevailed," and the SPLA killed or won over the "separatists, reactionaries, and opportunists."[9] The remaining Anya Nya II received arms and funds from the government; Anya Nya II was a low-cost way to harass the SPLA.

Nonetheless, after prolonged negotiations, the SPLM appointed the most effective Anya Nya II commander, Gordon Kong Chuol, to the SPLM/SPLA high command in January 1988. He led operations in his home district, fighting the army garrisons that had previously funded him. Only remnants of Anya Nya II remained under government control in Upper Nile. Those Anya Nya II members who joined the SPLA felt that their immediate interests coincided with the rebellion, even though they were not interested in the comprehensive ideology espoused by Garang; they emphasized the special needs of the south, its African heritage, and the establishment of a federal system of rule.

The SPLM lost the advantage of structural unity, however, in August 1991 when three commanders split from Garang. Two officers with a civilian background were supported by former Anya Nya II Gordon Kong Chuol in demanding that Garang resign, the SPLM institute internal democracy, that civilian needs be given priority in the SPLM-ruled territories, and temporary partition be accepted if that were the only means to gain peace.[10] The SPLM high command denounced the dissidents but addressed the reformist demands by giving qualified endorsement to the concept of establishing civil administration. Clashes between the two sides during the fall of 1991 threatened to exacerbate political differences.

Foreign Relationships

In addition to consolidating their internal bases, SPLM leaders sought stable relations with foreign countries that could provide sanctuaries, material assistance, and diplomatic support. The government of Mengistu Haile Mariam provided the most substantial support. When the battalions from Bor and Pibor took sanctuary in Ethiopia, Mengistu was already hostile to Numairi, whom he accused of supporting Eritrean secessionists and antiregime forces in Tigre and Oromo. Mengistu preferred to support the SPLA rather than Anya Nya II since Garang rejected the concept of secession. That support continued during subsequent regimes in Khartoum, as the basic tension in Sudanese-Ethiopian relations remained.[11]

Mengistu allowed the SPLA to operate a radio station, which reported the SPLA's military campaigns, the outcome of meetings with Sudanese political groups, and the basic philosophy of the movement. The radio served as a vital means for the SPLM to transmit its message directly to the Sudanese public. Ethiopia served as a sanctuary for SPLA forces. They operated training camps, logistical centers, and a prison and POW camp—all beyond the reach of the Sudanese army. Moreover, by early 1991 more than 400,000 southern Sudanese crowded into refugee camps operated by the SRRA in western Ethiopia. Mengistu probably also provided military support in the form of transport planes, helicopters, and trucks that sometimes ferried SPLA forces and supplies among base camps in western Ethiopia and even into Sudanese territory. Ethiopian forces may have assisted the SPLA's attacks on certain border garrisons, since long-range artillery shelled the towns from Ethiopian territory. In 1987 reports surfaced that Cuban advisors to the Ethiopian army aided SPLA operations; in 1990 similar rumors spread that Israeli arms and advisors reached the SPLA through Ethiopia. Garang denied contact with Israel and argued that such rumors were designed to discredit the movement. He maintained that most SPLA weapons came from the Sudanese army itself, either captured in battle or seized when garrisons were overrun. Other weapons, he asserted, were purchased on the international market.

Mengistu facilitated contacts with Muammar Qadhafi, who eagerly supported any groups that opposed Numairi. Garang's visit to Tripoli in April 1984 secured substantial military aid, but the SPLM resisted Qadhafi's pan-Arab political agenda.

Cooperation ended abruptly when Numairi was overthrown and Qadhafi signed a military protocol with the transitional government. Tripoli subsequently provided not only sizeable arms deliveries but also Libyan-piloted planes that bombed SPLA positions on behalf of all three post-Numairi governments. (Nonetheless, the SPLM never publicly attacked Libya and continued to seek to reopen its office in Tripoli.)

The SPLM established significant relations with Egypt, a pivotal country in both Africa and the Arab world. President Husni Mubarak consistently sought a negotiated settlement between the SPLM and the government. He facilitated Garang's meeting with al-Sadiq al-Mahdi at the summit of the Organization of African Unity (OAU) in July 1986 and strongly supported the talks in 1987–88 between the SPLM and the DUP. Cairo also tried to arrange negotiations between the SPLM and the military government that seized power in 1989 but shifted toward the antiregime National Democratic Alliance (NDA) in spring 1990. Mubarak's tilt became more pronounced during the Gulf crisis that autumn in reaction to Khartoum's overt sympathy for Iraq.

Egypt provided credible diplomatic support to the SPLM as a neutral but vitally concerned state. The SPLA also diversified its territorial sanctuaries by the time Mengistu was overthrown. That proved invaluable, since the groups that seized power in Addis Ababa in May 1991 (assisted, not surprisingly, by Khartoum) closed the SPLM office. The SPLM hastily dismantled its radio station, and troops and refugees surged across the border into Sudan. By then, the movement controlled virtually all of Equatoria and had access to neighboring Kenya and Uganda, although no SPLA forces were stationed on their soil. Tentative contact had also been made with Zaire and the Central African Republic, to which perhaps 65,000 Sudanese refugees had fled during fighting in 1990–91. Moreover, an agreement was reached with Chad in 1990 for SPLA and Dar Fur dissident forces to receive support: the agreement was never implemented since the government fell in December.

Nairobi, in particular, supplemented Ethiopia as a conduit for military and relief supplies and a locale for political offices. By mid-1988 Sudan's government was so irritated at the high-profile SPLM and SRRA presence in Kenya that the foreign minister charged President Daniel Arap Moi with abetting the SPLM. Khartoum threatened to aid ethnic and religious groups inside Kenya to destabilize the regime. Nonetheless, Kenya let the SPLM and SRRA retain their offices, and the relationship assumed enhanced importance with the overthrow of Mengistu.

Relations became cordial with Uganda after Yoweri Museveni came to power in January 1986. Despite Kampala's protestations of neutrality, supplies crossed into Equatoria from Uganda. Khartoum attempted to respond by supporting his predecessor Gen Tito Okello, whose forces launched sporadic attacks across the border into northern Uganda.[12] The Khartoum and Kampala governments managed, however, to avoid diplomatic crises, and Museveni (as well as President Mobutu of Zaire during his one-year term as OAU chair) tried unsuccessfully to arrange meetings between Garang and the Sudanese ruler.

Military Operations

The SPLA developed a five-pronged strategy to undermine the government and the armed forces in the south by mounting protracted operations that would wear them down. The guerrilla forces sought to cripple major economic projects; block communications routes; surround, isolate, and overrun army garrisons; seize and administer towns; and expand the fighting to the north. SPLA operations moved beyond purely guerrilla tactics to include limited positional warfare and rule over a vast territory.

First, the SPLA focused on destroying development projects in the opening months of the civil war. The Water Buffalo (jamus) Battalion, a combined SPLA/Anya Nya II force, compelled a French contracting company to stop digging the Jonglei Canal in November 1983. That ambitious and costly project was intended to increase the amount of water for agricultural projects in the south as well as in the north and in Egypt. Another combined operation in February 1984, led by Anya Nya II's Oil Battalion and the new SPLA Tiger and

Crocodile (tumsah) Battalions, forced Chevron to suspend drilling for oil at Bentiu—another vital project that Numairi had hoped would make Sudan self-sufficient in its energy resources. Garang commented that those two operations were designed "to achieve maximum shock and embarrassment" while Numairi was negotiating with creditors in Paris.[13] SPLA leaders believed that they could force Numairi's hand by demonstrating the economic price he paid by adhering to his policies. The actions also alerted Egypt and the United States to the Sudanese crisis, since they underwrote the Jonglei and Chevron projects, respectively.

Second, the SPLA attempted to block communication routes within the south as well as between north and south. That included boats and barges on the Nile River, the railway from Kosti via Babanusa and from Aweil to Wau, the unpaved roads that crisscrossed the region, and the airports in the provincial capitals. SPLA's ability to blow up railway bridges, ambush truck and rail convoys, mine roads, and sink barges devastated the armed forces and economic life. Rebuilding bridges and railway lines drained the government's budget. By the end of his rule, Numairi conceded that the army could not protect communication routes in the south. Even resupplying garrisons was generally impossible from April through October because the summer rains made land routes impassable, and cloud cover made flying hazardous. In February 1988, for example, the SPLA sabotaged river barges approaching Malakal, delayed a Juba-Bor overland convoy, and ambushed a convoy from Juba to Torit; no additional supplies reached Torit before it fell to the SPLA the next year.[14]

As a result of the SPLA's effective grip on communication routes, the government dropped supplies by air to garrisons that lacked airports. But the SPLA used portable SAM-7 missiles to attack civilian and military planes as they landed or took off from airports in the south. Garang claimed in March 1986 that the SPLA had, in just two years, downed 13 military helicopters, two transport planes, and three of the air force's five F-5 fighters. In December 1987 and December 1988 Libya lost three of the four MiG-23s supplied to Sudan and appealed to the SPLA to return captured pilots.[15] Civilians died in several attacks: the most widely publicized case involved the death of 60 people on a civilian plane that was hit after it took off from Malakal airport on 16 August 1986. That cost the SPLA considerable public sympathy and gave the government an excuse to freeze diplomatic contact. The political damage outweighed the short-term value to the SPLA of forcing a three month's halt of flights to the south. Garang, however, insisted that blame lay with the government, which frequently used civilian planes to ferry troops and supplies and therefore made them potential targets.

Third, the SPLA sought to isolate strategically located towns and army garrisons, initially in Upper Nile and Bahr al-Ghazal provinces. By controlling the surrounding countryside, the SPLA prevented soldiers from entering or leaving the garrisons and disrupted civilian life with artillery shelling. The SPLA also periodically occupied garrison towns on the Ethiopian border, notably Nasir and Kurmuk, to catch the army off balance and seize heavy weapons from its military supply bases.

Fourth, the SPLA seized and held populated towns once it controlled sizeable areas and established reasonably secure lines of communication into Ethiopia. The SPLA took over Yirol and Tonj, for example, in eastern Bahr al-Ghazal during 1985 and gained dominant influence in Upper Nile by 1986–87. (Access remained limited in central Bahr al-Ghazal, where the government armed a Fertit militia against the SPLA, playing on Fertit resentment of the populous Dinka.) Moreover, SPLA's Locust (jarad) Battalion rapidly consolidated its hold in eastern Equatoria in the winter of 1987–88, which was signaled by the capture of the strategic town of Kapoeta astride the only land route from Juba and Torit to Kenya. As local militias began to switch to the SPLA, bringing their government-supplied arms with them, the guerrilla forces swelled to nearly 40,000, according to Sudanese military commanders. In their ultimatum to Mahdi in February 1989, the officers complained of a serious disequilibrium in the balance of forces. That imbalance enabled the SPLA in early 1989 to operate throughout the south and overrun not only Torit (near Kapoeta) but also Nasir (Upper Nile on the Ethiopian border), Nimule (near Uganda), and Jummayzah, north of Juba. By May 1989, when the dry season ended, the SPLA had gained control of the north-south land route and had seized three more towns, including Bor, where the 1983 mutiny had touched off the revolt. Only then did the SPLA accept a UN-mediated cease-fire, which the government had desperately sought throughout the spring.

That cease-fire ended in October 1989, when the three-month-old military government launched a dry-season offensive southward from Damazin. But the SPLA ambushed those troops and pinned down government forces in the hilly southern Blue Nile province. With an additional 50,000 government soldiers bottled up in Juba, the capital of Equatoria, the SPLA mounted its own offensive in western Equatoria. By March 1991 the SPLA had captured more than 16 garrisons and six important towns, including Yambio and Maridi. The SPLA controlled all of Equatoria except Juba and Yei. Juba had to be supplied by air from Khartoum, and the only convoy that reached Yei from Juba in the winter of 1990–91 took three weeks to travel that 100-mile route.[16]

The seizure of western Equatoria completed phase three of the SPLA operations in the south, code named Bright Star. In the winter of 1990–91, the fourth phase (intended to be the final phase) began as SPLA forces moved into western Bahr al-Ghazal and intensified operations in central Bahr al-Ghazal around Wau and along the railway near Aweil. The SPLA controlled an area larger than Uganda or Ghana but remained vulnerable. Aerial bombardment damaged the towns and complicated the task of caring for civilian residents. Moreover, the military government tried to prevent international food and medical aid from reaching SPLA-controlled areas by delaying the signing of agreements with the United Nations that would permit emergency airlifts and even by bombing authorized relief flights and distribution centers. In 1990 the UN estimated that 3.5 million of the 5.5 million southern residents had fled their homes during the seven years of war; entire areas were depopulated, with cattle and crops lost.[17] The SRRA lacked resources to cope with that overwhelming dislocation. Coupled with the SPLM/SPLA's apparently weak administrative structures, those difficulties indicated the problems that the movement faced in moving from a purely guerrilla struggle to static operations in which the SPLA would hold fixed positions, and the SPLM would bear responsibility for the well-being of the local population.

Fifth, the SPLA tried to expand the fighting to the north, notably southern Blue Nile, southern Kordofan, and southern Dar Fur. The Kurmuk operations signaled that the SPLA could operate on the fringes of the northern provinces since Kurmuk was located in the Ingessana Hills of southern Blue Nile province. The loss of that province's vital hydroelectric installations and agroindustrial areas would black out Khartoum and destroy sugar and cotton projects in which the government had invested substantial resources. After the SPLA seized Kurmuk in November 1987 and hundreds of civilians and wounded soldiers fled to Damazin, Mahdi mobilized frantically to recapture the town. The army suffered heavy casualties in the effort and remained vulnerable to ambushes. Moreover, the SPLA made inroads among the Ingessana people who were already angry at the government and private businessmen for establishing agricultural projects near Damazin that drove them off their traditional lands.

The SPLA also capitalized on grievances among the peoples in the Nuba Mountains of southern Kordofan.[18] Long-standing tensions between Nuba and Arab tribes over water and grazing rights were exacerbated both by the alienation of land to private mechanized agricultural schemes and by drought, which forced herders to move into the Nuba-populated hills. Nuba political groups supported the aims of the SPLM and decried the government's use of Nuba foot soldiers against the SPLA. By mid-1985 the transitional government responded by arming Arab villages, initially for self-protection. Mahdi increased support for those militias, which raided Nuba villages and Dinka, Shilluk, and Nuer areas in Bahr al-Ghazal and Upper Nile. Militia raids and SPLA operations devastated the rural areas. Civilians who fled to southern Kordofan to escape the fighting were sometimes killed by revenge-driven Arab militias, as in the massacres of Dinka in Daien in 1987 and of Shilluk in Jabalayn in 1989. Only after the SPLM and the NDA aligned in 1990 were there serious attempts to reconcile the peoples: Mahdi's Umma party reversed its prior support for the Arab militias, and the NDA reached an accord with the SPLA for Nuba to train and fight jointly against the military government's forces.

That accord was extended to political movements in the westernmost region of Dar Fur where many had long resented control from Khartoum. They were also angry that central governments allowed Chadian rebels to camp in Dar Fur, thereby provoking retaliation by the Chad army. Intra-Chadian battles destroyed villages and livestock in Dar Fur, particularly in 1989–90. Nonetheless, relations were limited between the SPLA and the Dar Fur groups because of the vast distance separating them. Guerrilla warfare was hampered in the stark terrain of Dar Fur and southern Kordofan by the lack of cover and a limited rainy season. The air force bombed and burned villages in periodic scorched-earth retaliations, and the SPLA could not supply its guerrilla outposts in the west systematically.

By 1992 the SPLA controlled nearly all of the south and had probed into the north. But the loss of its sanctuary in Ethiopia set back its operations and helped to precipitate a serious split within SPLA ranks in late August 1991. With eastern Upper Nile under the dissidents' control, intra-SPLA fighting weakened the movement and eased the government's task. The sharp decrease in financial and arms support to the SPLA, just as it faced the increased costs of providing for refugees fleeing Ethiopia and internal displaced persons, risked overwhelming its rudimentary administrative structure. Moreover, the SPLA lacked the capacity to move north toward Khartoum. Despite its control over nearly one-third of the country, the SPLA realized the fighting had reached a stalemate.

Perspectives from Khartoum

Government officials and politicians in Khartoum reacted to the SPLM/SPLA according to their own perceptions of the political realities in Sudan. The dominant political groups—notably Mahdi's Umma party, DUP, and NIF, which together controlled 83 percent of the seats in the parliament elected in 1986—viewed Sudan as a predominantly Arab and

Muslim country, within which the south comprised a discrete entity. The geographically peripheral areas appeared politically marginal: political life revolved around the needs and perspectives of the Nile Valley core. Moreover, those politicians assumed that the Muslim majority had the right to institute Islamic legal codes concerning not only their own personal matters but also political, economic, and social life. The non-Muslim minority would have to accept that reality but would retain freedom of worship and certain other legal protections.

In light of those perspectives, officials and politicians from the dominant parties tended either to stress the insubstantiality of the SPLM as a political movement or to argue that its values were so threatening that accommodation was impossible. If the rebellion was merely "Garang's movement," the SPLA need not be taken seriously, and the leader's personal ambitions could be satisfied by an honorary post in Khartoum. If the SPLM were an extension of the secessionist Anya Nya movement, restoring the Addis Ababa accord and offering economic aid to the south could end the uprising. (Only in the fall of 1991 did it appear that the dissident commanders of the SPLA might accept separation, at least as a temporary measure.) Another delegitimizing tactic was taken by politicians who argued that the SPLM was not the sole voice representing the south by arguing that several southern parties won seats in the elections of April 1986 and could represent the region in a constitutional conference. Other politicians claimed that the rebellion would collapse without support from Israel or communist states. The NIF military government sought to mobilize Arab support by arguing that Israel aided Ethiopia and, by extension, the SPLA to control the Red Sea and encircle the Arab world. Mahdi frequently called Garang a puppet of Mengistu and argued that Ethiopia sought to impose a Marxist regime on Sudan, with help from Moscow and Havana.

Primarily, however, political leaders argued that the SPLM was fundamentally anti-Arab and anti-Muslim in mounting a challenge to the structure of power in Khartoum and therefore had to be defeated at all cost. They appealed to Libya, Iran, and the Gulf states for financial and military aid on the grounds that Garang would impose an African identity on the Muslim Arab majority in Sudan.[19] Islamic-oriented politicians viewed the SPLM's insistence on a secular constitution as proof of its anti-Muslim orientation, rather than as a reflection of its concern for upholding the multireligious dimensions of Sudan. NIF adherents blamed western Christian churches which, they argued, sought to thwart the natural spread of Islam in Africa.[20] A movement that embodied African racism, Christian missionary influence, Zionist designs, and/or Marxist dogma could be delegitimized. The speaker proved that negotiations were impractical and impossible and reinforced the image of a zero-sum conflict.

Nevertheless, some political forces in Khartoum addressed Sudan's problems from perspectives that were compatible with the SPLM and thereby prevented the conflict from becoming entirely zero sum. The charter drawn up by the professional and trade union groups that underpinned the uprising in April 1985 supported the secular constitution of 1956 (as amended in 1964). In March 1986 a broad range of political parties and unions met with the SPLM in Ethiopia and issued the joint Koka Dam Declaration that resolved to create the New Sudan "free from racism, tribalism, sectarianism and all causes of discrimination and disparity."[21] The declaration asserted that the "basic problems, not the so-called southern problem" must be addressed and that the government must repeal the Islamic laws of September 1983 and promulgate the constitution of 1956. Those resolutions indicated that bases for dialogue and agreement existed between the SPLM and potentially influential political and intellectual groups. And yet most parties that signed the Koka Dam Declaration lacked representation in the parliament elected the next month. The Communist party, Nuba-based Sudan National party, and coalition of African parties totalled barely 17 percent of the MPs. On their own, they could not transform the conflict from zero sum to positive sum, although their efforts could contribute to that transformation.

Thus, only a minority of the political forces in Khartoum sought to construct a constitutional system that would incorporate the political perspectives of the SPLM. The majority in the government and parliament tried to contain the movement by force and avoid making political concessions. Numairi viewed the rebellion as a mutiny that could be stamped out, even though his generals warned him that the strife would drain his resources. The TMC sent missions to Arab countries in search of military and financial aid so that the armed forces could quell the SPLM or negotiate from a position of strength.

Mahdi built up the armed forces and sought to persuade Ethiopia to stop assisting the SPLA. His heightened rhetoric after the SPLA shot down the civilian plane in Malakal (August 1986) and overran Kurmuk (November 1987) helped win extra Arab armaments but could not turn the tide militarily. Mahdi relied on direct Libyan air support to help recapture Kurmuk: by February 1988 Tripoli provided more than one-half of the military aid received in Khartoum. NIF encouraged Mahdi to look to Iran for support, once the Iran-Iraq war ended in August 1988. Ironically, Iraq also armed Sudan, supplying jet fighters, transport planes, rockets, and heavy guns. Despite Iraqi and Libyan airlifts of supplies, towns and garrisons fell to the SPLA at an accelerating pace. By winter 1988–89, Jordan, Saudi Arabia, and Oman rejected Sudanese appeals for arms, disillusioned by the possibility of a military solution to the civil war. (Iraq also briefly withheld arms, protesting pro-Iranian statements by NIF.) The government paid a substantial political price for military aid. Libya, in particular, expanded its influence in Dar Fur and pressed politicians to support unity between the two countries.

The military regime that seized power in 1989 depended heavily on arms from Libya, Iraq, and Iran. Qadhafi gained free access to Dar Fur, which served as a sanctuary for the Chadian rebels under Idris Deby and as the vital launching pad for the rebels' thrust into Chad to overthrow the government in December 1990. Qadhafi also won an integration charter in 1990, the establishment of Libyan-style popular committees (jamahiri) in Sudan, and joint pledges to disseminate the Arabic language and culture throughout the country. Saddam Hussein received his payoff when Bashir supported Iraq's stance against the international coalition; Sudan thereby antagonized Egypt and the Gulf states. When Iraq was defeated, NIF turned to Iran, whose diplomats declared that the Sudanese civil strife was a war of "Islam versus blasphemy."[22] Iran viewed the Sudan as a means to promote Islam throughout Africa and provided substantial funds to NIF to purchase Chinese weapons for the government. Chinese teams also trained Sudanese pilots, following the signing of an accord in March 1991. Meanwhile, Khartoum succeeded in helping to overthrow Mengistu in May 1991. Bashir supported several opposition groups, but especially favored an Islamic-oriented group among the Oromo people who attacked SPLA forces and refugees in western Ethiopia. The long-term aim of eliminating the SPLA's presence in Ethiopia finally succeeded.

The governments paid heavy prices financially and in national sovereignty to prosecute the war. Under Numairi, the government admitted that the war cost $1 million a day; under Bashir, the cost skyrocketed to $3 million daily. Bashir's government conceded, at its conference on peace prospects in September 1989, that 4,593 soldiers and officers had died since 1984, as well as 2,700 policemen. (The conference also stated that the total military casualties reached 340,000; 27,000 members of the SPLA had died; 62,000 civilians had died from the violence; and another 260,000 civilians had died from disease and starvation.)[23]

The economic cost escalated. Numairi accumulated a $9-billion debt by 1985, which soared to $13 billion by 1990. Bashir admitted that one-half of that debt was due to economic projects that were cancelled in the south.[24] The fighting wrecked plans to diversify cash crops and reduce dependence on foreign oil. The cost of living soared as the government frantically printed money and raised prices; severe food shortages spread in the towns as well as the countryside. Foreign donors withheld assistance, and the International Monetary Fund (IMF) froze loans to Sudan in February 1986. In the fall of 1990 the IMF issued a formal Declaration of Non-Cooperation that isolated Khartoum from virtually all international donors except China, Iran, and Libya.

Economic collapse was an important reason for governments to seek a way out of the conflict. The finance minister argued in April 1987 that negotiations with the SPLM were essential: since the war consumed most of the general budget, neither balanced budgets nor economic reforms were possible as long as the war continued.[25] The economic crisis was a factor in the decision of DUP to initiate talks with the SPLM a year later. A senior party member stated:

> We are in a terrible and deteriorating economic situation and threatened to lose our democracy, indeed our independence. We will never be able to stand on our feet unless there is stability, which can only be achieved by stopping the war—which is why the initiative was made.[26]

In contrast, Bashir's government refused to let economic collapse and even widespread famine deter its efforts to prosecute the fighting. The government sold its entire reserve stock of grain to Libya and Iraq and Europe in 1990 to finance the war. The government did not admit that the country lacked food until October 1990, despite reports that the grain crops had failed in the west, east, and south.

Obstacles to Conflict Resolution

In view of the polarity between the dominant political forces in Khartoum and the SPLM, creating mechanisms to overcome the mutual distrust and resolve the conflict was a laborious process. Nonetheless, major obstacles to negotiations—if not to ultimate agreement—were overcome by mid-1989. The coup d'etat cancelled four years of effort and deepened the divide.[27]

Until the Koka Dam Declaration of March 1986, northern political forces did not realize fully that the SPLM would refuse a political settlement based on the Addis Ababa accord and would insist on adhering to its comprehensive program to transform the government in the center. For the north, according limited self-rule to the south was easier than countenancing a fundamental shift in power in Khartoum. By November 1985, however, the transitional government accepted Garang's idea of convening a constitutional conference to establish agreed upon legal bases for the political system. But the government was not willing to annul the September laws prior to the conference, as the SPLA demanded. Both sides jockeyed for advantage on the ground, believing that the other side would not make significant political concessions until it was hurt militarily. The continued fighting, in turn, made each side more suspicious of the other's intentions.

If the Koka Dam meeting had convened in January 1986, as its organizers had hoped, its resolutions could have encouraged the promulgation of a new, secular constitution and cease-fire prior to the April elections. In fact, the conference met barely two weeks before the elections and had no impact on their outcome. Thus, even though terms of the conference met virtually all of the SPLM's requirements—in particular repealing the September laws and adopting the 1956 constitution—the new prime minister, al-Sadiq al-Mahdi, ignored these terms. If he had pledged to implement them, Mahdi could have ended the guerrilla warfare after only three years, with relatively limited destruction of the social fabric and economy and with relatively limited political polarization within the Sudan as a whole. Instead, Mahdi signed a coalition agreement with DUP that endorsed the Islamic-oriented draft constitution of 1968 and indicated that he sought modified Islamic laws that would protect religious minorities. His room for maneuver was also limited by NIF, which formed a vocal opposition bloc in the assembly to prevent any dilution of Islamic law.

Attempts at personal diplomacy between Mahdi and Garang failed to bridge the gap: Mahdi tried to bypass Koka Dam, which Garang insisted was the only legitimate framework of negotiations.[28] Mahdi claimed that Garang had negotiated in bad faith when the SPLA subsequently shot down a civilian airplane. Garang claimed that Mahdi intended to accelerate the war by turning to Libya for additional bombers and shooting down the peace process itself by refusing to engage in further meetings with him.

Nine months later, when the government's dry-season offensive ended inconclusively in April 1987, both sides resumed their tentative contact. Garang noted that dialogue was necessary since neither side could win militarily, and Mahdi responded with a proposal partly based on Koka Dam.[29] The September laws would "be replaced by a legal position based on diversity in a way which satisfies the aspirations of the Muslims in areas of Muslim majority and the aspirations of non-Muslims in other areas." The SPLM rejected Mahdi's proposal and argued that Sudan was multiracial and multireligious; therefore, Islamic law must be abolished rather than reinforced. The SPLM questioned the sincerity of Mahdi's commitment to end the war since the army had launched a military offensive two weeks after he sent the letter and since the government declared an enhanced state of emergency in late July.

The political groups that had organized the Koka Dam meeting were scattered by then and could not take any effective political initiative to counter Mahdi's approach. Rather, the African (Nuba and southern) parties represented in parliament took the lead. They held three meetings with the SPLM during August and September 1987 that restated support for the Koka Dam Declaration. They also persuaded Umma and DUP (but not NIF) to sign a transitional Sudan charter on 10 January 1988 that stressed the dual Arab and African identity of Sudan and the importance of sharing power and resources equitably within the country. Nonetheless, Mahdi used the excuse of SPLA military victories to break off diplomatic contact and countered the growing pressure to negotiate by adding NIF to the cabinet in May 1988, on a platform that insisted on full-scale Islamic laws.

The African parties could not counter that Umma-NIF alliance. Rather, DUP broke the diplomatic impasse, not because it accepted SPLM terms but because its leaders feared political marginalization by Umma and NIF. The DUP-SPLM accord, signed on 16 November 1988, agreed to the SPLM's long-standing preconditions: pending the convening of the constitutional conference, the government would "freeze all the clauses related to the hudud [Islamic punishments] and other pertinent clauses included in the September 1983 laws. No laws with clauses that refer to the above clauses shall be enacted until the National Constitutional Conference is convened and the issue of the laws is finally settled."[30] They also agreed to abrogate military agreements that affected national sovereignty, end the state of emergency, and declare a cease-fire. A national preparatory committee would fix the date, venue, and agenda of the conference which, they hoped, would open on 31 December 1988.

In effect, each side agreed to reserve its substantive differences until the conference. The DUP leader noted: there will be "a long debate and a great deal of argument at the . . . conference [on these issues], but in the end we believe in the democratic course. . . . Through dialogue and democratic practice we shall be able to build a modern Sudan."[31] There was no guarantee that the outcome of the conference would be a secular constitution; instead, they agreed that debating rather than fighting was the appropriate means to resolve the country's deepest divisions.

The DUP-SPLM accord was blocked by NIF and Mahdi because it threatened their rule: DUP, SPLM, and the African parties suddenly emerged as an alternative ruling coalition. NIF leaders also feared that Islamic laws would never be promulgated once they were frozen. The multiparty fragmentation in Khartoum and relative power of Umma and NIF prevented DUP from gaining parliamentary support for the accord. The military high command had to step in to resolve the crisis with its ultimatum in February 1989. The officers forced Mahdi to form a broad-based government that excluded NIF and to make the DUP's chief negotiator with the SPLM the foreign minister in charge of finalizing arrangements for the constitutional conference.

The government and parliament formally endorsed the DUP-SPLM accord, and parliament voted to shelve debate on Islamic laws until the constitutional conference. The SPLA responded with a cease-fire on 1 May for Ramadan, the Islamic month of fasting, and both sides agreed that the state of emergency would end simultaneously with the beginning of a permanent cease-fire just before the constitutional conference would convene on 18 September. As a final step, on 29 June Mahdi initialed the draft law that would suspend the existing Islamic laws in preparation for a cabinet decision on 30 June and parliament's vote on 1 July. The stage was set for the government meeting with the SPLM on 4 July with all the preconditions satisfied for the conference. Garang commented later, "The peace process was at an advanced stage when the fifteen army officers seized power."[32]

With the establishment of Islamic-oriented military rule in Khartoum, prospects for negotiations vanished. The government immediately canceled the DUP-SPLM accord, reinstated the September laws, and reverted to viewing the conflict as a southern problem. The officers insisted that negotiations could not question the majority's right to Islamic laws. The nominally federation system promulgated in 1990 only allowed provinces with non-Muslim majorities to exempt themselves from certain Islamic personal-status laws. The SPLM countered with stiff terms of its own: the government must restore democracy and accept the DUP-SPLM accord before the war would end.

The two meetings between the junta and SPLM, held in August and December 1989, reached a deadlock. SPLM negotiators spoke of "an extremely wide gap" and an "agreement to disagree" after the first meeting.[33] The gathering in Nairobi, hosted by former US President Jimmy Carter on 1 December 1989, found itself presenting diametrically opposed positions. No mediator could bridge the differences; if anything, the meeting deepened the discord.[34] Subsequent mediation attempts by Mubarak and Mobutu through the OAU failed even to get the two parties to the table. Similarly, an American proposal to separate the two sides by an internationally monitored buffer zone in the south proved a nonstarter.

Each side geared up for a military showdown. The SPLM strengthened its ties with the NDA and former military high command in 1990 and gained control over virtually all the south. The government viewed substantive negotiations as ideological suicide and deepened its ties with radical Libya, Iran, and Iraq. The fall of Mengistu, however, affected the political balance by reducing material support for the SPLM. Some SPLA commanders calculated that, given the impossibility of achieving the SPLM's political aims, the protracted struggle was destroying the country rather than bringing about the anticipated transformation. They calculated that the cost of war outweighed the benefits: the SPLM must cut its losses and make peace, even on limited terms. Otherwise, fighting could continue indefinitely or until the south was destroyed. Whether the government would agree to the de facto or de jure separation of the south as the price for consolidating an Islamic regime in the north remained uncertain. That radical excision of the major non-Muslim area would benefit the government at the price of relinquishing a third of the country's territory.

Three opportunities emerged to negotiate an end to the prolonged war. The Koka Dam conference of 1986 and the DUP-initiated accord of 1988–89 could have resulted in an agreement that would restructure the political bases of power in Sudan to meet the needs of the peripheral areas that the SPLM claimed to represent. The fragmented nature of the Sudanese political system, the weakness of the central government, and the lack of political consensus about the appropriate form of rule, however, enabled powerful political forces to block the implementation of those accords. The third opportunity, expressed by support for territorial partition, approached the problem from a radically different angle, based on the failure to reach a mutually acceptable accommodation and a reversion to the view that the conflict is a zero-sum game.

The civil war in Sudan since 1983 thus represents a classic prolonged war. The stakes have been so high that a negotiated resolution is extremely difficult. Each side has perceived itself as defending its core identity and ensuring the survival of its political community. At key moments, one side or the other has perceived the benefits from conflict as outweighing the benefits of peace and has hoped that a clear military victory will obviate the need for negotiated compromises. The struggle has therefore been prolonged far longer than anyone anticipated or sought and has caused profound suffering to the people. Those living in the war zone have been affected directly, but the economic and social life of the country as a whole has been damaged by the nearly decade-long strife. The collapse of the political agreement in 1989 meant that the civil war would continue until another political transformation occurred or until both sides accepted partition. The prolonged conflict was destroying the people and society that the political forces sought to preserve.

Notes

1. For background, see Mohammed Omer Beshir, *Revolution and Nationalism in the Sudan* (London: Rex Collings, 1974); P. M. Holt and M. W. Daly, *The History of the Sudan* (Boulder, Colo.: Westview Press, 1979); and Peter Woodward, *Sudan, 1898–1989: The Unstable State* (Boulder, Colo.: Lynne Rienner, 1990).

2. Quoted in full in *Perspective on the South* (Khartoum: Ministry of Guidance and National Information, 1983), 21–22. For subsequent developments in the south under Numairi, see Mom K. N. Arou and B. Yongo-Bure, eds., *North-South Relations in the Sudan Since the Addis Ababa Agreement* (Khartoum: University of Khartoum, 1988); and Raphael Koba Badal, *Oil and Regional Sentiment in Southern Sudan* (Syracuse: Department of Geography, Syracuse University, 1983).

3. Abel Alier, *Southern Sudan: Too Many Agreements Dishonoured* (Exeter: Ithaca Press, 1990), 244–47. Analyses of the causes and initial period of renewed civil war include the author's "Rebellion in the Southern Sudan," *UFSI Report* 8 (1985); and "Confrontation in the Southern Sudan," *The Middle East Journal* 40, no. 3 (Summer 1986).

4. For details on 1985–1986, see the author's "The Fall of Numairi," *UFSI Report* 9 (1985); "Transition in the Sudan," *UFSI Report* 20 (1985); and "Party Politics in the Sudan," *UFSI Report* 9 (1986).

5. All quotations in this paragraph come from Garang's address on 3 March 1984 and are quoted in full in Mansour Khalid, ed., *John Garang Speaks* (London: Kegan Paul, 1987), 21–23.

6. Ibid., 26.

7. Ibid., 112.

8. Commanders who led negotiations included Col Kerubino Kuanyin Bol (the former commander at Bor) at Koka Dam, March 1986; Chief of Staff Lt Col William Nyuan Bany Deng (who led Ayod troops to the bush on 6 June 1983) with African parties, August 1987; Lt Col Lual Deng Wol with African parties, September 1987; and Dr Lam Akol (former professor of engineering, Khartoum University, who joined the SPLM in winter 1985–1986); Yusuf Kuwa (Nuba officer, former member of the Kordofan regional assembly) negotiated with DUP in August and October 1988. Akol also headed the delegation at the meetings with the foreign minister on 10–12 June 1989 that finalized arrangements for the constitutional conference and met the military government officials in the summer and fall of 1989.

9. Khalid, 54. Concerning relations with Anya Nya II, see Alier, 251–55; *Africa Confidential*, 30 October 1985; SPLA radio, 7 August 1985 and 28 January 1988; Omdurman (Sudan) radio, 17 August 1985; and Sudan News Agency, 6 February 1988.

10. Declaration on 28 August 1991 in Nasir (Upper Nile) by "SPLA Two," headed by Riek Mashar, Lam Akol, and Gordon Kong; reprinted in *Sudan Update* 3, no. 4 (24 September 1991). The SPLM high command met in Torit 6–11 September. See *Sudan Monitor* 2, no. 4 (October 1991): 6.

11. For a discussion of relations with Ethiopia in the context of Sudan's foreign policy, see the author's "A View from Khartoum," *Foreign Affairs* 65, no. 4 (Spring 1987): 807–26. For details on foreign meddling in the civil war, see the author's "External Involvement in the Sudanese Civil War," in "Foreign Intervention in Sub-Saharan Africa: Making War and Waging Peace," ed. David R. Smock (US Institute of Peace, forthcoming 1992).

12. When SPLA forces overran western Equatoria in the fall of 1989, they captured Okello's bases and virtually imprisoned them. The Sudan air force then bombed Ugandan villages in November 1989 in which international

relief supplies and personnel were located. See John Garang, speech in Khalid, 20 March 1986, 139; Omdurman radio, 31 January 1986.

13. Khalid, 45, 55. In a speech on 22 March 1985, Garang claimed that the government spent $12 million to rebuild the Lol River railway bridge alone.

14. SPLA radio, 20–24 February 1988 and 2, 7 March 1988; Omdurman radio, 21 February 1988 and 2, 3 March 1988; SUNA, 27 February 1988.

15. SPLA radio, 18, 25 December 1987, claimed that one MiG-23 was shot down by the SPLA and that the other crashed. On Libya's appeals to release the two pilots, the SPLA captured the previous year, see SUNA, 15 January 1989, and Middle East News Agency (MENA), Cairo, 20 January 1989. SPLA radio, 21 January 1989, claimed that the third MiG-23 was shot down near Nasir within the previous six weeks. Earlier references to military losses come from *Ittihad*, 5 March 1986; SUNA, 13 and 16 October 1985; and SPLA radio, 14, 16 October 1985.

16. Sudan radio, 11 February 1991. *Africa Confidential* 3, no. 22, discussed the difficulty of supplying Juba and Yei. See *Africa Confidential* 3, no. 22 (9 November 1990). See *Sudan Monitor* 3, no. 4 (October 1990). *Sudan Monitor* commented that 200,000 displaced persons had crowded into Juba, in addition to the 100,000 normal residents, and the army was preventing anyone from leaving.

17. UN and US figures cited in *Sudan Update*, 25 September 1990.

18. Alier, 255–57.

19. For an overview of Sudanese foreign policy, see the author's "Sudan's Foreign Policy: In Search of Arms, Aid, and Allies," in *Sudan: State and Society in Crisis*, ed. John O. Voll (Bloomington: Indiana University Press, 1991), 43–70.

20. See the arguments in Hassan Makki Mohamed Ahmed, *Sudan: The Christian Design* (London: The Islamic Foundation, 1989).

21. Khalid, 145.

22. Quoted in *Africa Confidential* 3, no. 22 (9 November 1990). For information on Chinese military aid to Sudan via Iran, see *Sudan Democratic Gazette* 19 (December 1991): 1, 3; and on the political relationship with Iran and Libya's reaction, see *Sudan Democratic Gazette* 20 (January 1992): 1, 3, 7.

23. MENA, 21 October 1989; and *al-Watan al-Arabi*, 9 February 1990.

24. *Al-Sharq al-Awsat*, 16 July 1990.

25. Dr Bashir Umar Fadlallah, former economics professor and member of Umma, whose views on negotiations did not coincide with Mahdi, SUNA, 6 April 1987.

26. Sid Ahmad Husayn, DUP's chief negotiator with the SPLM, SUNA, 10 December 1988.

27. Details on the conflict resolution process are provided in the author's "Negotiations in the Sudan," in "Foreign Intervention in Sub-Saharan Africa."

28. They met for nine hours in Addis Ababa on 31 July 1986; the author interviewed Mahdi, 21 September 1986; and interviews with Garang in al-Musawwar, 14 August 1987 and on SPLA radio, 9 November 1987.

29. Text of the letter broadcast on SPLA radio, 20 April 1987, the day after the SPLM received it. Garang's comments on SPLA radio, 6 April 1987; and SPLM reaction to the letter on SPLA radio, 8 August 1987.

30. Text found in MENA, 17 November 1988.

31. Comments at their joint press conference, SPLA radio, 18–19 November 1988.

32. SPLA radio, 14 August 1989; on the coup d'état, see the author's "Khartoum Diary," *Middle East Report* 162 (November 1989): 36–38.

33. AFP, 19 August 1989; and SPLA radio, 20, 22 August 1989.

34. SPLA radio, 2, 4, 6 December 1989; *al-Sharq al-Awsat,* 5 December 1989; SUNA, 7 December 1989; and Omdurman radio, 7 December 1989.

Fire in the Horn
Prolonged War in Ethiopia and Eritrea

Cobie Harris

The tragedy of Ethiopia is really an African tragedy because a generation of Africa's most precious resource, its young people, have been sacrificed on the altar of war, famine, refugee camps, and exile in a prolonged war that should have been avoided. This study examines the circumstances that contributed to Ethiopia's descent into an interminable fog of conflict and prolonged war.

In Africa's modern history, Ethiopia is probably the most important nation-state because it is the only African country to defeat in war a European power, the Italians. In the last 30 years, however, Ethiopia has moved from a symbol of African freedom and dignity to a country plagued by famine and internal wars and is now teetering on the verge of disintegration. Presently, Ethiopia is the only African country where a subnationalist group has effectively fought the central government, and is now on the verge of seceding from the republic.[1]

Five major factors contributed to the prolonged war in Ethiopia. First, the intervention in Africa by Italian imperialism eventually led to the creation of an enclave called Eritrea. Second, the existence of the kingdoms of Tigray and Shoa, two Christian islands in a sea of Islamic states, forced these kingdoms, for their own survival, to pursue a regional imperialist policy as a form of insulation against Islamic domination. Third, after World War II, the rise of Israel as a nation-state exacerbated and highlighted the strategic importance of the presence of a powerful Christian state in a mostly Islamic region, transforming the problem between Ethiopia and Eritrea into a regional conflict. Fourth, the rivalry amongst the superpowers to establish hegemony over the Horn and Red Sea areas predisposed Ethiopia to pursue a military rather than a political resolution to the problem since the superpowers had flooded the area with weapons.[2] Fifth, after its collapse and eventual overthrow, the feudal monarchy was replaced with a military dictatorship based on repression and organized around Leninist principles.

European imperialism radically transformed African society because it brought the revolutionary idea of the nation-state to the African continent; this was a revolutionary concept in that it compelled the radical transformation of the geopolitical landscape. Imperialism created states in a heretofore stateless environment and introduced the idea of the nation-state to Africa for the first time. The uneven penetration of European imperialism, however, led to the differential effects of nationalism and state formation on the nascent Ethiopian state.

Europe's profound impact on the region was most evident by the fact that neither Ethiopia nor Eritrea existed before European penetration. Although independent kingdoms had existed in the regions for thousands of years, the Horn region was devoid of a self-conscious nationalism associated with any particular state boundary. Thus, one dimension of the modern civil war in Ethiopia was the simultaneous emergence of two competing nationalisms within the same geographical area. This paper focuses on the intersection and conflict of ethnicity, feudalism, and foreign interests on prolonging the war in the region.

Italian Interests

In the 1870s, when the Italians first attempted to colonize the northeastern part of Ethiopia now called Eritrea, they met stiff resistance from the indigenous people there. Later, when the Italians attempted to seize the area, King Yohannes incorrectly assumed that the British would protect his flank against encroachment by another European power. Instead, the British encouraged the Italians to seize the coastal lowlands from King Yohannes and permitted them to advance on the highland plateau area in Eritrea.[3] Once the Italians realized that they had the consent of the British, they decided to conquer all of the Eritrean territory under the control of King Yohannes.

During this period another Christian kingdom rose in the center of Ethiopia. Under the leadership of King Menelik, a new Amhara kingdom became the dominant force in the region by incorporating the Oromo people and gaining their fertile agricultural lands. When the Italians began to encounter fierce resistance from northern Tigrean forces, they embarked on a program to exacerbate and intensify the latent cleavages between the Amhara and Tigrean kingdoms. While fighting the Tigreans from Tigray and Eritrea, the Italians also signed a nonaggression treaty with King Menelik who, in exchange, received approximately 5,000 guns from the Italians.[4]

In 1887 King Yohannes was killed while fighting the Mahdist's forces from Sudan. His death created a power vacuum in the northern Christian kingdoms which intensified competition for the title of King of Kings formerly possessed by King Yohannes. The death of King Yohannes started an intense rivalry between his two sons, Alula and Mengesha, and King Menelik over control of the northern Christian areas. In pursuit of this goal, King Menelik enlisted the Italians, who were quite eager to assist him because this alliance would allow Italian control over Eritrea. King Menelik conceded this expedient alliance with the Italians to gain time to build a modern military force without antagonizing them. King Menelik made this alliance even though it meant he would accept Italian control over Eritrea.[5]

By 1893 a conflict emerged between the Italian and Ethiopian interpretation of the Treaty of Uccialli, which stated that the Ethiopians might use the Italians as their intermediary with other European powers. The actual controversy between the Italians and Ethiopians revolved around the issue of whether Ethiopia was a dependent Italian protectorate.

To establish Ethiopia as a protectorship, the Italians attempted to colonize the region by military conquest. Instead, the Italians suffered one of their worst military defeats ever at the battle of Adowa, marking this military feat as the first victory ever by non-Europeans over Europeans in Africa. However, the fateful decision of Menelik not to drive the Italians completely from Ethiopian soil eventually led to the creation of the Eritrean colony.

After their defeat, the Italians consolidated their power within their Eritrean colony. In short, Italian rule stimulated and nurtured Eritrean national consciousness. On a political level, Italian colonization served as a capitalist-battering ram to destroy or distort noncapitalist formations in Eritrea.[6] Italian colonization of Eritrea transformed their feudal subsistence/peasant economy into a market economy. On the one hand, the insertion of capitalist components into Eritrea created working and commercial class fragments, whose struggles to protect and advance their interests made their society more liberal and free than feudal Ethiopia. On the other hand, King Menelik's victory over the Italians essentially froze their feudal social structure until the overthrow of the monarchy in 1974. Hence, the Italian transformation of Eritrean society is particularly noteworthy because it planted the seed for Eritrean nationalism which made the fusion between semicaptialist Eritrea with feudal Ethiopia inherently unstable. The simultaneous uneven penetration of Italian imperialism and the rise of the Ethiopian nation-state set the stage for the drama of the longest war in postcolonial Africa.[7]

The Social Historical Context

The Ethiopian and Eritrean conflagration highlights one of the central problems of African politics: the question of legitimacy and the relevance of nation-states created directly or indirectly by European imperialism. Ironically, the only principle African states agree on today is that the colonial boundaries are legitimate and any attempt to overturn these boundaries violently is illegitimate; such an idea is one of the charter principles of the Organization of African Unity.

African political discourse on the postcolonial state is grounded on the sanctity of colonial boundaries as a non-negotiable principle. Since African states are multinational and multicultural, any boundary dispute could lead to the disintegration of the nation-state in Africa. Thus, most African states consider the Eritrean struggle illegitimate. Nonetheless, Eritreans believe that Italian colonization legitimated their claim for independence because only sovereign nations can be colonized and colonization presupposes a sovereign state. They further contend that the difference between oppression and colonial subjugation is that the resolution of the nationalist question is justice, while the resolution of the colonial question is independence. Of course, Ethiopians reject the Eritreans' formulation of the problem in this manner. To Eritreans, Ethiopia's annexation of Eritrea symbolized an act of aggression against a sovereign nation by another African country.

United Nations and British Perspectives

After the 1936 Italian invasion of Ethiopia and their final defeat in 1941 by the British forces, Eritrea fell under the control of the commander in chief of the British forces in East Africa. The power vacuum left by the unconditional surrender of Italy compelled the British to design a plan for the future of the Eritrean territory. Proposals put forth by the British ranged from forming a unity between Tigray and Eritrea to allowing Ethiopia to absorb the area.

In either case, Britain did not support independence for these states because it could adversely affect their geopolitical hegemony over the vital Red Sea shipping lanes. In addition, Ethiopia asserted their own "natural right" to control the future of Eritrea for two main reasons: first, Eritrea provided Ethiopia's only access to the sea; and second, the idea of separation between the two countries was artificial, since they were one culture and one people.

On 2 December 1950, the United Nations General Assembly voted for a federal solution as the best way to resolve the Eritrean question. The resolution recommended the following:

1. Eritrea shall constitute an autonomous unit federated with Ethiopia under the sovereignty of the Ethiopian state.
2. The Eritrean government shall possess legislative, executive, and judicial powers in the field of domestic affairs.[8]

The federal constitution developed by the UN provided that the emperor would be represented by a governor general in Eritrea. The federal charter also established an autonomous legislative chamber for Eritrea and elected a president who was responsible for electing permanent secretaries to staff the national civil service departments. In addition, the president would appoint judges and an independent judicial branch of regional government. The federal constitution enabled Eritrea to establish Tigrinya and Arabic as their national languages. The UN designed these measures to protect the autonomy of Eritrea from total subordination to Ethiopia.

The first election to the national assembly of Eritrea was held in 1952.[9] What is noteworthy about the first free election ever held in Eritrea was that the prounion and federation supporters garnered over two-thirds of the votes cast in the election. The election results clearly indicated that a majority of the Eritrean people favored unification and/or federation with Ethiopia. Perhaps what is even more significant was how the decrepit feudal regime managed to convert profederation Eritrean sentiment into volatile antifederation and anti-Ethiopian feelings.

Within 10 years, Haile Selassie fundamentally transformed the relationship between Eritrea and Ethiopia and forcibly annexed Eritrea in 1962. Ethiopia's unilateral decision to annex Eritrea abrogated the UN provisions, which stated that only the UN General Assembly could amend the federal relationship. Ethiopia justified the annexation of Eritrea by referring to the Eritrean General Assembly's majority vote for the union in 1962. Eritrean nationalists, however, discounted the General Assembly's vote by alleging that the Ethiopian regime had packed the chamber with its supporters and destroyed the integrity of the National Assembly. In 1960, in the conclusion of his book, G. K. N. Trevaskis prophetically stated:

> The temptation to subject Eritrea firmly under her [Ethiopian] control will always be great. Should she try to do so, she will risk Eritrean discontent and eventual revolt, which, with foreign sympathy and support, might well disrupt both Eritrea and Ethiopia herself.[10]

War Clouds

To understand how the conflict between Ethiopia and Eritrea degenerated into Africa's longest war requires a brief analysis of the major political, geographic, and religious divisions. Such divisions within Eritrea transformed the war against Ethiopia into two separate wars; the Eritrean Liberation Front waged the first war, and the Eritrean People's Liberation Front waged the second.

The major internal political cleavage within Eritrea is between Christianity and Islam. The most contentious issue in the formation of Eritrean nationalism focuses on whether Islam or Christianity will become the hegemonic power and organizing principle of the new state. The fact that Eritreans are evenly divided between Christians and Muslims further exacerbates this religious cleavage. Moreover, each religion dominates a particular region: Christians comprise the majority in the core highland area, while Muslims dominate the coastal regions and the western plateau area.[11]

Other major factors which led to the nationalist revolt included the Ethiopian *ancien regime*'s backward and repressive strategy toward Eritrea, the rise of Pan-Arabism, and an Islamic revival. Emperor Haile Selassie's regime's relentless implementation of policies that alienated and intensified Eritrean secession sentiment fueled the fires of war even more.

The federation of Eritrea and Ethiopia was complicated by Eritrean's indirect participation in democratic government under colonialism in such areas as freedom of association, independent political parties, trade unions, and an independent legislative branch, whose function was to elect an Eritrean chief executive and to legislate for the state. In Ethiopia, on the other hand, once the emperor returned to the throne after the defeat of the Italians, the government granted none of these rights to the Ethiopian people.

In 1956 the Ethiopian government began to destroy the democratic features of the Eritrean state. The Ethiopians banned independent political parties and muzzled the press. It forced opponents of annexation to go into exile while the office of the chief executive was put under the control of the government's representative, the *enderase*. In 1958 the disembodied Eritrean National Assembly voted to rescind their right to fly an Eritrean flag and later, in 1959 when Ethiopian law was imposed on Eritrea, it virtually destroyed the Eritrean

National Assembly's legitimacy and power to act independently. The final act of Eritrean annexation was the National Assembly's announcement of the legalization of the annexation.

What is most remarkable about this tragedy is that before the Ethiopian government pursued the fateful path of annexation, the majority of Christian Eritreans supported some level of affiliation with Ethiopia. An ethnic dimension further reinforced the solidarity between these two Christian communities, since the Christian Eritreans, who were Tigres, comprised the same linguistic and ethnic group as the Tigres of Ethiopia. What should have been a natural unification between two nationalities was destroyed by the repressive policies of the Ethiopian regime.

Tension between Christians and Muslims began to increase in the late 1950s and early 1960s, when Nasserism, Pan-Arabism, and Islamic solidarity began to rise. In Egypt Nasser's emergence to power was based on a Pan-Arab platform, which called for the unity of Arabic and Islamic people and exhorted Muslims to form a political union that transcended artificial colonial boundaries. More importantly, it provided a fertile political climate for nurturing the rise of Eritrean nationalists.

The initial Eritrean Liberation Movement (ELM) was formed in the Port of Sudan in November 1958. Origins of this movement are somewhat unclear because of the obscure background of the early founders. What is definite, however, is that the founders were Islamic emigrants from the Keren and Sahel coastal region who flocked to the Port of Sudan.

The principal leaders of the movement were Saleh Ahmed Iyay, Yasin el-Gade, and Mohammed Said Nawid.[12] Mohammed Said Nawid was politicized by the Sudanese Communist party, the largest and best organized Communist party on the African continent. The ELM was a Leninist party, predicated on the vanguard party and the principle that it was the party's responsibility to bring revolutionary consciousness to the masses. Following this recipe for revolution, the vanguard of Eritrean Islamic workers and intellectuals attempted to galvanize support for their movement against the Ethiopian state. They attempted to return to their area and use

Sudanese passports to organize cells within the Muslim areas located in the Asmara and Massawa regions. In addition, they mobilized Christian workers and students who had been alienated by Ethiopian repression. To the surprise of these organizers, they saw their overtures well received by segments of the Christian community.

However, like most Communist movements that were organized along the Leninist model, the ELM was immobilized because of two conflicting goals: one, the political education of the masses and the other, the desire to seize power during or after a popular revolt. The recent coup in Sudan and its apparent success had predisposed the ELM to consider a coup as the best way to wrest power away from Ethiopia. ELM's ambivalence on these questions led other Eritrean nationalist movements to accuse them of passivity.

While the ELM struggled for power, another Eritrean nationalist movement arose outside of Eritrea by Egypt. Egypt supported the Eritrean movement because Ethiopia had sided with the Western powers against Egypt and the rest of the Arabic states in the 1956 Suez Canal crisis. During the late 1950s, characterized as the apex of Pan-Arabism and Islamic solidarity, Nasser provided free scholarships and room and board to Eritrean students and prominent exiles. In retaliation for Ethiopia's support or the West against Egypt in 1956, Nasser asked the most prominent Eritrean exile in Egypt to broadcast anti-Ethiopian programs into Eritrea. For this assignment, Nasser chose Woldeab Woldemariam, one of the most important exiles in Egypt at the time.[13]

A third political organization, called the Muslim League and led by Ibrahim Sultan, also emerged during this time to fight for Eritrean independence. Ibrahim Sultan was joined by Idris Mohammed Adam, a leader of a splinter faction from the Muslim League called the Muslim League of Western Eritrea. These leaders embarked on journeys to the Middle East to secure money from the expatriate Eritrean community to raise funds for a resistance movement against Ethiopian domination. The expatriate Eritrean community encouraged them to form an alliance with the ELM, who were already organizing resistance from their base at Port Sudan. Ibrahim and Idris refused, referring to the ELM as outsiders who had been contaminated by communism. Furthermore, they both distrusted the ELM because they had recruited into the movement Christians whom they considered traitors.[14]

Their strong rejection of the ELM set the stage for the development of other Egyptian-backed movements to resist Ethiopian domination. Idris Mohammed Adam, who came from a minority ethnic group within Eritrea, and Ibrahim Sultan decided to form a new liberation front and to call it the Eritrean Liberation Front (ELF). The Ethiopian security forces began to round up suspected leaders or any strong Eritrean nationalist. Inadvertently, their sweep frightened a Shifta leader, Hamid Idris Awate, who decided in 1961 to resist arrest. His shots, according to most accounts, were the ones that started Africa's longest war.[15]

Despite opposition, the ELF continued to organize and later formed a central committee of major figures who intended to dominate the liberation struggle of Eritrea for the decade of the 1960s. The three central figures who led the ELF were Idris Mohammed, Idris Osman Galedewos, and Osman Saleh Sabbe. After creating the political structure, they received arms from friendly Arab states, as any level of external support for a national liberation war required an armed presence in Eritrea.[16] By 1962 the ELF formed a motley armed force comprised of Shiftas, who were Eritrean deserters from the Ethiopian security forces and from the Sudanese army.

Once the Eritrean nationalists established an armed presence in the region, they became pawns in the regional conflict between the Arab states and Israel as well as the superpower competition for strategic hegemony over Middle Eastern objectives. In this volatile regional context, Ethiopia, because of its strategic position at the mouth of the Red Sea, became a target of international and regional concern and attention.[17] The internal conflicts of Ethiopia became international problems leaving Ethiopia's internal affairs subject to external manipulation. Thus, when the Islamic states in the region realized that Ethiopia's support of the West during the Suez Canal period was not an anomaly and that Ethiopia was actually a recipient of Israeli aid, they began to support an Eritrean secession movement fervently.

Eritrean Perspectives

In 1961, when Eritreans opposed to the Ethiopian annexation fired the first shots, the Eritrean movement was composed exclusively of Muslims. Even though Christians represented 50 percent of the Eritrean population, it was clear that the Islamic dominance of the nascent Eritrean nationalist movement limited the recruitment of Christians into the nationalist movement. Despite the compelling necessity to organize Christians into an effective liberation movement, the presence of Christians in the nationalist movement remained marginal in the first decade.

From 1962 to 1965 the small bands of ELF fighters were ineffective and still based mainly around the western lowlands. Pressure to expand, coupled with previous ineffective campaigns, compelled the ELF's executive committee to establish a more decentralized and effective fighting force.

The ELF adopted a guerrilla model for protracted struggle and predicated it on the Algerian model of autonomous zones. They divided Eritrea into four fighting zones that ranged from the western lowlands to the eastern seaboard. Although the zones were designed to increase the fighting capacity of the ELF's national liberation forces, they failed to address the most important issue of war: the sociopolitical configuration of power between the various ethnic, linguistic, and religious groups found within Eritrea. This dimension of their struggle was even more complicated by the fusion of ethnicity, class, religion, and territory.[18]

During this period, the ELF found itself divided into two types of cleavages. On the one side, internal divisions existed between Muslims based on ethnicity, regionalism, and the timeless dispute between pastoralists and agriculturalists. The other major cleavage began between the Christians and Muslims. The ELF's leadership and rank and file distrusted the Christian population, holding them responsible for the forced annexation of Eritrea because of Eritrea's prounionist stance. For example, even though the third zone, the Christian area, held over 50 percent of the population and was the breadbasket and economic engine of the region, it was militarily the weakest zone.

The Christians, in turn, distrusted the Muslims because of their historical position as an embattled minority in a region surrounded and dominated by Islamic states long before the states of Ethiopia and Eritrea were created. That the official language of the ELF was Arabic and that the ELF's major source of funding and training also came from the Arab states reinforced Christian fears of Islamic hegemony.

In spite of these divisions, the feudal Christian Ethiopian regime did not embrace the natural alienation of Eritrean Christians by the ELF. Even though a Christian prounionist party received the majority vote in the 1952 election, the Ethiopian regime lost its legitimacy within the Christian sector when it destroyed the Eritrean Christian-dominated trade unions and the party organization that had voted for unionization.

The Ethiopian regime could have regained its credibility with 50 percent of the Eritrean population easily if they had incorporated Eritrean Christians into the bureaucracy on an equal basis. Instead, the Amhara-dominated Ethiopian government limited the Eritreans' entrance into the national administration. Although prominent Eritrean Christians held key posts in the Ethiopian government (Bereket Selassie as attorney general and Gen Arman Andom), opportunity was basically limited for the vast majority of Eritreans. The tendency of individual Eritreans to attain high status did not alter Eritrean perceptions that their assignment to low-level staffing positions in the national and provincial administration occurred solely because they were Eritrean.

The ditcum that "war is the continuation of politics by other means" sums up how the failed politics of the Ethiopian regime pushed the Christian Eritreans into conflict. Christians not only constituted 50 percent of the Eritrean population but were also natural allies of the Ethiopian regime. This political failure also explains the inevitable limitation of the ELF's effectiveness because of its failure to mobilize the Christian sector.

The ELF realized later, however, that to advance their liberation struggle, the Christian sector had to be accommodated. Towards this end in 1967 they created another zone called the Christian zone, designed to facilitate the incorporation of the

Christian sector. Before the creation of the Christian zone, the Ethiopian government treated the insurrection in Eritrea as an Islamic/Arabic problem against a Christian/African country.[19] Since Ethiopians had framed the Eritrean question in these terms, they did not even consider the seriousness of the insurrection. For example, they installed only one brigade in the region. This circumstance changed dramatically when the regime discovered the creation of a Christian sector. The killing of Asrate Kassa, the governor of Eritrea, highlighted in a dramatic way the depth and seriousness of the Eritrean crisis.

The Rise of the Eritrean Peoples Liberation Front

The social forces that pushed the Eritrean conflict into a prolonged war, based on Marxist and Maoist principles, were generated, ironically, by the rise of the more Christian and secular Eritrean Peoples Liberation Front (EPLF) movement, which was coupled with the collapse of Emperor Haile Selassie's regime. The Italians had transformed Eritrea into a more capitalist and industrial society than the feudal and agriculturally based one established by Ethiopia. Despite the political alienation and disaffection of Eritrean Christians from the Ethiopian regime, the Eritrean Liberation Front could not mobilize the Christian sector, because the ELF lacked a coherent and integrative ideology. The paradox of ELF's national ideology was that they predicated it more on the religious idea of community, which by definition was more particularistic, than on the secular and universalistic idea of nationalism, which was based on citizenship and individualism. The ELF's more sacred-than-secular conception of nationalism precluded the equal incorporation of all sections of Eritrean society within the national struggle.

The failure of the ELF to develop a more secular ideology of nationalism, independent of Islamic principles, and to develop an effective strategy eventually paralyzed their military and political positions in Eritrea. For example, the ELF based their organizational structure and distribution of military personnel and supplies on four autonomous zones that operated independently from the strategic needs of the movement. The ELF functioned as a loose confederation of war lords against the feudal regime of Haile Selassie. For example, if Idris Mohammed acquired resources from the more militant Arab states, like Syria or Iraq, he would only use them in his zone.

In this context, the ELF found it unnecessary to mobilize the Christians, who were considered to have dual loyalty to Eritrea and Ethiopia due to their religious sentiment. They essentially excluded progressive nationalist Christians from the liberation struggle while Christians who joined the ELF operated under a cloud of suspicion, since the ELF viewed them as potential spies for the Ethiopian regime.

A great deal of hostility and distrust also existed between the Muslims and Christians on the battlefield. These cleavages became even more acute when the Ethiopian government decided to wage a full-scale assault on these nascent forces, inflicting heavy casualties around the western and highland zone. The Ethiopian regime's first major offensive against the ELF sought to limit or destroy the effectiveness of the new Christian zone. The Ethiopian army offensive proved successful, since the ELF fighters did not have the organizational or military capacity to withstand a frontal assault by a professionally trained and seasoned Ethiopian army. Nonetheless, the ELF attributed their major defeat to the ineffectual fighting of the Christian contingents and to treachery. One commander, Osman Hishal, accused and executed 27 fighters for not fighting properly, which compelled some Christian leaders to defect to the Ethiopian side.[20] These defections only reinforced the suspicions of Islamic groups, who already viewed the Christians as traitors and union supporters.

Although the summary executions constituted a heinous crime against the Christian soldiers, in particular, and against the Christian community, in general, they did not deter Christians from continuing to join the Eritrean independence struggle. In 1970 the continuing hostility between Christians and Muslims led to a massacre of about 300 Christian Eritreans, who had previously left Addis Ababa to join the ELF. The massacre at Barka created an irreparable breech, prompting the remaining Christian elements in the liberation struggle to abandon the ELF and to form a new liberation front, the Eritrean Peoples Liberation Front.[21]

The core group of the ELF escaped to Ala, a Christian area, located in the desolate hills on a highland plateau 50 miles from Asmara. Isayas Afeworq, Mesfin Hagos, Tewolde Eyob, Solomon Wolde Mariam, and Asmeron Gebre Egzhiabher, the core of the group, together formed the central committee of the new movement. It was called the Peoples Liberation Front and was eventually transformed into the EPLF. The founding principles of EPLF rejected Arabism, denounced the use of Arabic as the official language of the liberation struggle, and argued against the subordination of Eritrean culture to Arabic culture. The EPLF also rejected Muslim sectarianism and the ethnic and regional divisions that dominated the ELF. Finally, the EPLF leadership rejected the ELF's lack of a secular political ideology. In this regard, the EPLF consciously sought to model their liberation struggle and movement on the "peoples war" theory advanced by Mao. In fact, the central committee of the EPLF and especially Isayas Afeworq developed this ideological orientation after their training in China during the midsixties, the highpoint of the Chinese cultural revolution. The EPLF's Christian ideological orientation actually facilitated its acceptance of Marxism, since Christianity, unlike Islam, divided society into two distinct realms, sacred and secular, from which came the social conditions for the emergence of civil society and revolutionary politics.

Of course, the ELF rejected the formation of the EPLF and issued a number of directives to eliminate the EPLF. These directives signaled the beginning of a brutal civil war between the EPLF and ELF as well as the demise of the ELF as a fighting and political force in the Eritrean national movement. The ensuing civil war lasted until mid-1975. Only the overthrow of Haile Selassie prompted the formation of tactical unity between the two movements. Even still, the movements were never able to join forces as a united front against the Ethiopian military government. On balance, the EPLF was the net beneficiary of the Eritrean civil war, since they survived attacks by the ELF and the Derg which, in turn, transformed the EPLF into a more effective fighting force.

The EPLF also incorporated in its struggle the ideological fusion of Leninist and Maoist precepts. From Mao, they took the idea that peasants could become a revolutionary force and the concept of a "new democratic revolution," which advocated the unity of all progressive forces in society. Following another of Mao's tenets, the EPLF proved themselves extremely self-reliant by acquiring most of their weapons from the Ethiopian army. They also established their own hospitals, collected taxes within their sphere of control, distributed food, and used overseas contributions from Eritreans abroad to buy needed supplies. The EPLF, unlike the ELF, distributed arms and military supplies on the basis of need and strategic necessity rather than for religious, regional, or ethnic reasons.

From Lenin, the EPLF derived the idea that professional revolutionaries organized in a vanguard party had the right to monopolize political and military activities in society. Hence, the EPLF included in their movement the dissident ELF followers, including deserters from the Ethiopian army and progressive Christians. In 1978 it was estimated that the ELF forces declined to 7,000 troops, while the EPLF increased to 30,000.[22] By the early 1980s, the ELF was neither an effective fighting force nor an effective political organization.

In sum, the progression of the Eritrean struggle from a sacred and religious struggle to a secular movement led by a Leninist party with a systematic fighting force transformed the Eritrean conflict into a protracted war. The fusion of the EPLF's vanguard organizational structure with Maoist ideological underpinnings enabled them to develop a more comprehensive military and political strategy than the more religiously and regionally based ELF.

The Ethiopian Response to the Crisis

Analysts can divide the Ethiopian response into two periods. The first period lasted from the undermining of the federation in 1956 until the overthrow of the emperor in 1974. To eliminate Eritrean opposition to the annexation, Haile Selassie used the strategy of incorporating antagonistic elites into the regime by giving them titles and land. Other tactics included incorporating Christians in his regime, excluding the Muslims, dismissing the ELF as bandits, and using a minimal amount of force. Selassie based the core of his strategy on a combination of coercion and cooperation.

However, famine and economic stagnation, accelerated by the dramatic rise in oil prices after the Yom Kippur War in 1973, paralyzed the regime of Haile Selassie. As popular protests became more widespread and intense throughout Ethiopia, Emperor Selassie's response of simply reshuffling his cabinet proved inadequate to deter demands for the overthrow of the *ancien regime*. Just when the regime was teetering and on the brink of failure, the military officers, observing the creation of a power vacuum, purloined the revolution from the popular movements. When the military officers established the Derg, they presumed that military rule could substitute itself for the *ancien regime*. However, the power struggles within the Derg proved that the military council as it was constituted could neither establish legitimacy nor sustain political stability.

The second period began in 1974 after Gen Aman Andom's initial attempts to achieve a negotiated peace with the Eritrean rebels failed, and the Derg assumed power. The failure of these talks fueled Mengistu Haile-Mariam's ascendancy to power and his adamant opposition to a nonmilitary solution to the conflict. Mengistu's elevation signaled the rise of military hardliners who wanted to achieve an unconditional military victory over the rebels. The Derg's exclusive reliance on coercion to resolve the issues created a zero-sum relationship between the Derg and the EPLF: either the Derg would destroy the insurgency or be defeated by the rebels. Under Mengistu, a political settlement of the Eritrean problem was unthinkable.

Mengistu's predisposition to use violence rather than negotiation was largely due to the social forces that enveloped him and which he had to overcome to gain power. First, the unrelenting power struggle between different radical, moderate, and conservative factions within the Derg led to bloody purges in the military as one faction attempted to gain hegemonic control over the Derg. Starting with General Andom, an Eritrean by origin who had favored a political settlement within the state, Mengistu eliminated all of the leading figures. General Andom was killed at his house in 1974 by a group of radicals and moderates. In February 1977 Mengistu also killed Brig Gen Terferi Banti, leader of the second division. In November of that same year, Mengistu executed his last rival for power, Atnafu Abate. By the time of Abate's execution, Mengistu had virtually eliminated all of the Derg's senior military officers.[23]

Second, the motley coalition of such urban groups as students, professors, taxicab drivers, and trade unionists that had initiated the overthrow of Haile Selassie, lost their revolution to the military because of disorganization. After the Derg seized power, these various groups could not decide if they should support or reject the Derg's leadership after the revolution. Haile Fida led *Meison,* one of the major popular groups, and called for an alliance with the progressive elements of the military. These groups advocated simultaneously politicizing the military and using it to advance the revolutionary transformation of Ethiopia.[24]

Another group, the Ethiopian Peoples Revolutionary Party (EPRP), under the leadership of Berhane Meskel, advocated a total break between progressive organizations and the military. EPRP tried to overthrow the Derg and the progressive organizations that sided with the Derg. Their widespread and popular base came largely from the working class and basically from students at different educational levels. In 1978, when they declared war on the Derg, then under the exclusive control of Mengistu, a bloody urban battle erupted, which led to the death of several thousand students. To silence the students, the military used the *kebelle* system, patterned after the "Committee to Defend the Revolution," a system the Cubans used to repress opposition. Ironically, the Marxist Meison had been a leading advocate for this type of repression, which eventually led to their own destruction.

After destroying EPRP, Mengistu destroyed the Meison, the only remaining independent political organization in Ethiopia. Mengistu executed Meison's leader and dismantled their organization, leaving himself in control of the entire state because of his virtual monopoly over the means of violence.

Nonetheless, the other factions in Ethiopia viewed Mengistu's exercise of political power as illegitimate. Even the military realized Mengistu's rise to power resulted solely from his execution of senior officers. Mass movements, which had overthrown Haile Selassie and rejected military rule, had to change their position when Mengistu unleashed *kebelle* terror in the urban areas. The only area where Mengistu could mobilize a sector in Ethiopian society effectively was the Oromo peasantry of southern Ethiopia.

Mengistu mobilized this sector around two issues. First, he either gave land or promised land to the tillers due to the inequality of feudal land holdings suffered by their region. Second, he exploited southern Oromo nationalism, which had stemmed from the Oromo's opposition to the Somali's claim of their lands.

Mengistu's two-pronged strategy enabled his regime to mobilize huge peasant armies successfully. His peasant army, reputed to be the largest in Africa, was sadly an army in name and uniform only; they were neither trained nor understood the idea of guerrilla warfare. Hence, thousands upon thousands of peasants lost their lives when they confronted the smaller seasoned EPLF army. Mengistu also used the peasant militia to control and repress any dissent in the urban areas. His skillful fusion of nationalism with socialist fragments, such as land to the tiller, allowed Ethiopia to become lost in the fog of a prolonged war against Eritrean secession.

In sum, Mengistu incorrectly believed that he could successfully substitute for both the Derg and the emperor. Since the military had usurped its power from the *ancien regime* as well as from the popular movements, Mengistu felt compelled to destroy the civilian opposition in the urban areas by using what he called "red terror," the indiscriminate killing of young students, and mass arrests. Mengistu then attempted to substitute the peasant militia for the regular army, since in his efforts to protect his power he had already decimated the senior officer corp through bloody purges. After Mengistu eliminated all effective opposition, he remained alone at the apex of power in Ethiopia, his regime illegitimate and unstable.

In his final attempt to salvage his regime, Mengistu declared himself the last of the true nationalists and offered his prolonged war with Eritrea as evidence. He garnered support for this war by becoming a pawn in the geopolitical superpower rivalry for hegemonic control over the Middle East region and by courting the United States, Israel, and the former Soviet Union to resupply his war machine. Yet as long as these countries continued to provide military support, Mengistu continued to destroy both Ethiopia and Eritrea. Only through the destruction of both countries could Mengistu's illegitimate rule achieve security.

Tigrai Liberation Movement

When the Derg replaced Emperor Haile Selassie's regime, Tigrai attitudes crystallized against Amhara chauvinism and the centralized government. Insurrection in the Tigray province was further stimulated by the lack of economic development, the imposition of Amharigna as the dominant language of administration, commerce and education, and the general resentment concerning Amhara political hegemony over Tigray.

The political vacuum created in Ethiopia by the collapse of the central government led to the emergence of two major opposition forces in Tigray: the Ethiopian Democratic Union (EDU) and the Tigrai Peoples Liberation Front (TPLF). The EDU was led by nobleman Ras Mengesha Seyoum, one of the few nobles who had escaped execution by the Derg. Seyoum sought to resurrect the *ancien regime* and restore Haile Selassie's son to the throne. Ironically, the EDU was destroyed by the combined efforts of the Derg and the TPLF, who repeatedly attacked the EDU from different levels.[25]

The TPLF, on the other hand, received inspiration from the EPLF because the constituent elements of the TPLF were Christian and Tigrigna speakers. The EPLF party was comprised of dissident members of the Tigrai intelligentsia, who emerged from Addis Ababa University, and segments of the Tigray peasanty, who had historically resisted the taxation policies of the central government. Both groups benefited from their mutual association. The EPLF benefited because the TPLF provided yet another front for the Derg to fight, while the TPLF gained because the EPLF offered tactical and strategic military support. The only significant difference between the EPLF and the TPLF was that the TPLF wanted to remain part of Ethiopia, based on a federal model, whereas the EPLF considered the Derg fascists and wanted to secede from Ethiopia.

The TPLF also engaged in a bitter struggle with the EPRP mainly because of their disagreements on the "national question." The EPRP's position emphasized that since all regions were oppressed by Haile Selassi's regime and then the Derg, a unitary party system, composed of regional and subnational entities, was required to create a new government structure. The TPLF, on the other hand, believed that every nationality had the right to organize separately and the right to self-determination. Eventually, the TPLF's forces routed the demoralized forces of EPRP, whose forces had been brutally repressed in Addis Ababa during the Derg's "red terror" campaign.

Because of its organic connection with the EPLF, the TPLF eventually found itself at odds with another liberation movement in the North, the Eritrean Liberation Front (ELF). The TPLF had to traverse through ELF territory when it procured supplies from the Sudan. Such circumstances made a conflict between TPLF and ELF inevitable. Eventually, the combined forces of the EPLF and the TPLF defeated the ELF which, in turn, made the TPLF more dependent on the EPLF than before the conflict.[26]

After the fall of the Derg and the dissolution of the Ethiopian armed forces, the TPLF with the support of the EPLF formed a coalition government called the Ethiopian Peoples Revolutionary Democratic Front (EPRDF). The leader of this front was Meles Zenawi, also the leader of the TPLF. Observers of the Ethiopian transitional period believe that the EPRDF was only a facade; the dominant force in the transitional Ethiopian government, Ethiopia was really the TPLF. However, the monopolization of power by TPLF during the transitional period is presently generating considerable resentment and fear of Tigray domination of the Oromo, Amhara, and Gurage and other smaller ethnic groups.

Conclusion

As the sunlight of Glasnost and Perestroika pierced the fog of the cold war, Mengistu's effort to support his regime continued in a self-destructive search for a military solution to the Eritrean problem, which already had destroyed the socioeconomic structure of Ethiopian society. Mengistu's capacity to raise huge peasant militias became extremely difficult. Seventeen years of war, famine, and Leninist economic policies had become an unmitigated disaster. To solve the country's political problems, Mengistu essentially had mortgaged his country's future to buy weapons. The officious nature of Mengistu's declining regime was clearly indicated by his sale of Falashas (Ethiopian Jews) to Israel for money and military supplies.

As Mengistu's reign began to fall apart, evidenced by increasing troop defections in 1990, continued economic stagnation, and the failure of his economic reforms to stop the hemorrhaging of his regime, the fragility and hollowness of Mengistu's dictatorship unraveled. These social forces contributed to a major revolt by the same generals Mengistu had appointed to conduct the war in May 1989. The failed coup of these personal appointees, however, further alienated Mengistu from his troops, the main pillar of his support. Because Mengistu's government had been maintained exclusively by military power, his rule became untenable when his troops began to feel alienated from his regime. Two years later, in June 1991, as the last fragments of his motley armed forces disintegrated or refused to fight, Mengistu fled the country, and the EPLF assumed control of Eritrea.

In sum, observers can attribute the success of the EPLF to wage a successful protracted struggle to a combination of propitious internal circumstances. First, the Christian orientation of the leadership of the EPLF allowed them to divide society into sacred and secular spheres, which in turn allowed the mobilization of women and Muslims. The division of society into sacred and civil spheres gave rise to what Weber calls "instrumental rationality" (a cost-benefit analysis of the relationship between means and ends). Such reasoning legitimated the mobilization of segments of Eritrean society because the goal was liberation, not religious salvation.[27]

Since Islam does not divide society into sacred and secular realms, it considers heretical all political behavior which is inconsistent with Sharia law or Islamic theology. Hence, in the Islamic context "instrumental rationality" is precluded because the principles of Islam are immutable and not subject to mitigation. On this view, Islamic theology constrains the conduct of a major prolonged war because it cannot mobilize all segments of society. In Afghanistan, questions of ideological purity also have divided the guerrilla forces which, in turn, limited their capability to wage and win a prolonged war.

Moreover, it is highly unlikely if someone can organize women or other nonbelievers as fighters in an Islamic jihad, while Christians as long as they do not have to denounce their faith can rationalize fighting with non-Christians and others. One significant effect dividing society into a secular part is that it facilitates the rise of Marxism-Leninism, the quintessential rational ideology of revolution. The rise of Marxism-Leninism as the dominant ideology of insurrection movements is important because it leads to the development of an "organization," "strategy," and "tactics," which are indispensable features for a successful protracted war.

Second, Eritrea's inaccessible terrain allowed the guerrillas to maintain a permanent presence inside the country. Most successful protracted wars in this century have occurred in environmental regions that contain large areas of inaccessible terrain. For Mao, it was the Yenan area and for Ho Chi Minh, the mountains and the jungle area. The presence of the EPLF in all areas of the Eritrean countryside created a symbiotic relationship between the armed forces and the people, which helped to reinforce their strategic position.

Third, the illegitimate nature of the Ethiopian government contributed to the EPLF's success because the will to fight was missing in the Ethiopian troops. Even through Mengistu had assembled one of the largest armies of men and equipment in Africa, the people's will to fight was absent.

Finally, Mengistu's slaughter of the urban intelligentsia, students, workers, and dissident forces created an ideological and political vacuum that constrained his ability to generate the popular support needed to actively defeat the EPLF's protracted war strategy. Hence, circumstances eventually forced Mengistu to rely on air power as the only weapon to prevent an outright EPLF military victory. After several coup attempts, he began to question even the loyalty of his generals and officers. It became only a matter of time before his regime collapsed. The rise of Gorbachev and the ending of the cold war only accelerated the process.

Presently in Ethiopia, the Eritrean problem has been settled with regard to Ethiopian domination. However, the question of how the remaining nationalities can or will construct a new nation-state, predicated on democracy and liberty, remains the leading challenge of the newly reconstituted states of Ethiopia and Eritrea.[28]

Notes

1. See Richard Greenfield, *Ethiopia: A New Political History* (New York: Praeger, 1965); Donald Levine, *Wax and Gold* (Chicago: University of Chicago Press, 1965); and Dame Margie Perham, *The Government of Ethiopia* (London: Faber and Faber, 1969).

2. See Matthies Volker, "The Horn of Africa and International Relations," *Intereconomics* 12 (December 1974); and T. B. Millar, *The Indian and Pacific Oceans: Some Strategic Considerations*, Adelphi Papers no. 57 (London: The International Institute for Strategic Studies).

3. G. K. N. Trevaskis, *Eritrea: A Colony in Transition* (London: Oxford University Press, 1960), 8.

4. Dielbo Getahun, "Historical Origins and Development of the Eritrean Problem 1889–1962," *Current Bibliography on African Affairs* 7 (Summer 1974): 226–31. See Stephen Longrigg, *A Short History of Eritrea* (Oxford: Clarendon Press, 1945).

5. Terrefe Woldestadik, "The Unification of Ethiopia 1880–1935," *Journal of Ethiopian Studies* 6, no. 1 (1968).

6. Richard Sherman, *Eritrea: The Unfinished Revolution* (New York: Praeger Publishers, 1980), 20–30.

7. Eric J. Hobsvawam, *Nations and Nationalism Since 1780* (Great Britain: Cambridge University Press, 1990), 15–30.

8. Resolution 390 (V) Eritrea—Report of the United Nations Commission for Eritrea, "Report of Interim Committee of the General Assembly on the Report of the United Nations Committee for Eritrea from the Fifth United Nations"; and General Assembly, 316th Plenary Meeting 1950—Official Records of the General Assembly, 5th sess. supplements, no. 8, 1c.

9. Trevaskis, 120; and Araia Tseggai, "The Case of Eritrean National Independence," *Black Scholar* 7 (June 1976): 20–27.

10. Trevaskis, 131.

11. Mesfin Wolde Mariam, *An Introductory Geography of Ethiopia* (Addis Ababa: 1972).

12. John Markakis, *National and Class Conflict in the Horn of Africa* (London: Zed Books, 1990), 174–75.

13. Haggai Erlich, *The Struggle over Eritrea 1962–1978* (Palo Alto: Hoover International Studies, 1982), 21–22.

14. See Sherman, 40–45; and Tom J. Farer, *War Clouds on the Horn of Africa: The Widening Storm*, 2d ed. (New York: Carnegie Endowment for International Peace, 1979), 35–45.

15. Gebre Medhin Yordanos, "Eritrea Background to Revolution," *Monthly Review Press* 28 (September 1976): 52–61.

16. See Spencer J. Trimingham, *Islam in Ethiopia* (London: Oxford University Press, 1952); and Eritrean Liberation Front, *The Eritrean Revolution: Sixteen Years of Armed Struggle* (Beirut: ELF Foreign Information Center, 1977).

17. Erlich, 25–35.

18. Lloyd Ellingson, "The Origins and Development of the Eritrean Liberation Movement" in Robert Hess ed., *Proceedings of the Fifth International Conference on Ethiopian Studies* (Chicago: 1978), sess. B.

19. Markakis, 122.

20. Ibid., 126.

21. Erlich, 88–96.

22. Eritrean Peoples Liberation Front, *In Defense of the Eritrean Revolution* (New York: 1978), 145.

23. See John Markakis and Nega Ayele, *Class and Revolution in Ethiopia* (Nottingham: Spokesman, 1978), chap. 1.

24. Ibid.

25. Max Peberdy, *Tigray: Ethiopia's Untold Story* (Rest Publications, 1985), 15; and Dawit Wolde Giorgis, *Red Tears* (New Jersey: Red Sea Press, 1989), 35–40.

26. Fred Halliday, *The Ethiopian Revolution* (London: New Left Books, 1980), 15; and David and Mariana Ottaway, *Afro-Communism* (New York: Holmes & Meirer, 1981), 128–30.

27. See Wolfgang Schlucter, *The Rise of Western Rationalism: Max Weber's Developmental History* (Berkley: University of California Press, 1981).

28. Christopher Clapham, *Revolutionary Ethiopia* (Great Britain: Cambridge University Press, 1990).

Chad
The Apparent Permanence of Ethno-Regional Conflict

Frédérick Belle Torimiro

Until the mid-1980s Chad was unknown to most of the world, even though it is the fifth largest country in Africa (four-fifths the size of Alaska). Like many other African countries, the brusque and arbitrary manner in which Chad's colonial boundaries were drawn reflected more the power politics of European states than local interests. The outcome was the creation of artificial borders compelling entirely dissimilar ethnic groups to live together. Chad's population, now estimated at 5.2 million, has come to represent what Samuel Decalo calls a "huge ethnic mosaic" of over 100 different languages.[1] The religious distribution is approximately 50 percent Muslim, 43 percent animist, and almost 7 percent Christian (primarily Catholics). The northern part commonly referred to as BET (Borkou, Ennedi, Tibesti) occupies about one-third of the national territory, although it is sparsely inhabited with about 6 percent of the population. By contrast, the southern area, or *Le Tchad utile* (useful Chad), has the bulk of the population and occupies only one-tenth of the total territory.

What has been termed the "north-south dialectic" is obviously an important facet of Chad's political development or decay.[2] Such a dichotomization does not however "capture the contextual complexities of ethno-regional conflicts over time."[3] It obscures the competitive and conflictual interplay of ethnic groups that share similar geographical boundaries or primordial elements. The prevailing assumption in the north-south dialectic is that the combative foes are easily identifiable and may be neatly separated into two distinct camps. However, the composition of the Chadian population makes it difficult to analyze the character of the conflict from a simple north-south perspective. In the north, for example, there are significant ethno-regional groups which are fragmented into subcultures, even though Chadic Arabic remains the *lingua franca*. The region's politics is therefore obliged to reflect the competing, if not conflicting, interests of the seminomadic, reticent, and fiercely self-reliant Toubou, who are subdivided into the Teda of Tibesti and the Daza of Borkou and Ennedi; the "sahelian" population of semisedentary groups with traditional strongholds in the eastern regions of Quaddai, Biltine, and the sultanates of Baguirmi in the West; the Arabs; and the Hausas and Fulanis, who are modern arrivals to the territory.

The multiplicity of ethnic groups and interests also poses a persistent challenge to the ideas that Islam is useful in institutionalizing sociocultural and political cooperation in the north. It does not always offer the Muslims an adequate *modus vivendi* or permanently provide the incentive to muster a strong and collective opposition against the non-Muslims. As René Lemarchand points out:

> One needs only to recall the bloody feuds that have periodically pitted Arabs against Arabs, as happened in 1947 when the Missiryes turned against the Rattatinine, resulting in 180 deaths, and again in 1972 when the same Missiryes slaughtered 120 "rebels" affiliated with the Front de Libération National Tchadien (Frolinat). Similarly, the deadly struggle between Hissene Habré and Goukouni Weddeye is also a trial of strength between two separate segments of the Toubou cluster, that is, Annakaza versus Tomagra.[4]

These events reinforce the assertion that ethnic and regional cleavages are intrinsic components of Chad's sociopolitical reality. At the very least, they expose the shaky relationships between the various factions in the north.

In terms of the south, the cultural and linguistic dominance of the Sara conjures the image of homogeneity. This perception fails to recognize that there are different clans within the larger Sara community. For instance, it distorts the cultural differences among the Ngambaye, Madjingaye, and Mbaye, just to name a few. What becomes most apparent is that intra-Sara unity is linked to their desire to politically control Chadian society. This means that any useful analysis of the strength and fragility of Sara cohesion must take into account the distributive or allocative capacity of the country's political leadership. In this context, the competition and distribution of political and economic fortunes or "spoils" provide a basis for sociocultural solidarity or fragmentation.

Equally important, Chadian ethno-regional fragmentation is related to its colonial experience. Although the BET remained under French military administration until 1965, and the East was periodically suppressed, there was some tolerance for local autonomy in the north. By adopting a policy of indirect rule or more accurately "omission" for political and administrative convenience, the locals were able to maintain their traditional authority. The sultans in Quaddai, Baguirmi, and Kanem, for example, ruled their people, even though ultimate political control still remained in French hands. This administrative option permitted the French to concentrate on the south (*Tchad utile*), where they implemented an assimilation policy. The incorporation of the southerners into a westernized culture was enhanced by their access to French education and upward mobility in the administration, military, and political life of Chad. Thus, by the time of independence, the westernized south had established its "beachhead" in the battle to gain political supremacy in the country. The point here is that instead of creating coparcenaries of Chad's political future, the colonial legacy fostered ethnic and regional animosity and distrust.

This chapter analyzes elements which might be brought to bear on the conflicts in Chad. It also assesses the pattern of political decay and any potential national reconstruction in terms of the dynamic character of ethno-regional animosities, personal ambitions, and external influence. What is implied here is that the scope of Chad's sociopolitical instability must take into consideration the impact of endogenous and exogenous factors. It is important to determine if these factors contribute to the prolongation of the conflicts in Chad and indispensably to any resolution scheme.

A Theoretical Foundation

One of the important concerns of international relations is the explanation of state behavioral ideals and attributes which leads to various levels and intensity of conflict. The concept of conflict is understood as the outcome of structural and perceptual incompatibilities that yield mutually exclusive and overt behaviors that may be violent. Put differently, it is a condition in which disputes over goals, interests, resources, and values between two or more identifiable groups may be of such magnitude that any resolution will adversely affect one of the contending parties. As Charles Lockhart suggests, "Conflict episodes can follow a variety of different patterns as they develop."[5] The implication is that the manipulation of domestic instruments of power and the influence of the international environment may be salient to any discussion of conflict prolongation in Chad.

What K. J. Holsti terms the "issue field," which is the "subject of contention between the parties and includes the positions they are attempting to achieve," may be complex and difficult to resolve.[6] The conflict becomes prolonged as civilian and military interventions fail to halt the rising tensions or guarantee a complete victory. The prolongation of the conflict becomes a deliberate means used by the competing forces to destroy the will of the other. The relevance of this distinction in the case of Chad is that it encourages a closer look at whether or not the ethno-regional hostilities have been strategically designed to persist. The following analyses therefore seek to determine if the persistence of conflict in Chad since its independence is a result of careful planning by the competing political actors and interests. These analyses also attempt to establish any possible connection between the character of the Chadian conflicts and the external environment. In other words, can it be said that the external influence was a deliberate and calculated effort to maintain a climate of perpetual violence in Chad? Or, has the dynamics of Chad's internal politics and the attraction of external actors resulted in the prolongation and not deliberate protraction of conflict?

The Internal Metamorphoses of Conflict Prolongation

A significant aspect of this study is analyzing the impact of endogenous factors in explaining conflict prolongation in

Chad. This method of analysis recognizes that the development of a modern state apparatus is challenged by

> raw power struggles among competing pressure groups (often communal or regional in character) striving for control of the political machine, for a greater share of economic rewards, for status, and for privilege. These rivalries sometimes occur between new interest groups or classes, between established and ambitious ethnic communities, or between new challenges and those wishing to defend a previous dominant position in a particular area.[7]

Or, as James O'Connell states, "Dissent and conflict in society are centered on the control of political authority, because it is seen as the main source of the allocation of rewards in the form of status, roles, and wealth."[8] In this sense, the absence of well-developed "national linkages and national identity" results in group fragmentation and resistance against any singular effort to expand the scope of "national authority."[9] The major concern rests on the question of who should govern, and by focusing on the following internal factors, it is possible to determine if conflict prolongation is a consequence of institutional dysfunction or structural weakness of the sociopolitical and economic system.

In the Chadian case, the idea of cooperating for a "common good" has been adversely affected by the pursuits of narrowly defined interests and goals. For over 44 years, party leaders have established and maintained their spheres of influence by bonding their political machinery to specific groups. This political orientation was exemplified by the failure of the Parti Progressite Tchadien (PPT) (formed by Gabriel Lisette) to emerge as a genuine nationwide party.[10] The PPT was perceived by the northern leaders as the "political wagon" of the non-Muslim population and challenged by northern-based party organizations. These organizations included the Union Démocratique Indépendante du Tchad, and the Groupement des Indépendants et Ruraux Tchadiens.

The formation of these splinter parties also explained the prevalence of incompatible objectives among the leaders. A variety of political organizations was created as personalities clashed or as party leaders built new alliances designed to maximize their political ends. The political "marriage of convenience" provided additional avenues for the leaders to broaden their support base or to gain access to the reins of government. For example, the political union between the Union Nationale Tchadienne of Marxist Ibrahim Abatcha and the Mouvement National de Libération du Tchad of the conservative Muslim Ahmed Moussa resulted in the organization of the Front Pour la Libération du Tchad (FROLINAT). This was a political marriage aimed at countering François Tombalbaye's push toward one-partyism and Sara domination of the government. However, the creation of FROLINAT did not help to eliminate the differences in long-term goals between the partners. It was therefore a union predicated on the purpose of correcting the political and social injustices in Chad but lacking a consensus on the definitions of social justice.[11] More specifically, a fierce internal struggle ensued for control among the various leaders in FROLINAT. In one of its loosely linked member groups, called the Second Army or the Forces Armées du Nord (FAN), the internal squabbles led to a split in 1976 between Hissène Habré and Goukouni Oueddei.

The Consequences of the Goukouni-Habré Rift

Unquestionably, the Goukouni-Habré split revealed profound disagreements among those seeking to reconstruct the political landscape of Chad. The immediate result of this rift was that it spawned new warring factions, including Habré's FAN, the Forces Armées Populaires (FAP), which was the outcome of a merger of Goukouni's forces, and the Volcan army of Ahmat Acyl, which had replaced the First Liberation Army, and Aboubakar Abdelrahmane's Third Army or the Populaire pour la Libération du Tchad. These rival warlords sought to expand their spheres of influence by squabbling with each other and by undermining the leadership of Gen Felix Malloum, who had ousted the postindependence regime of Tombalbaye. This crisis of political legitimacy intensified as Libya extended military and humanitarian aid to Goukouni and Acyl. Libyan support and the outbreak of fighting in July 1977 boosted Goukouni's position. Not only was Malloum's Forces Armées Tchadiennes (FAT) defeated, but they were forced to abandon the northern towns of Bardai, Zouar, and Ouri.

Malloum was compelled to initiate a rapprochement with Habré, due to the military threat posed by Goukouni's forces. Habré was perceived as a likely ally since he had shared Malloum's rejection of Libya's occupation of the Aouzou Strip and had built his military capability, with Egyptian and Sudanese support, in the eastern towns of Biltine and Abéché to approximately 1,000 men. However, any outcome of great consequence was preempted by an offensive launched by Goukouni's forces in January 1978. This military campaign resulted in the capture of Ounianga-Kebir, Fada, and Faya-Largeau by the end of February. The significance of these victories was that Goukouni controlled much of BET.[12] It also brought Malloum to cease-fire conferences involving Chad, Niger, Libya, and Sudan in Sebha and Benghazi, Libya. In the Sebha talks of February 1978, Malloum conceded to abandon his charge against Libya over the Aouzou Strip. Similarly, the Benghazi peace agreement in March compelled the Malloum regime to recognize FROLINAT as the political organization which symbolized the true will of the people in Chad.[13]

The fragility of these agreements was made evident by renewed fighting seemingly encouraged by Acyl's forces in April. French intervention with ground and air support was nevertheless crucial, since Malloum's FAT repelled FROLINAT advances in battles at Ati and Djedaa. More specifically, 1,500 men of the French expeditionary military force, 10 Jaguar fighter aircraft, troop transportation, and refuelling and electronic surveillance planes were used to stave off Acyl's attacks.[14] The immediate fallout of the French assistance was the break up of the FAP. The Goukouni-Acyl schism allowed Habré and Malloum to sign an agreement in August 1978. The resulting Fundamental Charter received the full support of France and also paved the way for a new national government. Malloum retained the presidency, while Habré was appointed prime minister. Of course, what became a serious weakness of the charter was the exclusion of Goukouni in the political system. This attempt at power sharing also was handicapped by distrust and eventually collapsed because of Habré's plot to forcibly overthrow Malloum. The upshot was the eruption of a civil war in N'Djamena between Habré's and Malloum's forces in February 1979 as well as a renewed Goukouni offensive in the north.

The First Battle of N'Djamena

The temporary collaboration between the FAN and the FAP tilted the balance of power in favor of the northerners. Malloum's southern-controlled government fell apart, and the FAT was forced to withdraw farther into the south. The tragic outcome of the first Battle of N'Djamena was that it aroused a new wave of communal violence, particularly between Muslims and non-Muslims. There were massive killings of southerners as the FAN established a stronghold in N'Djamena and at Abéché and Biltine. This was subsequently followed by a brutal revenge on thousands of Muslim civilians in the southern prefectures. It is estimated that roughly from 5,000 to 10,000 Muslims were killed in the south.[15]

Arguably, conflict prolongation in Chad may be a consequence of ineffective or the lack of political leadership. The attempt to fill the leadership vacuum resulting from the military campaign in N'Djamena was marked by the peace conference of March 1979 in Kano, Nigeria. The Kano talks produced a pact which sought to mend the rift between Habré and Goukouni. In the eight-man Provisional Council of State in which Goukouni became president, Col Wadal Abdelkader Kamougue (a southerner) and Habré were designated to serve as vice president and minister of defense, respectively. It was also agreed that the French would withdraw their forces under a pledge sanctioned by Cameroon, Central African Empire, Libya, Niger, Nigeria, and Sudan. Needless to say, the political legitimacy of the coalition government was strongly challenged by the rival warlords excluded from Kano I. Acyl's and Abba Siddick's insistence on recognition as members of the Council of State precipitated the failure of Kano II, held in April 1979. A similar demand of recognition came from Mohammed Abba Said of the People's Liberation Army and Adoum Dana of the First Army. The result was the resumption of a violent struggle for power.

It stands to reason that neither Goukouni nor Habré vigorously searched for legitimacy by pledging a government based on consensus. The political-military maneuverability of these warlords and the other rival groups was geared toward expanding their territorial claims within Chad. Notably, the Goukouni-Habré rift and the first Battle of N'Djamena lend credence to the following words of Gen Olesegun Obasanjo:

> Most of Africa's inheritors of political independence spent inordinate time not only "establishing" themselves to ensure personal and political survival, but also hunting down and dealing with "enemies," real and imagined. The kind-hearted allowed their opposition to go into exile or put them in prison; others put theirs under the soil.[16]

The efforts at coalition politics only demonstrated the spirit of distrust and violence among the various factions. Quite simply, it would appear that the prolongation of the Goukouni-Habré squabble for power and the challenges from the other forces were aided by the lack of a workable and long-lasting politics of accommodation.

The External Stimuli of the Chadian Conflict

The Chadian conflict assumed an international character as various external forces penetrated its domestic realm. The focus is not simply on the local warlords but also on the global and regional actors, who were eager to influence the events in the country. It is apparent that the internationalization of the conflict stresses the extent to which the local competing groups are receptive to outside assistance. What is pointed out is that any

> struggle for power can turn violent when opposition groups feel a strong sense of exclusion from the political system, a deep fear of domination by a major communal group, and bitter grievances about regional neglect. Violence is likely to occur when such groups consider themselves strong enough to resist—and even more likely when they feel they can attract external support.[17]

The rival groups seek political leverage by attaching themselves to foreign interests. The number of these outside parties increased as the conflict was transformed from a domestic to an international phenomenon. They found it essential to intervene in the internal affairs of Chad even though their objectives were not strongly connected to a durable peaceful solution.

The Libyan Nexus

The most significant and highly controversial aspect of Libyan-African interaction has been that nation's pursuit of interventionist policies. In broad descriptive terms, Libyan adventurism in Chad has often conjured images of neocolonialism and irredentism. Col Muammar Qaddafi's activities have had a political and a military dimension. From a politico-military standpoint, Qaddafi used Islam to link Libya's foreign policy interests with the struggle of the Muslim population in Chad to secure a greater share of political power and to reduce the intrusive effects of westernization. As maintained by Qaddafi, Chadians "are Muslims of Arab origin. They are subjected to plots, divisions and minority rule."[18] It is no wonder, then, that the Libyan leadership perceived the various conflicts in Chad as part of its anti-Western campaign.

Although Qaddafi's effort to include the conflict into the broader objectives of Dar-al-Islam was not clearly defined, his mere involvement delayed its resolution. The prolongation of the conflict was boosted by Qaddafi's ability to use the principle of "divide and rule." The pursuit of this interventionist policy was most evident with the annexation in 1973 of the uranium-rich Aouzou Strip on the northern end of the BET. It demonstrated Qaddafi's ambition to expand his "Islamic empire" deep into Africa as well as accounted for the rift between Goukouni and Habré. While Goukouni agreed to cooperate with Libya, Habré vehemently opposed such profound external control of a piece of northern Chad. Perhaps what may be recognized here is that what René Lemarchand refers to as "crass opportunism" sharpened the edges of factional politics in Chad.[19] Qaddafi saw the opportunity to feed and benefit from the political ambitions and personal greed of the rival forces. By the same token, some of the warlords gravitated toward Libya to foster their own political agenda. The openness of the factions to Libyan or other external financial and military support suggests that they had become clients in search of a patron. Libya became one of the patrons whose activities intensified the post-1980 conflicts in Chad.

It is generally acknowledged that Libya's invasion in 1980 was at the invitation of Goukouni. The military defense pact signed between Qaddafi and Goukouni allowed about 5,000 Libyan troops to occupy northern Chad and to be strategically positioned within 40 miles of N'Djamena. The Second Battle of N'Djamena in December 1980 was therefore marked by the presence of Qaddafi's Islamic Legion which assisted Goukouni's FAP in its violent struggle for power against Habré's FAN. The Islamic Legion was equipped with US Chinook helicopters, Soviet multiple rocket launchers, 81-mm mortars, and roughly 60 Soviet-made T-54 and T-55 tanks.[20] The participation of the well-armed Libyans in the push for N'Djamena as well as the military support received from Kamougue's FAT guaranteed Goukouni's victory. The triumph of Goukouni forced a retreat of Habré's troops to Cameroon and to Sudan.

Goukouni's control of N'Djamena eventually led to the establishment of the Gouvernement d'Union Nationale de Transition (GUNT). The GUNT was faced with the herculean task of rebuilding a war-torn capital, a paralyzed economy, and a country fettered by ethno-regional animosities as well as persistent personal ambitions. At the same time, the Libyan invasion had aroused the suspicion and disdain of neighboring states like Nigeria, Sudan, and Egypt and such patrons as France and the United States. The announcement of an agreement to merge both countries further heightened concern about Libya's interventionist goals in Chad. This action propelled the Organization of African Unity's (OAU) Ad Hoc Committee on Chad to convene in Lomé on 14 January 1981. The Lomé conference rejected the planned union and called for Libyan withdrawal from Chad.[21] It was agreed that the OAU would sponsor the creation of a peacekeeping force.

Unfortunately, the replacement of Libyan troops by an OAU peacekeeping force failed to bring peace to Chad.[22] Apart from Goukouni's inability to consolidate his power, Habré had rebuilt the FAN with Sudanese and American support. The revitalized FAN successfully launched an offensive from its station at Abéché to regain control of the BET and such places as Oum Hadjer and Ati. By June 1982 Habré had pushed into N'Djamena with hardly any resistance from the GUNT. The recapture of the capital by Habré's forces caused Goukouni to flee to Cameroon. At this juncture, Habré emerged as the new Chadian leader and unified the FAN and the FAT to form a national army or the Forces Armées Nationales Tchadiennes (FANT). However, Habré's ascendancy to power was seriously challenged by Goukouni, who used Libyan support in his absence from N'Djamena to revamp his military capabilities. From the northern town of Bardai and assisted by 2,000 Libyan troops and MiG fighters, Goukouni took control of Faya-Largeau on 24 June 1983.[23] GUNT hegemony was also established in most of northern and eastern (including Abéché) Chad. This imminent threat to Habré's leadership prompted an increased military assistance from the United States, France, Egypt, and Sudan and a paratroop unit from Zaire. The injection of this new life into the FANT resulted in the recapture of Faya-Largeau on 30 June, although it was lost again to the GUNT two months later.

The Character of French Military Involvement

In 1976 Chad signed a military agreement with France. It is therefore not surprising that the internal squabbles for power and Libya's intervention drew attention to the readiness of the French to meet the demands of its security commitment. This was particularly essential since Chad was more of a strategic than an economic interest to the French. Chad served as a "buffer state, partly shielding other French protected states (most immediately, Cameroon, Niger, and Central African Republic) from invasion or subversion from territories beyond French influence."[24] Libyan interference in the conflict should have provided the French with the opportunity to forcefully establish its presence in Chad. David Yost estimates that between 1960 and 1973 Chad received about 30 percent of all French military assistance to black African states.[25] The French even assumed a brief fighting role when their assistance was sought by Tombalbaye and Malloum. Nevertheless, the French have been uncertain of their responsibility toward Chad and played the role of a reluctant partner. They remained neutral in much of the fighting among the various factions. Perhaps the logical assertion is that the lack of a full-scale French military operation against Libya or the warring factions caused the prolongation of the conflict.

The cautionary disposition of France toward the conflict in Chad was essentially maintained by François Mitterrand and the Socialist party. This orientation provided a sharp contrast to the interventionist outlook of the Gaullists and the Giscardians. It would appear that the Socialists were concerned about Chad's sovereignty and the increasingly negative public opinion in France to any military adventurism in Africa. The escalation of Goukouni's Libyan-backed insurgency lessened the reluctance of the French to become embroiled in the Chadian conflict. In the so-called Operation Manta (Stingray), 3,000 French troops supported with Jaguar and Mirage fighter planes were sent to stymie the Libyan assault. Although the operation failed to prevent the fall of Faya-Largeau, it was nonetheless considered to be France's largest military campaign since Algeria.[26] The French troops established an east-west defensive ("red") line at the 15th parallel that stretched from the towns of Abéché, Arada, and Biltine in the east to Salal and Moussoro in the west. The French and 2,000 Zairian troops created a buffer between Habré and Goukouni forces.[27] Equally important, the cessation of fighting raised the expectations of a proposed OAU national reconciliation conference intended to bring an end to the conflict.

As it turned out, a controversy over protocol matters preempted the Addis Ababa talks scheduled for 9 January 1983.[28] Habré refused to participate in the conference because of what he perceived as the "presidential" treatment Goukouni received from the Ethiopian leader Mariam Mengistu. In return, Goukouni rejected the idea of meeting with Tahar Guinasson, the minister of interior and security, who was appointed by Habré to lead the government's delegation. The eventual cancellation of the reconciliation talks led to the outbreak of fighting. In effect, the renewed fighting was precipitated by the attack of Goukouni's Libyan-backed forces on the French military post of Ziguey in northern Kanem. The real significance of the assault is that it drove the French to extend the defensive line 60 miles north to the 16th parallel (along the Koro Toro-Oum Chalouba line).[29] In short, Chad was literally divided with Habré's French-supported forces in the south and the Libyan-backed GUNT in the north. Operation Manta had helped to establish a military stalemate.

The United States and the "Libyan Element"

It is a fundamental assumption that United States involvement in Chad is intimately related to the Libyan factor in the conflict. Much of the impetus in the US for support of Habré's forces has been connected to Qaddafi's anti-Western stance. The Reagan administration saw Libya as a violently aggressive "outlaw state" whose activities had to be halted.[30] The responsiveness of the US to Habré's request for military and financial assistance was demonstrative of the United States' desire to undermine the "empire building" attitude of Qaddafi. To support the war campaign of Habré, the Reagan administration in July 1983 provided him with $10 million in nonlethal military supplies, which included food, fuel, vehicles, clothing, and tents. It was expected that a portion of the aid would be allocated to the Zairian troops in support of Habré. The amity between Reagan and Mobutu Sese Seko of Zaire seemed to be based on their similar view on the threat to African stability posed by Qaddafi's aggression against Chad.

United States interest in the Chadian conflict was most evident in the battle for Faya-Largeau in August 1983. To counter the escalation of air attacks by the Libyans, the US offered antiaircraft weapons, trucks, guns, and ammunition to the Habré government. The Chadian army received 30 US Redeye and Stinger shoulder-fired antiaircraft missiles and military advisers. The Reagan administration also sent two airborne warning and control system surveillance (AWACS) planes, eight F-15 jet fighters, and approximately 550 US personnel to Sudan. The implication here is that the US was not eager to directly intervene in the fighting. As President Reagan stated, Chad "is not our (US) primary sphere of influence."[31] In fact, the US took the position that the AWACS in Sudan would only be used if Mitterrand were willing to upgrade French military involvement in the conflict. It is even claimed that by organizing Operation Manta, the US may have indirectly, if not directly, pressured France to take a stronger stance in Chad's destiny.

Collapse of the Military Stalemate

The military stalemate by the first half of 1984 was enhanced by the mutual agreement by France and Libya to withdraw their forces from Chad. Although the French troops pulled out by the end of 1984, Libya reneged on its pledge to leave by retaining about 5,000 soldiers in northern Chad. However, the most powerful jolt to the stalemate occurred when Libyan-backed GUNT forces and a unit of the Islamic Legion attacked FANT positions south of the 16th parallel. The offensive involved the towns of Oum Chalouba, Ziguey, and Kouba Olanga and was successfully repelled by Habré forces. The French responded by increasing their military shipment to FANT and by sending a squadron of Jaguar fighter-bombers stationed in the Central African Republic to raid the Libya airfield at Wadi Doum in northern Chad. What would appear to be a recurring trend, the Libyans retaliated by using a Soviet-made Tupolev 22 bomber on a raid at N'Djamena's airport.

Of great importance, the Libyan counterattack exposed the permeability of the redline. It also led to the redeployment of 1,200 French forces in Chad under the so-called Operation Epervier (sparrowhawk). But more important was the capture of the Libyan air base at Ouadi Doum by FANT forces in March 1987. This was a worthwhile military action, since Libyan and GUNT forces were compelled to withdraw from Faya-Largeau. The retreating forces also left behind military equipment valued at $1 billion.[32] Habré was now in control of northern Chad with the exception of the Aouzou Strip. One can therefore argue that his push into Aouzou in August was designed to completely liberate all of Chad's territory. The failure to achieve this goal was perpetuated by the refusal of France to support Habré. The French were uneasy about Habré's push north, since their troops could be drawn into the battle. By contrast, the US perceived the attack as an opportunity for Chad to humiliate Libya's forces.

The notion that the internal unrest in Chad persists is reaffirmed by the rise of Gen Idriss Deby to power on 4 December 1990.[33] The conflict between Habré and Deby was sparked by the 1989 abortive coup, which pushed the latter to flee to Sudan's Darfur province. What ensued were cross-border attacks that rekindled the civil war in Chad. Deby's Mouvement Populaire du Salut (Patriotic Salvation Movement—MPS) was aided by Libya and Sudan. It is estimated that the Libyans provided the MPS with 200 Toyota desert cruisers armed with 23-mm Soviet-made cannons and Brazilian-built six-wheel armored vehicles with 90-mm guns.[34] Habré did not only have to face a Libyan-equipped rebel force but also a decision by France not to allow its 500 soldiers in Abéché to intervene. The French described the fight between Habré and Deby as an internal matter and directed the efforts of their 700 additional troops toward protecting their 1,250 nationals.[35] Thus, the success of Deby's insurgency was facilitated by the entente between Deby and Qaddafi as well as the refusal of France to assist Habré.

Admittedly, the MPS is now faced with the ultimate challenge of altering the fabric of government authority in Chad. The MPS's call for multipartyism and democracy, probably inspired by the growing demand for democratization programs in several Françophone African countries, is now put to a real test. The guerrilla activities of the Movement for Democracy and Development (MDD), including the January 1991 attack on N'Djamena, has increased Deby's dependence on France. Ironically, the French government responded to Deby's call for help by sending in 450 paratroopers. The justification of this decision was linked to Deby's democratization effort.[36]

Toward a Conclusion

On the whole, this study has centered on the complexities of the conflict in Chad. The examination of the various battles sought to explain the prolongation of the Chadian conflict. In the first place, the absence of political leadership is notable in explaining the prolongation of the conflict. This emphasis reveals that conflict resolution might depend on political leaders who build successful coalitions or advance the politics of compromise among the rival warlords. The connectedness of leadership performance and conflict prolongation in Chad is one that has demanded a careful appraisal of the rivalries between Habré, Goukouni, Malloum, Deby, and the remaining rival leaders. Their strong desire to gain control of government at any cost meant that Chad would be infected with lengthy ethno-regional fighting. In other words, only a decisive victory in which the rivals and their forces are completely eliminated would the likelihood of political stability prevail in Chad. However, the remnants of a rival faction willing to fight for control of Chad's political systems make the search for peace a remote possibility.

A slightly different and equally important analytical focus is the internalization of the Chadian conflict. It is assumed that conflict prolongation in Chad is in part a function of the external political environment. The internationalization of the fighting was maximized by the presence of France, Libya and its wide variety of sophisticated Soviet-made weapons, the US, and the several African states (including Nigeria, Sudan, Egypt, Senegal, Zaire, and Cameroon). The external stimuli underscored the extent to which rival factions sought to win. The disposition of the external actors and their conflicting prescriptions to ending the Chadian conflict suggest that what occurs in Africa has an extraneous symbiosis. Even more significant, the internalization of the conflict exposed the weaknesses of the state to turn inward and simply provided the rival lords with different periods at which they exercised leverage.

It is clear that the government of Deby must also face the task of rebuilding a political system while fighting off the MDD. The random arrests and killings of Deby's political opponents reveal the fragility of his leadership. What is apparent is that the proliferation of ambitious and disaffected Chadians may hamper attempts to promote a final chapter to the conflict. In addition, the MPS's chances for success are not necessarily predictable, because it is difficult to accurately gauge the disposition of France, Libya, and the US. It is no longer easy to determine the willingness of France to serve as

Françophone Africa's policeman. The friendship between Deby and Qaddafi is also uncertain, since there is still a major disagreement over the Aouzou Strip. It is likely that Qaddafi may be drawn to support another rebel force with the goal of displacing yet another government if Deby maintains his resolve to liberate the mineral-rich area in northern Chad. Lastly, the US involvement by proxy seems to be influenced by Washington's antipathy toward Qaddafi. This raises the possibility that US military and financial assistance will end if Qaddafi is out of the picture. It may be, in fact, that the resolution of the Chadian conflict must emanate from the people themselves.

Notes

1. Samuel Decalo, "Regionalism, Political Decay, and Civil Strife in Chad," *Journal of Modern African Studies* 18 (1980): 26.
2. Robert Buijtenhusi, "La Dialectique Nord-Sud dans l'Histoire Tchadienne," *African Perspectives* 2 (1977): 43–62.
3. René Lemarchand, "Chad: The Misadventures of the North-South Dialectic," *African Studies Review* 29 (September 1986): 28.
4. René Lemarchand, "Chad: The Roots of Chaos," *Current History* 80 (December 1981): 415.
5. Charles Lockhart, "Flexibility and Commitment in International Conflicts," *International Studies Quarterly* 22 (December 1978): 551.
6. K. J. Holsti, *International Politics: A Framework for Analysis* (Englewood Cliffs, N.J.: Prentice-Hall, 1983), 400.
7. Colin Legum et al., *Africa in the 1980s: A Continent in Crisis* (New York: McGraw-Hill, 1979), 30.
8. James O'Connell, "The Inevitability of Instability," *Journal of Modern African Studies* 5 (1967): 186.
9. Donald G. Morrison, *Understanding Black Africa: Data and Analysis of Social Change and Nation Building* (New York: Paragon House, 1989), 18.
10. Gabriel Lisette, a black colonial administrator of Guadeloupian descent, was elected to represent Chad in the French National Assembly. Lisette served Chad in this capacity from 1946 to 1951 and from 1957 to 1959, and rose to the rank of prime minister, although he was eventually replaced by François Tombalbaye. Lisette's PPT was the indigenous version of the *Rassemblement Démocratique Africaine,* which was fully operational in Guinea and Ivory Coast.
11. Ali A. Mazrui and Michael Tidy, *Nationalism and New States in Africa: From About 1935–Present* (London: Heinemann, 1984), 208.
12. Kaye Whiteman, *Chad* (London: The Minority Group, 1988), 10.
13. Arthur S. Banks, ed., *Political Handbook of the World 1990* (New York: CSA Publications, 1990), 114.
14. Mazrui and Tidy, 209.
15. Whiteman, 11.

16. Olesegun Obasanjo, *Africa in Perspective: Myths and Realities* (New York: Council on Foreign Relations, 1987), 8. Obasanjo served as Nigeria's head of state from 1976 to 1979 at which time the military handed over the reins of power to civilians.

17. Legum et al., 32.

18. John Wright, *Libya, Chad and the Central Sahara* (Totowa, N.J.: Barnes and Noble Books, 1989), 139. According to Qaddafi, the Muslims in Chad were the obvious majority.

19. Lemarchand, "Misadventures," 38.

20. Mazrui and Tidy, 209; Lemarchand, "Chad: The Roots of Chaos," 414.

21. It is claimed that Tripoli pressured Goukouni into signing the merger agreement. The proposed union was also strongly criticized by Vice President Kamougue, who called it an "impossible marriage" and by Dr Abba Siddick (the minister of health), who subsequently resigned and fled to Sudan.

22. The OAU peacekeeping force was comprised of 1,500 Nigerians, 900 Zairians, and 600 Senegalese.

23. Banks, 115.

24. David S. Yost, "French Policy in Chad and the Libyan Challenge," *Orbis* 26 (Winter 1983): 966.

25. Ibid., 968.

26. Franziska James, "On the Battlefield," *African Report* 31 (July–August 1986): 81. It is estimated that the military operation cost the French $500,000 each day.

27. Whiteman, 13; Martin Meredith, *The First Dance of Freedom: Black Africa in the Postwar Era* (New York: Harper and Row, 1984), 279; René Lemarchand, "Putting the Pieces Back Together Again," *Africa Report* 29 (November–December 1984): 65.

28. There was another failed attempt to foster Chadian political reconciliation in Brazzaville, Congo, in July 1983.

29. Whiteman, 14.

30. Franziska James, "Habré's Hour of Glory," *Africa Report* 32 (September–October 1987): 21.

31. *Facts on File*, 12 August 1983, 597.

32. Banks, 116.

33. Idriss Deby, a French-trained professional soldier who served as Habré's head of the military, masterminded the 1982 coup which brought Habré to power. A Muslim from the north, Deby was eventually forced to flee Chad after the April 1989 abortive coup.

34. *Time*, 17 December 1990, 40.

35. *West Africa*, 10–16 December 1990, 2990; *The Economist*, 1 December 1990, 42.

36. Assane Diop, "Plus Ça Change," *Africa Report* 37 (March–April 1992): 26.

Liberia's Conflict

Prolongation through Regional Intervention

Karl P. Magyar

Gus Liebenow's comprehensive review of Liberia, published in 1987, concludes with quotations from Rev Thomas Hayden, who observed that if the Samuel K. Doe government and the opposition cannot "respond to the legitimate needs of the people, spontaneous violence might erupt and an entirely new leadership might evolve." He also expected that "soon there will be a changing of the guard in Liberia"—and that it may come by "Quadafi-backed Liberian exiles."[1]

His judgment proved to be chillingly prescient. But those who have been following Liberia's tumultuous affairs realize that a prediction of Liberia's catastrophe was safe to make. Since Doe, as a master sergeant, ousted President William R. Tolbert and his Americo-Liberian elite from power in 1980, that country had been on a steady course toward self-destruction. Whatever wrongs that coup sought to correct, everything in fact only deteriorated further, and it soon became a classic case of history repeating itself. Numerous coup attempts followed; existing social divisions polarized; the already fragile economy deteriorated further; and relations with the international community plummeted. Under these circumstances, either a successful coup, or the commencement of an insurgency, could have been expected. The latter transpired in 1989.

Liberia had enjoyed a unique history in comparison to that of other West African states. The territory had been settled by freed slaves from other areas of West and Central Africa, but more significantly, by numerous former slaves from the US who returned, starting in 1822. This latter group soon formed the sociopolitical elite, as they extended their influence and control inland over the territory's indigenous inhabitants. In 1847 the Americo-Liberian elite declared its formal "independence"—though it had not been a colony of any external power. The elite introduced a constitution and currency, all patterned on the American standard; this repatriated elite soon built fine antebellum houses on Southern-style plantations and worked them with indigenous slaves well past the period of America's Civil War.

Rule by the Americo-Liberian elite was authoritarian with the power elite being characterized as an ethnic oligarchy, rent by nepotism, and ruled through paternalistic authority over a largely traditional and impoverished native population, which comprised some 95 percent of the country's inhabitants. They in turn were divided into 116 ethnic groups, each occupying distinct geographic zones which, significantly, range across Liberia's borders, well into neighboring Ivory Coast, Guinea, and Sierra Leone.[2] The ruling elite, as the 17th group, dominated the capital city of Monrovia—grandly named for America's President James Monroe, who facilitated the return of the ex-slaves.

Few in the external world knew of the plight of Liberia's rural populations and few cared. A handful of foreign enterprises were lodged in Liberia, extracting mostly iron ore, diamonds, and rubber—which was dominated by America's Firestone Company. These, plus Roberts Airfield, a major refueling stop for international flights, comprised the bulk of the external world's interest in Liberia. The US, because of its historic ties to the ruling elite who had guaranteed long-term stability, also had installed some major international navigation, radio, and communications facilities in Liberia, and this tie added to America's interest in the country.

Over the years, the Americo-Liberian elite had grown too complacent, which was not justified in view of the tumultuous events associated with the decolonization process throughout Sub-Saharan Africa. As colonial powers terminated their rule in Africa, starting with Ghana in 1957, William V. S. Tubman held forth as president of Liberia from 1944 until his death in office in 1975.

Tubman was succeeded by President William R. Tolbert, who ruled until his brutal ouster by MSgt Samuel K. Doe in 1980. Apparent was the ignorance of continental developments on the part of both rulers. In fact, their regimes approximated and were viewed by the indigenous population as a foreign colonial force, replete with a foreign language and other colonial accoutrements of power. Others saw it as black-on-black apartheid. When Doe took over in his grisly coup, he was but another youthful African military man who would challenge a corrupt, stagnant, and entrenched elite. Certain historical facts concerning Liberia were indeed unique, but the general trend of these historic developments now comprised somewhat routine Africawide political practices which had engulfed the continent during the previous two decades.

Doe led the fatal putsch, after which he and his armed forces-based People's Redemption Council undertook to rule the country under the usual austere and emergency conditions, but ultimately at the point of increasingly undisciplined guns. President Tolbert was the victim of grisly atrocities—which became the fate of Doe himself a decade later. The international community was horrified to learn of the summary executions of 13 top officials of the Tolbert government. These executions were carried out in public along Monrovia's beach. Many of the ecstatic and cheering people in attendance at the executions would in turn become victims of violence during the events of 1990–91. The offending Tolbert government officials had been cited with, among other crimes, disrespecting human rights. Treating them so brutally in turn set the tone for the Doe regime's modus operandi.

Little would change—least of all the drift towards disaster—but at least a representative of the indigenous population, along with mostly his own tribal-originated junta, had replaced this unique "colonial government." Rather than arrest the drift towards disaster, Doe only expedited the process. And that process would deteriorate rapidly. Where previously social relations among the indigenous tribes had been tranquil, they now would become polarized—a process in modern Africa which, when initiated, has rarely been reversed.

Africa's Conflict Environment

It is tempting to speculate that the longer the process of social disintegration, the more likely it is that an ensuing civil conflict will continue. In the case of Liberia, tensions in the social fabric emerged over several decades and were influenced especially by the liberation processes in neighboring countries. Doe may well have been only the spark that set off the final vortex of the next decade.

Explaining the reasons or causes for this incident of disintegration could easily occupy an entire volume in itself and not yield a definitive statement. Readers receive an enhanced understanding of the process of conflict within Liberia when they view it in a greater African perspective; that is, a perspective where similar conflicts have been experienced during the last four decades. However, we must remember Liberia's unique situation in that the ruling Americo-Liberian elite was in fact not a transient colonial force but a permanent, foreign-originated, social oligarchy much like the Afrikaners in South Africa. At issue in both instances is the question of majority rule (in Liberia, implying the indigenous natives) rather than simple liberation.

Africa has experienced and is presently undergoing several distinct types of conflicts. They may be categorized as wars for independence (e.g., Algeria); wars which preceded and which continued after independence had been attained (e.g., Angola); wars starting after independence (Uganda); transnational wars (Somalia and Ethiopia); wars of secession (Biafra); military coups and countercoups; and massive internal disturbances.[3] Most of these examples contrast with conflicts in the developed world which, if and when they break out, tend to be almost exclusively transnational wars. In this respect, Africa's conflicts may be collectively characterized as "conflicts of consolidation," which suggests that such conflicts are endemic to most states in their early histories as independent nations.

In the preindependence phase, resistance and militarism is introduced as these colonial or subject territories become politicized. The indigenous population identified the colonial regime as the enemy, but in most of Africa the colonial power departed without a fight. The most notable exception were three Portuguese colonies, where the anticolonial struggles had been standard protracted conflicts and turned into complex civil wars. Often, competing insurgent groups, each with different ethnic and leadership bases, opposed the enemy. These groups did not easily overcome their separate identities after independence had been attained. But the process was different where no substantial anticolonial wars were fought—as was true for most of West Africa. There, the initial security forces were more integrated into the ruling circles until opposition emerged from within the single security apparatus, led usually by younger military men. These men did not perceive the struggle as one against competing ethnic-based social groups but against an entrenched, corrupt, and stagnant ruling elite. This latter tendency, introduced by the Dergue in Ethiopia, by Jerry Rawlings in Ghana, and Thomas Sankara in Upper Volta, inspired the ascension to power by Doe in Liberia.

Whether peaceful or violent, the abrupt replacement of an entire power elite by another group based on different social classes amounts to a revolution and introduces a new consolidative period. This period is the most volatile and disruptive in the history of most nations. In these instances, law and order breaks down; social cleavages polarize; new leaders emerge who are bent more on achieving power than on restoring peace; economies are devastated; and external interventions often do more to exacerbate problems rather than to resolve them. In essence, the major thrust of political efforts in this period concern the attempts to establish legitimacy and social tranquility to restart the process of socioeconomic development. But with the established social infrastructure having been destroyed, and considering that the general welfare level had never been advanced, new aspirants to power face formidable challenges indeed. These characteristics describe Liberia's predicament quite well.

Whether protracted or prolonged or conflicts or outright wars, organized hostilities by African governments in power or by their opponents tend to get bogged down when neither side can eliminate the other in a short, sharp strike. The conflict rages, but as the active combatants come to respect each other, each in turn inflicts great havoc on the other's society. The embattled government's only concern becomes sheer survival, while administrative programs are placed in abeyance due to the absence of stability and the lack of domestic and foreign capital formation. Opponents, on the other hand, rarely constitute in the initial stage more than a motley group of armed opportunists seeking to legitimize themselves by embellishing their cause with terminology borrowed from external guerrilla and insurgent groups. Their lack of materiel and financial resources is soon overcome by the unscrupulous exploitation of the rural masses among which they move, but in time they gain some respectability. They become, in Mao Tse Tung's formulation, the fish in a sea of people. Their claims to be taken seriously is enhanced when they receive overt support from external neighboring or distant benefactors. In the cold war days, this support was obtained easily from the Soviet or the American camps. Today, external support may come from Libya, Israel, Iraq, Iran, Saudi Arabia, or India or old, established west European sources. Africa has not lost its attraction to external meddlers, although most of the continent has been marginalized as a global player.

There have been respectably managed protracted wars in Africa. For the most part, however, other insurgencies in Africa have lacked sophisticated formulations of protracted warfare at the outset, although several, such as those movements led by Savimbi in Angola, Dhlakama in Mozambique, and Garang in Sudan, developed them once they were under way. Most other conflicts in Africa have been prolonged wars. Conflicts such as those in Chad or Sudan or the western Sahara have gone on interminably, and as is the case of Ethiopia, Angola, and Sudan, they raged on and became "wars without results."

Liberia's civil war early demonstrated the classic tenets of prolongation as the proliferation of competing opposition groups, changing objectives, expanding battleground, and extending the conflict into neighboring states. These tenets also included divisions among regional supporters, external intervention, devastation of the social infrastructure, and large numbers of direct and indirect civilian casualties. Although not planned developments, these tenets offer emerging evidence of progressive deterioration—the result of avoiding one massive, concentrated confrontation, which certainly would eliminate one of the protagonists decisively. However, it may indeed be likely that the ultimate damage done as the result of an initial decisive confrontation between the government and the insurgents may be less than the damage inflicted from the prolonged war and the absence of a clear victory. External interests, whose intervention almost invariably leads to the prolongation of conflicts, should have had a prime concern in this proposition. Certainly the examples of Chad, Ethiopia, Angola, Mozambique, and, presently, Liberia, attest to this.

The Evolution of Liberia's Prolonged War

Progressive deterioration, social tensions, and authoritarian domination have marked the reign of President Tolbert. The "rice riots" of 14 April 1979 offered evidence of the Liberian military's reorientation from traditional support for the government. For example, when Doe ascended to power in the 12 April 1980 coup, he faced not only a declining social situation but also a deteriorating economy that Tolbert's civilian government had failed to stem. As the subsequent decade of Doe's own rule progressed, however, there were no signs to feel good about the country's fate. An unsuccessful coup attempt by Gen Thomas Quiwonkpa in 1985 had a remarkably prescient tinge in that the coup had wide popular support and engaged the neighboring countries of Sierra Leone and the Ivory Coast, and to an undetermined degree, Israel and the United States.[4] That coup started the active demise of Doe's regime.[5] His decade in power had been marked by at least one-half dozen coup attempts, but it took a full-scale civil war to bring him down—after which the war continued.

On 24 December 1989 Liberia's devastating conflict commenced when Charles Taylor and his small band—variously estimated to range between a few dozen to well over a hundred strong—crossed into Liberia from the Ivory Coast. Its composition represented acknowledged dissidents, participants at Quiwonkpa's coup, and an undetermined number who may have received training in Libya. Allegedly, included in this latter group were Charles Taylor and Prince Johnson, who split from Taylor's movement a few months later.

This modest team of insurgents were led by Charles Taylor, a man who must have assessed two important factors correctly: that Doe's troops, the Armed Forces of Liberia (AFL), were poorly prepared for combat, and that his own ethnic-based rural insurgents would advance quickly as the AFL would pose only a challenge in the capital city. This weakness afforded Taylor the opportunity to prepare for a quasi-protracted war while extending his control over the rural areas, which in turn also would generate additional recruits for his force. These strategies went according to plan, and after a few successfully quick strikes at minor government installations and garrisons in Nimba county—adjacent to the Ivory Coast from where he entered—Taylor moved rapidly towards the coastal towns. His early intentions were to seize the key economic and infrastructural grids while avoiding a premature attack on Monrovia. This latter objective would require a much larger and trained force and the prior alliance with most of the country's rural inhabitants.

Taylor's successes came rather easily and were abetted by Doe's widely perceived ineptitude and by the seemingly total incompetence of the AFL, which quickly resorted to pillaging and wanton slaughter of villagers. D. Ellwood Dunn and S. Byron Tarr note that Liberia was severely "tribalized" during the 1980s, and these writers expected a severe backlash against Doe's Krahn kinsman, whose members dominated the government and military.[6] This backlash may have been anticipated by Taylor, who may have incorporated the existing tribal animosities into his insurgent strategy. If he did so, he once again calculated correctly that this would weaken the government and would enable him to gain the wide backing of other Liberians who had come to resent the Kahn ascendancy.

The question of ethnicity and its role in civil wars holds special interest to students of African conflicts. In Liberia ethnic relations were generally peaceful during the long period of domination by the Americo-Liberian elite. The abrupt termination of that oligarchic rule and the wider indigenization of political participation may have constituted a proper step towards democratization, but only the naive should have failed to anticipate the ensuing social conflict. While observers may accuse Tubman and Tolbert of not learning from recent African history, they may level the same charge at Doe and his supporters (and in the distant country of South Africa, the same phenomenon may be underway). Once activated, ethnic politics cannot be easily turned off, and the distant repercussions will likely plague the region in ways not initially foreseen by the perpetrators. Thus, in Liberia not long after Taylor's initial forays, the Gio and Mano of mostly Nimba county supported Taylor's efforts. In contrast, a Mandingo and Krahn coalition—the latter being the privileged kinsmen of Doe—opposed him. By this time the Americo-Liberians had ceased to be a major factor, although one may argue that they had served as the functional peacekeepers while they dominated power.

Doe's government troops failed to mount a credible offensive against Taylor's National Patriotic Front of Liberia (NPFL) forces; hence, they assumed a mostly defensive posture by digging in around key points in the capital. Before a year went by, Taylor dominated 95 percent of the country, but he remained frustrated in not being able to go in for the final kill. (Some observers argue that he could have done so, but he chose not to as this would have entailed massive civilian casualties.) Again, this tactic is a frequent feature in African wars where the main strategic objective always involves the difficult-to-conquer capital city, despite the loss of control over the vast rural areas. Luanda in Angola and Maputo in Mozambique are such examples. The failure of the NPFL to take Monrovia may be attributed, as is the case in Luanda, to foreign intervention. Cubans defended Angola's capital, while a combined force, the Economic Community of West African States Cease-Fire Monitoring Group (ECOMOG), introduced a mixed force that was comprised in mid-1990 of five West African countries in defense of Monrovia.

ECOMOG's active intervention, starting in 1990, secured the capital city but in the process had introduced a new fighting force. Remnants of Doe's army numbered about 1,000; Taylor's NPFL troops numbered several thousand; Prince Johnson, who left the NPFL and organized his Independent National Patriotic Front of Liberia (INPFL), commanded fewer than 1,000 troops, but they were highly aggressive; and ECOMOG began with about 3,000 troops, which, after a slow start, doubled a year later. ECOMOG's mission began ostensibly as a peace-keeping effort, to separate the sides in the conflict while allowing the political process to resolve the dispute. However, Taylor's hostile attitude towards ECOMOG led ECOMOG partially to align itself with Johnson's INPFL. This alignment allowed Johnson's men to capture Doe and his contingent of palace guards while Doe was on a visit to ECOMOG headquarters in September 1990. The terms of this capture have not yet been cleared up, but the subsequent mutilation of Doe by Johnson and his cohorts cast a terminal pall over the INPFL as an internationally approved participant in the future affairs of Liberia. Subsequent atrocities committed by the INPFL have only verified the innately perverse nature of that group as personified by its leaders.[7]

External interventionists cannot escape the task of becoming kingmakers—and subsequently being burdened with supporting them in power. The US experienced this situation in Vietnam and in Panama; the Soviets in Eastern Europe and in Afghanistan; and in Liberia, ECOMOG had little choice but to support the interim government of Amos Sawyer. As Taylor had no hand in that appointment, ECOMOG then inherited the obligation to defend Sawyer in a country where 95 percent of the territory was out of his control, thus, ensuring a classic tenet of prolonged wars. Liberia would become ECOWAS's "Vietnam."

Prolonged wars tend to engage various efforts at intervention by the international community, and this is likewise the case in Liberia. Some observers believe Taylor received training and financial support from Libya; most of this support was channeled through Burkina Faso. Taylor operated from the Ivory Coast, where he made his initial incursions. Unsubstantiated reports note that Burkino Faso even had some active troops supporting Taylor's efforts.[8] As Taylor had Libyan connections, and his program and leadership capabilities were uncertain, relations between Taylor and the US were cautious. Taylor in turn castigated the US for its enthusiastic backing of ECOMOG's effort and its support of Amos Sawyer.

The US had backed Doe during his initial years in power but gradually reduced that support after failing to note positive gains. Once hostilities against Doe commenced, the US moved quickly to guard American facilities, especially its embassy, and a substantial contingent of US Marines was stationed on warships offshore. They eventually did have to secure the US

Embassy and evacuated American and other foreign personnel from Liberia. As peace talks broke down, the US offered to fly Doe to a destination of his choice, but he declined the offer.

Throughout the war, all opposing factions acknowledged America's long interests and future role in Liberia, but the factions cast a weary eye at America's involvement and occasionally accused the US of siding openly with opponents. In the cold war days, this alignment may have led to more acute problems if the Soviets had chosen to enter the fray, but now Taylor had to make do with only limited external support. It appears that initially Taylor had lived mostly off the land, however ruthlessly, and that he did gain much of his military supplies from captured government stores. This strategy is also a feature of many prolonged wars in Africa. But it is hardly conceivable that he could have succeeded so much had he not secured external support initially from Burkina Faso and the Ivory Coast. He also realized that he would receive continued Libyan backing and substantial profits from diverse French commercial interests during the course of the conflict.

Prolonged wars tend to attract participation by external benefactors, and they soon involve neighboring countries. Indeed, Liberia's civil war actually spilled directly into Sierra Leone, whose government was subsequently overthrown in a coup in 1992. Motives for external intervention or involvement are difficult to isolate, and we cannot generalize about them, as each participant has his own agenda. Qadhafi's objective may well be his penchant to intervene wherever possible or to capitalize on opportunities afforded while world attention is focused on another combat theater. However, he has made no significant or permanent gains in black Africa. The motives of Burkina Faso may well represent little more than the personal ambition of President B. Campaore to exert influence in a wider region. Houphouet Boigny of the Ivory Coast is rumored to have been related to President Tolbert through marriage, and he found Doe difficult to work with.

Sierra Leone may have feared the conflict spreading across her borders and hence gambled by offering a contingent of troops to ECOMOG to help stabilize Liberia. This act, however, led Taylor to support a Sierra Leonean dissident faction against that government, which then expanded it to the level of a quasi-civil war. Most accounts by the print media agree that Taylor's NPFL forces entered deep into Sierra Leone's territory with the support of Libya and Burkina Faso. That action by the NPFL did not lead to Sierra Leone's pulling out of the ECOMOG force, and Guinea and Nigeria had to come to Sierra Leone's rescue to beat back the incursions.[9]

The other foreign participants include members of the ECOMOG force led by Nigeria, which subsidizes the largest portion of the bill and offers the largest contingent of troops, and later Ghana, Gambia, Guinea, Sierra Leone, and Senegal later. Again, each participant had different objectives for intervention, most of which have not been articulated openly. Those with contiguous borders generally fear the expansion of the conflict into their areas, while the more distant participants present a mix of political, or "imperial," ambitions. According to Kenneth B. Noble, Campaore justified his country's participation "because he believed foreign military intervention would only exacerbate the situation."[10] His words, however, were overtly corroborated by his own actions.

A Lutta Continua

"The battle continues" embodies the classic expression of Africa's struggle against colonial domination. But in the case of Liberia, this phrase introduces an updated dimension which implies that the battle is becoming prolonged. The elements of the war which normally lead towards the "bogging down phenomenon" were manifest early in Liberia's civil war, and the war-related activities after the first two years only expanded on the established complexity but added little that was new. A detailed listing of intervening events would in effect yield a history of the conflict—which is not the purpose of the present analysis. However, several developments may be presented as they promote an understanding of prolonged wars, particularly Liberia's prolonged war.

On the international front several conferences were convened, with or without Charles Taylor's participation. The most notable one was Yamousoukro IV (October 1991), at which time the convening nations agreed to open the roads throughout Liberia, to begin disarming the warring factions and encamp the troops, and to prepare for elections. Little of this was realized as Taylor would not accept a situation in which his perceived just rewards would not be ensured. Instead, Taylor began to sense an emerging split in the ranks of ECOWAS and its ECOMOG peacekeeping force.

Nigeria, which supplied the leadership and the greatest number of personnel and finances for the ECOMOG contingent, also began to show doubts about the entire affair. Indeed, some in Nigeria began their own domestic "Vietnam" debate. Military operations in distant lands depleted fragile treasuries quickly. ECOMOG's troop strength escalated to around 11,000, and by mid-1992 Ghanaian and Nigerian airplanes were engaged in bombing sorties to destroy Taylor's headquarters. All this promised a long and escalating involvement.

Whereas West Africa had traditionally seen a pervasive political split between Anglo and Francophone states, regional concern over the Liberian affair introduced a further split among the Francophone ranks. Guinea, a contributor of forces to ECOMOG, feared that the war was spilling into her territory—as it had done in Sierra Leone, while Senegal offered a moderate-sized contingent, perhaps in response to external requests as there was no problem of territorial contiguity. These two states and Mali, another supporter of ECOMOG, were opposed by Burkina Faso and the Ivory Coast. In 1992 the US recalled its ambassador from Burkina Faso in response to continued active support of Taylor. Houphouet Boigny of the Ivory Coast on the other hand was under great pressure by his West African colleagues to end his support of Taylor. The Ivory Coast then began to assume a more ambiguous position. This split in the Francophone ranks, however, did not reduce its traditional suspicions of Nigeria and its unavoidable leadership role in West Africa and ECOMOG. Dominating the military operations in Liberia, it was feared, could whet Nigeria's appetite for a hegemonial role and Nigeria would gain valuable leadership and battlefield experience. Nigeria's long-established opposition to French initiatives in West Africa once again raised old questions about certain Francophone states serving French designs. French commercial interests had moved rapidly into the area controlled by Taylor, and the revenues thus generated served as Taylor's principal financial support for his insurgency.[11]

President Doe was killed in September 1990. Conferences at Banjul, Lome, and Monrovia then established an interim government led by Amos Sawyer. He had the backing of the ECOWAS and the US but remained opposed by Taylor, who either boycotted most conferences or refused to abide by their formulations. This strategy then introduced a common feature of prolonged wars, namely the de facto existence of two governments. Under these conditions, the insurgents develop their own political and administrative machinery: introduce at least the intent of providing social services, starting with schools and rudimentary clinics; and attempt economic revival by focusing on the exploitation of extractive commodities and subsistence agriculture. Inevitably, external commercial interests are available to capitalize on this high-risk, but profitable, opportunity. The besieged government, on the other hand, occupies the capital city and relies on external aid, military support, and resolution of the conflict at international conferences and distant negotiating tables. This procedure raises a curious dilemma, one which Liberia has to face as well: the government in the capital city relies on its legitimacy which stems from the diplomatic activities of external interests, while the insurgents work systematically to gain legitimacy from among the mostly rural population. This pattern surfaced in China, Cuba, Vietnam, Afghanistan, Guinea Bissau, and Angola and now characterizes Liberia. Whoever wins in the end will still face a formidable task in attracting a former adversary's support base.

Ongoing factionalism embodies another pervasive characteristic of prolonged wars. If insurgents fail to act quickly to oust the government, as a protracted strategy is then assumed, they will notice a rise in factionalism with the insurgent group or the introduction of new armed dissident groups. Prince Johnson and his followers broke away from Taylor's NPFL and pursued their own insurgent agenda in the suburbs of Monrovia. Then the United Liberation Movement of Liberia (ULIMO) emerged in March 1991, representing the Krahn-based Liberian United Democratic Front (LUDF), led by

Arma Youlu, and a mostly Mandingo force led by Alhaji Kromah. These new groups have the backing of neighboring Sierra Leone and Guinea and concentrate their energies more against the NPFL than on the interim government.[12] Their ultimate objectives, aside from servicing the needs of their benefactors, are unknown. And reportedly, before long, each group, including Taylor's NPFL, experienced an outbreak of bickering and outright hostilities among internal factions. Such continuous bifurcations virtually guaranteed the attendant prolonged public suffering and foreign meddling. This outcome was evident in Angola and continues to be the case in Liberia.

ECOMOG's role also underwent redefinitions. That organization was constituted as a peace-keeping force, but when it bombed certain strategic targets in areas controlled by Taylor, it acted as a combatant. ECOMOG backed the interim government of Amos Sawyer and came to appreciate the efforts of ULIMO to weaken the NPFL. But ECOMOG demonstrated notable weaknesses. Several—if not most—of the constituent national forces, such as Sierra Leone's and Guinea's, were hardly neutral in the conflict. Nigeria's massive commitment is, of course, also suspect. And persistent rumors hold that there is dissent within the ranks as soldiers perceive their "Vietnam-like" situation. Some soldiers seriously question their willingness to confront Taylor's force in a systematic ground campaign on his own territory.

While the conflict festers and its resolution is pursued at various negotiating tables, the atrocities continue, whether inflicted by war or by deprivation. The massacre of 600 Gios and Manos in St. Peter's Church in July 1990 provides a stark reminder of the inevitable dark side of a prolonged war.[13] Casualties in Africa's wars are not reliably reported, but observers generally assume that civilians account for 90 percent of the deaths in the continent's civil war. Another dimension concerns the long-term damage of such wars. The physical infrastructure throughout most of the country is soon destroyed, or it deteriorates through neglect. Domestic capital flees—as do many of the entrepreneurial managers—while traditional external trade and investment ties are abrogated. The mere cessation of war will hardly reinstitute these lost assets, hence the damage will be compounded for some time to come. Liberia has never been featured as the focus of major external economic interest; therefore, the reconstruction of the destroyed economic infrastructure will not be accomplished soon.

Conclusion

The prolonged Liberian conflict has been raging since the end of 1989. Some wars have been much longer, but scholars in the field do not define prolongation within a purely temporal context only. Rather, observers should focus on the failure of the conflict to come to a head in an early decisive battle—which would eliminate at least one of the major protagonists. In this regard, the conflict in Liberia, while still in its early stages, experienced most of those attributes associated with prolonged wars of much longer duration.

Charles Taylor never did develop an acknowledged sophisticated protracted war strategy. Instead, external forces and rival insurgent groups interdicted the NPFL's surprisingly rapid advance on Liberia's jugular vein. Certainly, the way in which Taylor built his force required elements of a protracted strategy, which also guided his systematic advance through the countryside. Gaining adherents in the rural areas in Africa is not, however, as significant as it may be in the Latin American or Southeast Asian contexts. Authoritative power in Africa is almost totally concentrated in the capital city, and if that city is secured with the aid of foreign forces, the classic protracted struggle, focusing on the domination of the rural areas, loses considerable relevance. In effect, the struggle continues with two "governments" in power: one dominating the formal accoutrements of state authority in the capital, which is embellished with claims to wide diplomatic recognition; the other gains incremental legitimacy in the countryside. Both however, rely on competing foreign interests to determine their fate. The insurgents rarely topple the government, but by retaining a credible capability to survive and by launching an occasional terrorist act or attack on a governmental installation (which are often little more than suicide missions), they determine their fortunes at international negotiation forums (e.g., SWAPO in Namibia), or they continue their fight sporadically and interminably and without resolution (e.g., PLO in Palestine and RENAMO in Mozambique).

Liberia has experienced virtually all the characteristics of a prolonged war in a compressed time frame. An insurgent group eliminated the head of state, but they could not capture state structures. The stated objectives of participants changed frequently, ranging from the desire for the mere ouster of Doe's corrupt government, ostensibly for altruistic reasons, to the unprincipled drive to become the new leader. Some external analysts have referred to the entire embroglio as little more than tribalism—a charge which cannot escape a pejorative connotation. Were this to be true, we would have to expect that all fighting in Africa would not cease until each tribe had its own state. Tribalism in Liberia's conflict is obvious, but as in other African conflicts, it is a symptom of the breakdown of an established order which did not possess wide legitimacy. External analysts will be mistaken if they focus solely on this issue—and miss the nuances of changing power relationships. At issue is progressive social deterioration and a contest of who shall prevail. Balancing or manipulating tribal cleavages is but one demonstration of power articulation. In Liberia, a fundamental revolution has already taken place, and the contest for power has moved out of the capital city and into the rural areas. The consequent resort to tribal reidentification as an expected reaction surprises no one. This resort also reflects a standard feature of prolonged wars throughout Africa, although it does not necessarily hold true in all other areas of the third world.

Negotiated settlements alone will not resolve Liberia's prolonged war. The traditionally placid rural population has been activated and polarized. Standard democratic institutions drawn from Western forms will find little relevance in Liberia's disrupted third world context. The most acute problems relate to socioeconomic standing and security matters. Analysts may argue that these needs cannot be addressed outside of a political context. But here they would do well to remember that most modern African states started with democratic structures but only a few of them survived. Certainly little realistic basis exists for expecting the economy to strengthen after a political settlement; hence, political and ethnic tensions may also be perpetuated. Ghana and Nigeria's alternations between democratic, authoritarian, and military governments reflect the fundamental problems associated with attempts to resolve poetical disputes without commensurate economic progress. In this regard, Liberia's offending head of state was removed from power, but the conflict has hardly ended. A negotiated settlement may represent little more than a reprieve before parties renew hostilities again. And with numerous external allies or benefactors available, the prospects for the prolongation of the war are greatly enhanced.

Because of the inherent nature of prolonged war, analysts find it difficult to predict when they will end. And again, the arrival of peace will only signify the start of new problems as the country begins to redevelop. In Liberia this latter task poses a social challenge just as profound as the civil war itself.

Notes

1. Quoted in J. Gus Liebenow, *Liberia: The Quest for Democracy* (Bloomington: Indiana University Press, 1987), 315.
2. D. Elwood Dunn and S. Byron Tarr, *Liberia: A National Polity in Transition* (Metuchen, N.J.: The Scarecrow Press, Ltd., 1988), chap. 2.
3. See Karl P. Magyar, "Low-Intensity Conflicts: The African Context," in Stephen Blank et al., *Responding to Low-Intensity Conflict Challenges* (Maxwell Air Force Base, Ala.: Air University Press, 1990), 19.
4. Liebenow, 300–303.
5. "Liberian Warfare Has Roots in 1985," *New York Times*, 5 June 1990.
6. Dunn and Tarr, 199–200.
7. "A Foe to Be Feared," New York Times, 1 September 1990; Foreign Broadcast Information Service (*FBIS*), "Daily Report, Sub-Saharan Africa," 5 August 1991, 30.
8. See "Civil War in Liberia Threatening to Divide West African Neighbors," *New York Times*, 29 September 1990.
9. *New York Times*, 16 April 1991.
10. Ibid., 29 August 1990.
11. S. Byron Tarr, "ECOMOG's Initiative in Liberia: A Liberian Perspective" (paper presented at the annual meeting of the African Studies Association, Seattle, 1992), 26.
12. *Africa Confidential* 32, no. 23 (22 November 1991): 6–7.
13. Janet Fleischman, "The Liberian Tragedy," *Reconstruction* 2, no. 1 (1992).

The Rhodesian Conflict 1966–79

Herbert M. Howe

All sides at least initially misjudged the war's length. The Rhodesian government, led by Prime Minister Ian Smith, issued its Unilateral Declaration of Independence (UDI) from Great Britain in November of 1965, convinced that the blacks could not mount a significant military threat and that white rule was thus assured. Among the blacks, the Zimbabwe African National Union (ZANU) and the Zimbabwe African Peoples Union (ZAPU) thought erroneously that significant guerrilla incursions from Zambia could trigger African uprisings against the whites. The British government, which had legal responsibility for its errant colony, at least publicly believed in 1965 that the Rhodesian cause would last only "months, if not weeks."[1] The United States government in its 1969 National Security Studies Memorandum (USSM)-39 reaffirmed its belief that whites in southern Africa would remain through the foreseeable future.

All sides were wrong. The prolonged war began in the mid-1960s, sputtered until the early 1970s, and flared increasingly in the mid-1970s until the Lancaster House agreement and the cessation of hostilities in 1979 and the subsequent election of ZANU's Robert Mugabe as prime minister in March 1980.

Between 1965 and 1972 the Rhodesian Defence Force (RDF) seemingly had won the military struggle; it had contained the initial guerrilla hostilities and had destroyed much of ZAPU's and ZANU's capabilities. Rhodesia used mostly military, and not political, means to counter the guerrillas. Its basic military doctrine was mobile, rather than area, defense. Both ZANU and ZAPU began hostilities believing that relatively large guerrilla incursions would secure immediate peasant support.

Between 1972 and 1976 the conflict started anew. By 1970 ZANU had shifted to protracted war with an emphasis on political mobilization, whereas ZAPU remain wedded to more conventional military strategy. ZAPU and ZANU benefitted from communist (Eastern bloc and Chinese) weaponry and European financial support. The independence of Angola and especially Mozambique (which granted sanctuary to ZANU), coupled with rising levels of East European and Chinese aid and rising Zimbabwean nationalism, helped to persuade white Rhodesia's two major hopes, the United States and South Africa, to work for black majority rule.[2]

By late 1979 the Rhodesian military remained capable of inflicting large losses on guerrilla forces and their regional allies, notably Mozambique, but the country's increasingly depleted economy, a declining white manpower pool, an almost inexhaustible manpower pool for guerrilla manpower recruitment, and rising guerrilla capabilities signalled the need for negotiations. The conflict claimed some 40,000 lives and affected postwar development by damaging Rhodesia's—and the region's—economy.

Events unforeseen in 1965 that contributed to the prolongation included: (1) the growth of large guerrilla forces capable of conducting a protracted war ("chimurenga")[3]; (2) Rhodesia's ability to withstand sanctions by developing alternative domestic industries and by circumventing United Nations sanctions; and (3) the accession to power of the Popular Movement For the Liberation of Angola (MPLA) and Front for the Liberation of Mozambique (FRELIMO). The massive Soviet/ Cuban support given to the MPLA and FRELIMO's support of ZANU guerrillas caused the US and South Africa, Rhodesia's two major hopes, to oppose continued white rule.

Black Grievances

Rhodesia was bound to resist African nationalism more than any other of the European colonies. Rhodesia had more settlers—some 200,000 by 1965—and these settlers had established a strong vested interest by developing Rhodesia's mineral and agricultural wealth. A self-governing colony since 1923, Rhodesia enjoyed almost total freedom from England, its legal but nominal ruler.

The settlers' strong will to control Rhodesia's economic and political future increasingly collided with external factors, most notably support for postwar decolonization and such internal considerations as a growing core of well-educated Africans, who saw continued white domination as barring the rights and privileges of 6 million other Africans.

Rhodesia's racial system, while not as pervasive or harsh as South Africa's, embittered blacks. Various land laws, including the Land Apportionment Act and the Land Tenure Act, gave one-half (and most of the best) of Rhodesia's land to whites. The Color Bar Act limited social mixing between the races. The Masters and Servants Act denied at least one-half of all African workers the right to form or join unions. The Unlawful Organizations Act authorized the government to restrict or ban political organizations.

Repression of African's initial protests and the government's UDI increasingly spurred a drive for violent opposition. ZAPU in 1961 and the more radical ZANU in 1963 replaced the moderate organizations, notably the (Rhodesian) African National Congress and the National Democratic Party of the late 1950s. African frustration turned to military insurgency by the mid-1960s when, after the government banned the recently formed ZANU and ZAPU in 1964, these two groups established their military wings, Zimbabwean African National Liberation Army (ZANLA) and Zimbabwean African Peoples Revolutionary Army (ZIPRA). Throughout the war, ZANU drew most of its support from the Shona, while ZAPU relied primarily on the smaller Ndebele population.

Major Actors

Ian Smith's Rhodesia Front government and the guerrilla forces of ZANU and ZAPU were the war's major combatants. Smith enjoyed overwhelming support from Rhodesia's whites (whose numbers peaked at 275,000 in 1972). His party never lost a seat in the 50-member parliament. The Rhodesian Front strongly opposed any significant political reforms. In talks with Britain in the late 1960s, Smith remained recalcitrant. As late as March 1976, Smith vowed that majority rule would not occur within the next one thousand years. The government believed the insurgency as externally directed and that Rhodesia's blacks, described by Smith in 1972 as "the happiest Africans in the world," would not willingly support the struggle.[4] The government had failed to understand the growing black anger towards white rule. In 1965, the year of UDI, the Rhodesian Defence Force was small and unable to fight a prolonged bush war. Yet, during the next 15 years, it forged a highly capable force that conducted highly successful internal and cross-border operations.

Joshua Nkomo, sometimes described as the "father of Zimbabwean nationalism," established ZAPU in 1961. While containing some Shonas, ZAPU had its greatest strength among the Ndebele, who comprised about 18 percent of all Rhodesians. Nkomo, unlike ZANU's Robert Mugabe, would sometimes meet with the white government and business establishment in hopes of a peaceful settlement. Yet the uncompromising Rhodesian Front government forced such moderates as Nkomo to use military pressure. Nkomo favored conventional military structure and tactics. Based in Zambia and relying on conventional Soviet support, equipment, and theory, ZAPU and its military wing ZIPRA posed only a limited threat to the Rhodesian government.

Reverend Ndabaningi Sithole founded ZANU in 1963 as a split off from ZAPU. But, following several leadership changes, Robert Mugabe by early 1975 had established himself as ZANU's leader. ZANU, and its military wing ZANLA, largely drew on the Shona, who comprised 80 percent of Zimbabwe's total population. Increasingly operating from Mozambique during the 1970s, ZANU supported Mao Tse-tung's three-staged protracted-war concept and opposed negotiations until achieving the clear military advantage. Mugabe's ZANU described itself as Marxist-Leninist, rather than simply nationalist, and received much of its weaponry from the People's Republic of China and from Eastern Europe.

At the war's outbreak, internal and regional relations greatly favored the Smith government. Opposition Rhodesian whites never posed any political challenge. In 1964 supporters of ZAPU and ZANU attacked each other over ideology ("moderate"

ZAPU versus "radical" ZAPU) and group/ethnic rivalry and caused both loss of life and property damage.

The military balance at the outset of UDI overwhelmingly favored the Rhodesian government. The two guerrilla "armies had perhaps two hundred poorly trained and equipped men." The Rhodesian army had one regular battalion of Rhodesian African Rifles, one battalion of Rhodesian Light Infantry, one Special Air Services squadron, and one armored car regiment. These forces could draw on the supporting corps—artillery, signals, and service units. Besides these regular units, Rhodesia had three white territorial battalions mobilizable with one day's notice and at least two more all-white reserve battalions that required about a 10-day mobilization period. The army, the uniformed police, the Special Branch, and the Department of Internal Affairs provided separate intelligence flows.

The air force had about one hundred aircraft, including Hunter fighters, Canberra bombers, Vampire jets, Alouette helicopters, Provost trainers, and Dakotas. Rhodesia started its counterinsurgency (COIN) operations with an officer corps that had effectively fought guerrillas in Malaya during the early 1950s.

South Africa, on Rhodesia's southern border, had militarily cooperated with its fellow white minority government in training exercises as early as 1961. To Rhodesia's east, Portuguese-ruled Mozambique adamantly opposed independence. Rhodesia's two black-ruled neighbors—Botswana to the west and Zambia to the north—opposed Rhodesia's minority rule. Zambia initially granted sanctuary to both ZANLA and ZIPRA but could offer only limited resources. Furthermore, the Zambezi River and Lake Kariba hindered guerrilla infiltration. To Rhodesia's west, Botswana would allow little help to the guerrillas other than transit permission.

Early Guerrilla Operations

Shortly after UDI had proven the futility of peaceful opposition, ZIPRA and ZANLA began guerrilla operations. ZIPRA received its earliest training from the Soviet Union,

Cuba, and Algeria. ZANLA's first trainers were from Ghana, the People's Republic of China, and Tanzania.

In April 1966 a 14-man ZANLA band moved from Zambia into Rhodesia and then split into three small teams. At the Battle of Sinoia (Chinhoyi), Rhodesian security forces in Operation Nickel wiped out a seven-man squad, including some guerrillas trained at Nanking Military College.

This early battle demonstrated the initial incompetence of both forces. The political commissar of this group apparently was an agent of the Rhodesian Central Intelligence Organization (CIO), and the group was to guide security forces to arms caches and chimurenga supporters. Yet government mistakes resulted in the group's inefficient slaughter: a Rhodesian air force history notes that "it was a very unconvincing and unprofessional action. . . . Fortunately for the police, the guerrillas were too confused to take advantage of the inexperience of the hunters."[5] The police armed themselves with single-action Lee-Enfield .303s and insisted then, and for several later operations, on wearing their blue uniforms while operating in the bush.

In August 1967 an 80-man joint ZIPRA-South African ANC team crossed the Zambezi River into the Wankie game preserve.[6] The force planned to establish base camps from which ZIPRA was to infiltrate into Matabeleland while the South Africans would move into the northern Transvaal. The Rhodesian African Rifles' response reportedly met "several nasty reverses," and the final tally was 30 guerrillas killed and 20 captured versus seven Rhodesians killed and 13 wounded.[7] Then captain Ron Reid-Daly (later commander of Rhodesia's Selous Scouts unit) described the government's casualty rates as "extremely high."[8] From March until the end of May 1968, Rhodesia conducted Cauldron, its first prolonged operation. Having reevaluated several of the mistakes from Nickel, the RDF killed 69 guerrillas and captured 50 out of a total of 125. According to Rhodesian figures, guerrillas killed only six Rhodesian troops. In another 1968 action, Operation Griffin, Rhodesian Provost T-52s dropped napalm and white phosphorous on guerrilla forces. By the end of 1968, only 12 security forces had been killed versus more than 160 insurgents.

Effects of Early Operations

These early, poorly conducted operations of ZANU and ZAPU had four significant effects: converting ZANLA to protracted war, reducing ZIPRA's effectiveness, deluding the Rhodesians about their invincibility, and prompting South Africa to insert security personnel.

Most importantly, around 1969 ZAPU and ZANU's military failures convinced ZANI (especially Mugabe) to embrace Mao's concept of protracted war. Previously, the scattered inhabitants of remote and mountainous northwest Zibabwe had greeted their liberators with suspicion, and the Rhodesian army was able to quickly locate, enclose, and then destroy such relatively large units. As Herbert Chitepo, ZANU's national chairman, noted, "These initial battles were fought in a social climate in which our people had not been given full political ideology and line." Chitepo concluded that "we cannot expect to wage guerrilla warfare successfully without mass support."[9] In December of 1972 the appointment of Chinese-trained Josiah Tongagara as chief of ZANLA and secretary of defense cemented ZANLA's conversion to Mao's three-stage concept of rural revolution.

Protraction of the struggle along Maoist lines initially involved organizing and consolidating safe bases, then expanding low-risk military operations, and finally engaging in more conventional military offensives. Protraction gave the guerrillas time to minimize their weaknesses while playing on the enemy's shortcomings. The white Rhodesians preferred a short war, whereas the Zimbabwean nationalists could win only a protracted war.

Protraction husbanded scarce resources, allowed guerrilla units to engage in low-risk operations (notably shoot-and-scoot ambushes, sabotage, and mine laying), and enabled ZANU to gain a rural base from which it could draw intelligence, sanctuary, food and water, and porters and recruits. Herbert Chitepo, ZANLA's operations chief, described the strategy's goal as "to attenuate the enemy forces by causing their deployment over the whole country. The subsequent mobilization of a large number of civilians from industry, business, and agriculture would cause serious economic problems. This would have a psychologically devastating effect on the morale of the Whites."[10] A protracted conflict would also damage white interests by giving UN-imposed sanctions more time to work and by grinding away at white morale.

Having taken the brunt of Rhodesian response to the northwest operations, ZIPRA witnessed "a collapse of morale and the withdrawal of ZIPRA from the conflict for a number of years."[11] As a result, ZIPRA turned down FRELIMO's offer to operate from Tete province into northeast Rhodesia. Perhaps because of its reliance on Soviet advisors or because it may have felt that it needed a conventional onslaught to defeat Rhodesia, ZIPRA did not shift to a Maoist-protracted strategy. ZAPU would later pay heavily for its early inactivity and decision not to engage heavily in internal guerrilla warfare.

The Rhodesian success in Operation Nickel gave a false sense of security to whites and a confirmation of African ineptitude. A limited military buildup and no political concessions appeared sufficient. Defense spending remained fairly constant, and the government became even more hardline in its social policies. Aided by a seemingly sanctions-proof economy and security complacency (which included a bordering South Africa), white immigration began growing to its peak of 275,000.

South Africa showed more prescience than the Rhodesian public. The ZAPU incursions, which had included South African ANC members, persuaded Pretoria to dispatch police units into Rhodesia's Zambezi valley. Numerous reasons explain South Africa's support. Pretoria considered Rhodesia and its some 150 miles of common border as South Africa's most important buffer. Twenty percent of white Rhodesians had Afrikaner roots, and the two nations benefitted from bonds built from race, tourism, and business as well as a common minority status in black southern Africa. South African right wingers had always seen the Zambezi as South Africa's real defense boundary. In 1961 (almost five years before UDI) the two nations participated in combined air exercises. By 1969 South Africa had deployed 2,700 paramilitary troops into Rhodesia—a total force of only one thousand fewer than the entire Rhodesian regular army—as well as the "V Troop" (an South African Defense Force (SADF) until that quickly cracked all of Zambia's radio codes). Later military assistance included helicopters and their air crews—vitally necessary for Rhodesia's COIN operations—and specialized flight and underwater demolition training in South Africa. Yet unforeseen in the late 1960s was that South Africa's regional goals increasingly would collide with those of Rhodesia.

Prelude to Chimurenga

Without much fanfare ZANU's several hundred supporters began consolidating their position within Mozambique's Tete province around 1970. While Rhodesia still focused much of its attention on a line running from Victoria Falls to the end of Lake Kariba (the infiltration site of previous operations), ZANU intensified its political buildup along the Mozambican border. The shared Shona language and culture of eastern Rhodesia and western Mozambique greatly aided ZANU.

ZANLA quickly turned to its greatest potential resource—the 6 million Zimbabeans—and began an effective process of politicization. ZANLA's adherence to the concept of people's war had ZANLA draw upon all possible Zimbabweans for military manpower, porterage, sanctuary, food, and intelligence. ZANLA for years would emphasize political recruitment (and reliance on the peasantry for support) rather than direct military conflict. David Martin and Phyllis Johnson note that until early 1978, ZANLA was not on the offensive but was engaged in defending the process of mass mobilization.[12] As one guerrilla recalled, "We were taught that we had come from the people and that we had to go to the people . . . our source of supplies, shelter, and security."[13]

The guerrillas sought out Tribal Trust Lands (TTL), where most of Rhodesia's Africans lived under subsistence conditions. Recruitment often occurred at a village gathering, or pungwe. Popular spirit mediums significantly aided recruitment. ZANU sought and gained the support of these popular mediums. Many, and probably most, Zimbabweans believed that departed spirits communicated through these living mediums. The mediums helped to legitimatize the new guerrilla organization in many nighttime pungwes and would advise on a wide range of combat-related issues. While ZANLA recruiters cited a wide range of African grievances, they sometimes benefitted from apolitical considerations. Many guerrillas, including Josiah Tongagara, later ZANLA's chief political commissar, thought of enlistment as leading to "probably an adventure."[14]

The guerrillas chose targets that denied the government's presence and legitimacy and, in time, installed ZANU administration. Achieving these two goals would secure guerrilla legitimacy among blacks while lowering white morale. ZANU's nonmilitary targets included white farmers and their farms, government personnel (who implemented governmental policies for the blacks), protected villages, tribal chiefs and headmen, and such strategic infrastructure as railway and power lines, roads, and bridges.

Beginning Chimurenga

The Rhodesians knew of ZANU's shift to politicization but apparently failed to gauge both its importance and its scope. The southern movement by FRELIMO opened up areas of western Mozambique from which ZANU increasingly infiltrated into Rhodesia. The rewards of this quiet campaigning appeared in late 1972. On 21 December small ZANLA units staged surprise attacks in Centenary and elsewhere in northeast Rhodesia and marked the beginning of the seven years of protracted guerrilla struggle.

ZANLA grew from one hundred fighters in 1964 to perhaps 40,000 by the late 1970s. It had both male and female fighters and noncombatant male auxiliaries (mujibas) and female auxiliaries (chimbwindows). Perhaps 50,000 mujibas supported the chimurenga effort.[15] Auxiliaries aided the guerrillas with intelligence, food, and psychological and physical support. By the early 1970s, "large-scale porterage groups numbering 100 or more were . . . covering the distance from the Zambezi to the Rhodesian border in under twenty-four hours. They carried hundreds of kilograms of ammunition, land mines, and weapons."[16]

The guerrillas' ordnance and training reflected ZANLA's predisposition towards protracted warfare. The fighters used light weaponry—rifles (AK-47 assault, AKM, and SKS carbine), light machine guns (DP, RPD, and RPK), rocket launchers (RPG-2 and RPG-7), grenades (fragmentation, percussion, and fragmentation rifle), mines (POMZ-2 antipersonnel and TM 48 and 48 antivehicle), 60-mm mortars, as well as Tokarev pistols and TNT blocks.

ZANLA's basic tactic was to hit the enemy with minimal risk and from some distance. Guerrillas attacked security outposts with mortars and RPG-7s, which allowed an attack base some 600 meters from the target. When hitting a convoy, guerrillas would often halt the convoy with a mine blast or RPG firing. Then a section (between 15 and 20 guerrillas) would fire AK-47s and several of their RPGs.

ZANLA implemented sabotage operations with two sections. The sabotage section had about seven combatants trained in explosives. A larger support unit numbered between 15 and 20 fighters with (at maximum strength) mortars, RPG-7s, and machine guns. ZANLA never used remote control explosive equipment.

Guerrillas often planted land mines—"the single most devastating and effective device"—according to Henrik Ellert.[17] ZANU and ZAPU usually had an abundance of mines, guerrillas (and mujibas) needed little training in their handling or placement, and mine-laying operations rarely encountered the enemy. The mine's weight—about 10 kilograms—proved their major disadvantage.

The government's central claim about the guerrillas' methods was that of terrorism. The government documented a number of cases where guerrillas, usually ZANLA, killed or mutilated Africans who refused to join the struggle or who supported the government. Such violence usually occurred at a pungwe, where residents would accuse other residents of collaboration and "the accused were then put on 'trial' and invariably found guilty. The penalty was usually death. However, amputation of hands, fingers, legs, toes . . . was regarded as a less severe form of 'punishment'."[18]

All combatant forces employed terrorism, or the use of physical force, against noncombatant/innocents for political ends. The government's security forces regularly used torture, especially against captured guerrillas or when seeking guerrilla locations during COIN operations. Terrorism had short-term tactical advantages but could be used politically later by the other side. The government's demonizing ZANU and ZAPU as terrorists aided Rhodesian military morale and intensified overall white refusal to negotiate with the guerrillas.

The Government's Response

Over the next two years, the government responded to ZANLA's growth with both typical and atypical counter-insurgency tactics. Most notable were protected villages and elite COIN units. The government's overall strategy was to control its internal population while stemming the insurgents' flow into Rhodesia.

In early 1973 Special Branch reported that ZANU enjoyed active support from the peasantry into the northeast. The resulting Operation Overlord worked to isolate the guerrillas from this support. Rhodesia established protected villages (i.e., strategic hamlets in Vietnam and "aldeamentos" in Mozambique) to remove the peasantry from the guerrillas. By August 1977 the government had placed one-half million Africans into such villages. These villages, often called keeps, were surrounded by barbed wire and had watchtowers. The government also created "free-fire" zones and increased patrols and checkpoints. Salisbury sometimes imposed collective punishment on villagers. Along the Mozambican border, the government created a cordon sanitaire by defoliating a strip and then seeding the area with antipersonnel mines. Begun in May 1974, the minefield by its completion in April 1976, ran about 120 miles, from the Musengedzi River to the Mazoe River. Manpower shortages prevented adequate patrolling. Some of the population controls, notably checkpoints and protected villages, did help to isolate the guerrilla from peasant support, but in the longer run, they created significant peasant antagonism towards the government.

By late 1973 the increased infiltration and politicization by ZANU within the TTL was drying up the government's intelligence sources. The Rhodesians responded with the Special Air Service (SAS) and two new groups, the Selous Scouts and RENAMO (Mozambican National Resistance Movement). Used mostly from external operations, the SAS was a small force; in the mid-1970s it had only about one hundred fighters. The SAS (and the Rhodesian Light Infantry) began covert, small-unit operations against ZANLA in Mozambique as early as 1969. SAS mastery included map reading, long-range tracking and infiltration, demolition, intelligence gathering, parachuting, and assault tactics. Security observers credit SAS as Rhodesia's most professional fighting unit.

No longer able to rely on chiefs and headmen for internal information, the RDF decided on a "pseudo" unit patterned loosely on units in colonial Kenya and Mozambique. Such a force would dress and act as guerrillas. In 1973 captain, later colonel, Ron Reid-Daly assumed control of these Selous Scouts, whose main function was to locate the enemy and then radio in a reaction force. A second role was that of "hunter-killer" groups. The Scouts received their funding from the CIO as well as from a South African Defence Force supplemental fund. The unit enjoyed unusual freedom from the normal chain of command. When at full force, the unit comprised about one thousand operators (combat personnel). Many were captured guerrillas whom the government had converted (turned).

The Scouts operated in "frozen" areas where no other RDF personnel were permitted: guerrilla impersonation could prove fatal to the Scouts if regular government troops sighted them. The Scouts, as pseudo guerrillas, often visited villages to gather intelligence. Initially, they fooled the villagers, who could not distinguish between vakomana (guerrillas) and the mazkuzapa (scouts). The villagers soon developed recognition codes which the Scouts would need to learn quickly—validation was essential. Hence, "turning" a captured guerrilla within 24 hours was crucial before codes changed.

The Scouts employed various means to force a guerrilla to support the government: emphasize that death was his only alternative, financial incentives that included a kill-bonus of $1,000, or intense political indoctrination. The Scouts employed physical torture, but it was done more for extracting information than for turning a guerrilla into a government supporter.

When a Scout's "call sign" (between two and 12 men) sighted a guerrilla band, it radioed the map coordinates to a quick-response unit called Fire Force. Comprised usually of the all-white Rhodesian Light Infantry in Dakotas (DC-3s) or helicopters (initially Alouettes and later Hueys), Fire Force would vertically envelop the guerrilla area. Overhead helicopters provided command instructions and cover fire.

The government claimed that the Scouts, largely through this function and only secondarily through direct combat, helped to account for 68 percent of all guerrilla deaths.

The Scouts also gained a reputation for "dirty tricks." They sometimes kidnapped or assassinated political undesirables, poised clothes (especially blue jeans), or booby-trapped radios, which they believed the guerrillas would later acquire. In 1974 a group of eight Scouts kidnapped four high-ranking ZIPRA personnel inside Botswana, an action which dealt ZIPRA a stunning loss. To convince a village that they were guerrillas, Scouts would sometimes flog a resident who villagers had accused as a sellout. Sometimes Scouts would deliberately mistreat villagers, hoping that the mistreated would blame ZANLA or ZIPRA.

A third new unit, which achieved notoriety following the Rhodesian conflict, emerged at the end of Portuguese rule. Former Portuguese businessmen and military personnel, as well as black Mozambicans, opposed FRELIMO's crossing into Rhodesia in 1974. Strapped both for effective manpower and the ability to gather current intelligence inside Mozambique, the Rhodesian military gathered these refugees—especially former special force Flecha soldiers—into the newly formed National Resistence Movement and tasked the SAS to train (and often lead) the unit.

During its Rhodesian days, RENAMO functioned more as long-range reconnaissance and as eyes and ears for the RDF;

it usually fought only when necessary. Its knowledge of Mozambique (i.e., geography, language, and customs) helped the Rhodesians in their spectacular raids during the mid-and late 1970s, although, comments Reid-Daly, "It came too late in the war to have any serious effect on the outcome."[19]

Rhodesia's ability to skirt international sanctions proved vital in sustaining the war effort. For the war's first decade, Rhodesia baffled the experts as its economy seemingly winked at sanctions. Rhodesians, working with South Africans, Portuguese, and businessmen from other countries, circumvented UN sanctions and obtained vital supplies, including oil, despite United Nations sanctions. Military imports included Siai Marchetti SF 260 trainers and 11 Augusta Bell 205 (Huey) helicopters. This last item proved invaluable to a government facing a theater, internally and extremely, of widening guerrilla operations. Rhodesia's commodity marketing boards (aided by favorable international commodity prices) helped to export dairy, cotton, grain, and beef products. The government and the private sector helped to manufacture a wide range of products, including such military items as "rat packs," uniforms, tents, radio sets, armored personnel carriers, and weaponry, including several assault pistols and 82-mm mortar tubes and bombs.

The United States played a secondary but important role in the Rhodesian conflict. It initially aided the Smith government through the Byrd amendment. Enacted in 1971 despite UN sanctions; the Byrd amendment allowed purchases of Rhodesian minerals if the Soviet Union were the only other major supplier. Until its repeal in 1977, the amendment accounted for trade amounting to $400 million in foreign exchange for the Rhodesian government. The US did not prosecute any of its approximately 500 citizens who fought in the Rhodesian Defence Force.

Divisions within Chimurenga

The government's population control measures, its elite units, and its securing of some international support, as well as divisions within guerrilla ranks, gave Rhodesia a short-term military boost which fooled some observers into believing that Rhodesia could win the war.

Relations between ZAPU and its now larger splitoff, ZANU, had never proven especially cordial. In the early 1960s, disagreement led to killings and property destruction. The differences were multilayered. Ethnically, ZAPU was largely Ndebele, while ZANU was predominantly Shona, but ethnic subdivisions existed within each of these groups, especially within ZANU.

To avoid ongoing competition, the frontline states (Zambia, Angola, Tanzania, Mozambique, and Botswana) and the Organization of African Unity's African Liberation Committee attempted to unify the armed struggle. In November 1975 ZANLA and ZIPRA contributed some forces to ZIPA (the Zimbabwean People's Army). Yet long-standing differences rose again, and as Dr Masipula Sithole notes, "Clashes ensued." After a few weeks of joint operations, the surviving ZAPU men withdrew from ZIPA in Mozambique and fled to Zambia.[20] The division of targets—ZANLA concentrating on eastern Rhodesia and ZIPRA focusing on western Rhodesia—lessened possible contacts and rivalries.

Guerrillas smarted against what many perceived to be a double standard: relative luxury in Zambia for high-ranking ZIPRA and ZANLA officials and intolerable supply and battlefield conditions for lower supporters. In November 1974 several ZANLA officers—in what became known as the Nhari rebellion—attempted a coup against their high command. While failing, the attempt caused some 60 deaths.

Kenneth Kaunda's anger at the guerrillas' disunity increased when, in March 1975, a bomb killed ZANU's national chairman, Herbert Chitepo, in Lusaka. Kaunda, believing that internecine ZANU conflict was getting out of control, jailed ZANLA's high command until mid-1976 and transferred ZANU's bases to Joshua Nkomo. "ZANU survivors fled to Mozambique," writes Paul Moorcraft, "but the military infrastructure had been emasculated."[21] Ironically, Rhodesia's Central Intelligence Organization may have planted the bomb.[22]

Government and guerrilla forces alike faced internal rivalries and jealousies which lessened their capabilities. The Nhari rebellion sidetracked ZANLA. On the Rhodesian side, divisiveness between the police and military and between the Selous Scouts and SAS lessened Salisbury's COIN abilities. Beginning in the mid-1960s, the Rhodesian police "jealously regarded" the maintenance of law and order as their preserve and an area where they would not allow the army to intrude. "It was an attitude," notes Reid-Daly, "that created much unnecessary friction between the police and the army and hindered cooperation between different branches of the security forces."[23]

Chimurenga Increases

But other foreign events aided the chimurenga (liberation struggle) forces. Without foreign support, the guerrillas could not have prolonged their struggle. Communist nations supplied military equipment and several west European countries and international organizations gave nonmilitary aid. The liberation of Angola and Mozambique proved most crucial.

"We need arms—arms—arms. We need training facilities. Give us these, please, we plead and we shall do the rest."[24] Mugabe's plea underlined ZANLA's overriding need. Most of its weaponry came from Communist states, either from the People's Republic of China or such east European nations as Czechoslovakia or Hungary. The Soviet Union gave little aid to ZANLA because of its Maoist doctrine and perhaps because it had gambled from the beginning on an eventual Nkomo victory.

Finland and Norway provided cash grants and the United Nations High Commission for Refugees, the United Nations Development Program, the International Committee of the Red Cross, and the World Council of Churches provided cash grants, relief supplies, and technical assistance. Such aid also provided obvious psychological support.

ZANU's ability to open political offices in foreign countries greatly assisted its ability to obtain external aid. By 1977 ZANU had offices in Australia, Botswana, Canada, Lesotho, Mozambique, Nigeria, Tanzania, the United Kingdom, and the

United States. The OAU regularly championed the chimurenga verbally but, given the largely empty coffers of its Liberation Committee, provided little financial assistance.

The April 1974 junior officer coup in Portugal installed a new government that quickly recognized the right of Portugal's colonies to independence. MPLA rule in Angola not only provided some military bases for ZIPRA, but its acceptance of Soviet and Cuban help prompted South Africa and the US to reconsider their Rhodesian policies. Newly independent Mozambique allowed ZANU bases all along the 650-mile border with Rhodesia. Rural black youth increasingly flocked to join ZANU, which soon began infiltrating into the southeast lowveld.

Another crucial external occurrence began in 1974 when South Africa sought a "deténte" with black, and especially southern, Africa. Detente could lessen both the international criticism of South Africa and the nascent international call for sanctions. Peaceful resolution by South Africa of the Rhodesian conflict could bolster South African prestige while preventing the war from washing over into South Africa.

In December 1974 South Africa obtained Rhodesia's acceptance to detente proposals which, inter alia, forced Rhodesia to release some leading nationalist politicians (including Mugabe, who had been in detention for 10 years) and to stop all offensive operations, including pursuit of guerrillas leaving the country. Smith's reluctance in early 1975 to seek a significant agreement with nationalist forces prompted South Africa to withdraw its 2,500 paramilitary police in Operation Nutcracker. Smith subsequently followed South Africa's wishes, but the peace talks failed. The cease-fire allowed the battered guerrilla forces to regroup and retrain. Furthermore, the cease-fire convinced many Rhodesians "that the guerrillas had won a victory similar to the one which the cease-fire had brought FRELIMO in Mozambique. All intelligence sources had dried up and the army's position on the ground was weaker than it had been since the beginning of the war in 1972."[25]

On 3 May 1976 Mozambique closed its border with Rhodesia. This action trapped one-sixth of Rhodesia's railway stock and significant exports. More importantly, the border closing forced Rhodesia—which would quickly construct the Rutenga-Beitridge railway line to South Africa—to depend almost totally on its southern neighbor for trade access to the outside world.

ZANLA infiltration by mid-1976 prompted Rhodesia to open two major theaters of operations (Repulse and Tangent). By mid-1976 ZANLA had over 1,500 internal members and a previous 11 to 1 kill ratio had temporarily slid to a 6 to 1 ratio. Defense costs were rising rapidly and so was white emigration (in March, only 40 whites had left the country, as compared to 817 the next month). Increasingly, the TTLS throughout eastern Rhodesia had become base areas that sponsored ZANLA attacks.

External Operations

ZANLA's larger base camps and increased infiltration from Mozambique prompted Salisbury to consider major "external operations." These incursions were the only major conflicts of the war: inside Rhodesia, contacts were usually between small units and involved slight casualties.

While militarily sensible, the cross-border operations exacted a political price and discredited Salisbury's belief that increased guerrilla casualties could end the war. By July 1976 Rhodesian security planners believed that upwards of 6,000 ZANLA were training for a rainy season offensive by perhaps 2,000 guerrillas. Some senior Rhodesian officers questioned whether their troops could contain such infiltration.

On 9 August a group of 84 Selous Scouts drove 25 miles into Mozambique to a camp at Nadzonya. Entering a ZANLA camp of perhaps 5,000 at daybreak, the Scouts, with no losses to themselves, rapidly killed some 1,200 ZANLA military and civilian supporters.

For the next three years, the RDF conducted a huge number of both large and small externals; in 1979 General Walls maintained that not a day passed without his troops operating inside at least one bordering country. The raids' initial objective was to inflict maximum casualties. As guerrilla encampments adjusted to the raids (smaller concentrations in larger areas, better antiaircraft and bunker configurations), the Rhodesians concentrated more on disrupting logistics.

"Wonderful" was Smith's description of Nyadzonya.[26] South Africa, however, feared that the raid could attract Soviet and Cuban personnel to South Africa's borders. To deter any future raids, South Africa instituted Operation Polo, which pulled out a significant number of military personnel and cut into half Rhodesia's air-strike capability. South Africa's "congesting" of its railway lines reduced ammunition supplies to only 12 days. Two days after the raid, South Africa for the first time publicly supported majority rule for Rhodesia.

Political Moves

External events continued to weaken Rhodesia's cause. In mid-1976 US Secretary of State Henry Kissinger moved to resolve the Rhodesian conflict. While he failed in his major goal, Kissinger did persuade Smith to announce that majority rule would come within two years. Kissinger also brought the combatants together in Geneva for a peace conference.

White morale understandably flagged with Smith's majority rule announcement as "in a single speech Smith had reversed the original war aim of the whites."[27] The war no longer was to preserve the white Rhodesian state. Now the war was only to buy Smith time to obtain a moderate black internal government.

Under pressure from their respective patrons—the frontline states and South Africa—the guerrilla forces and Rhodesia reluctantly attended the stillborn Geneva conference in late 1976. When Smith agreed to attend, Pretoria began supplying enough weaponry for Rhodesia to survive the growing infiltration. Subsequent Western diplomacy, led primarily by British Foreign Secretary David Owen and America's UN Ambassador Andrew Young, failed to achieve peace.

Yet Smith needed to change his government's image to counter growing South African and US hostility and to lessen growing internal support for chimurenga. In March 1978 the internal settlement headed by Smith, Bishop Muzorewa, and the docile Chief Jeremiah Chirau began to rule in Rhodesia. Nkomo refused to participate and ZIPRA activities increased, which in turn prompted increased Rhodesian strikes across the Zambezi. The settlement did not address the guerrillas' major concerns and provided whites with inordinate power (e.g., an effective veto for at least the next 10 years).

Beginning of the End

White Rhodesia was slowly coming to its knees, despite its clear combat superiority. The guerrillas' ability to prolong the struggle had provided them time to secure manpower, destroy some government infrastructure, gain military knowledge, and generally to wear down white morale and the economy. "Purely statistically," J. K. Cilliers writes, "the security situation was deteriorating with each passing month."[28] At least since 1976, ZANLA had faced difficulties finding enough facilities for its expanding numbers. The guerrillas effectively closed such government operations as clinics, schools, and cattle dips and increasingly attacked white farms and major transport routes.

The security forces countered with a series of dramatic and destructive external raids. In November 1977 a combined SAS, Scouts, and RLI unit attacked a ZANU encampment in Chimoio, Mozambique, and killed over one thousand ZANU soldiers and civilians. The government also recruited young supporters of Muzorewa and Sithole into an auxiliary force while offering safe return and amnesty to guerrillas.

But such operations were only a finger in a crumbling dike. ZANLA "simply punched too many terrs across the border for us to handle," recalls a former Selous Scout.[29] ZANLA infiltration went from about 300 in 1973–74 to about 5,000 by late 1977, then to 9,000 in December 1978.

Also, Mozambican-ZANLA resistance was stiffening against Rhodesian attacks. The SAS-RLI Operation Uric of 1979 involved "a very light infantry assault against a well-defined conventional position. For the first time in the history of the war, the Rhodesians had been stopped dead in their tracks."[30]

Martial law rose from 50 percent of Rhodesia in November 1978 to 95 percent in September 1979. Special courts helped to enforce dawn-to-dusk curfews. In December 1978 ZANLA guerrillas destroyed 25 million gallons of fuel at Salisbury's depot. The worsening battlefield conditions lowered morale. Henrik Ellert claims that "security force reaction had in some cases sunk to the levels of the once despised Portuguese army who merely reacted to such events by firing into the bush and getting out of the killing zone as quickly as possible with little thought of any follow-up."[31] The protracted guerrilla war and international sanctions bit hard at Rhodesia's economy. Rhodesia's gross domestic product for 1979 was about the same as in 1965—the year of UDI.

The final year of combat, 1979, saw dramatic examples of escalation and the continuance of two related trends: increased guerrilla infiltration and Rhodesian external operations. Between December 1978 and January 1979—less than two months—the number of insurgents increased from 8,954 to 11,183. In early 1979 a major Soviet military delegation began reorganizing ZIPRA's overall military strategy. Ominously for Salisbury, ZANLA reportedly was preparing for the final, conventional state of protracted war. In February ZIPRA shot down a second Viscount (the first occurred in September 1978) and an aging Canberras hit ZIPRA bases in central Angola. In April the SAS conducted a dramatic, but unsuccessful, assassination attempt against Joshua Nkomo in Lusaka.

Also in April, the government managed to mobilize 60,000 personnel to ensure undisturbed elections that brought Bishop Muorewa to power. As with the 1978 Internal Settlement, effective white retention of power and the guerrillas' insistence on a military solution foreclosed any hope for the new government.

Rhodesian external raids continued to exact an increasing toll on Mozambique's and Zambia's economies, while internally ZANLA and now ZIPRA increasingly gained effective control over much of rural Rhodesia. Officially recorded contacts with insurgents rose from under 600 in late 1978 to 1,706 in May 1979. A Rhodesian Intelligence Corps paper noted that "in some areas a well-organized and security-conscious [guerrilla] civil administration is working"[32] while an army briefing document in mid-1979 tacitly admitted defeat: "in classical COIN terms, this is a no-win or rather, a sure-lose equation."[33] Rhodesia was ready for peace.

Britain's foreign secretary, Lord Carrington, decided to resolve the rapidly unraveling conflict. Primarily speaking to the combat's patrons, South Africa, and the suffering frontline states, Carrington brought all combatants to Britain's Lancaster House. As the talks opened, Rhodesian troops destroyed important bridges in both Mozambique and Zambia. "Zambia was now almost totally dependent," J. K. Cilliers notes, "for all her imports and exports on the remaining link through the south" (i.e., Rhodesia).[34] Kaunda and especially Machel pressed their reluctant clients and all parties agreed to a 21 December cease-fire. A commonwealth monitoring force supervised the movement of Patriotic Front guerrillas to assembly points, and in April 1980 Robert Mugabe stunned most Western observers by gaining 63 percent of the vote and 57 of the 80 seats reserved for blacks. On 18 April Mugabe became Zimbabwe's first prime minister. In his first address to the nation, Prime Minister Mugabe stressed reconciliation. He then retained several whites for his cabinet and declined to nationalize the white industrial, service, and agricultural sectors. About 100,000 whites elected to remain under Mugabe's rule. Tensions between ZAPU and ZANU flared for several years, as former ZIPRA soldiers conduct banditry in Matabeleland, and the government employed harsh, and often arbitrary, countermeasures. But by the mid-1980s, the three former combatants—the Rhodesian Front, ZAPU, and ZANU—were living in peace.

Conclusion

Why did the Rhodesian war, initially between two seemingly mismatched opponents, persist for some 14 years? Internal decisions and actions by the combatants prolonged the war long enough for external events—UN sanctions, communist military support, and the independence of Angola and Mozambique (which changed US and South African policy)—to take effect.

The two most decisive internal factors were ZANLA's adoption of Mao's protracted war strategy and the Rhodesian government's continued insistence on using military means to combat these guerrillas. Maoist doctrine gave ZANLA time to muster political and manpower support from its greatest potential strength—the overwhelmingly and increasingly politicized black population.

Rhodesian military capabilities always remained strong enough to prevent a conventional guerrilla military victory. Yet Rhodesia could never win the war with strictly military means. ZANLA's politicization attracted increasing recruits, while Salisbury could not counter with more men or better equipment. A refusal to acknowledge African grievances and a lack of funds prevented Rhodesia from conducting a strong civic action campaign.

Salisbury underestimated the war's political dimension. It did not realize the depth (and perhaps even the existence) of African displeasure with its rule. At the outset of the 1972 Pearce Commission, a British delegation which documented black anger to a proposed British-Rhodesian accord, Smith had commented that Rhodesia's blacks were the happiest blacks in world. Even after the commission, Salisbury continued to believe that outside elements were confusing, deceiving, or intimidating its blacks. Therefore, the government only needed to militarily eliminate the troublemakers.

Some governmental officials knew of the discontent (and the guerrillas' appeal), yet Smith apparently refused to believe it. He spurned the more realistic intelligence appraisals of the Special Branch (composed largely of Englishmen) for the decidedly more rosy assumptions presented by Rhodesian-born Internal Affairs officials.

Salisbury's refusal to acknowledge the war's political content was illustrated when it sharply rebuked a respected white parliamentarian (and COIN expert), Alan Savory, for using the term "guerrilla" rather than "terrorist" and then for suggesting that a constitutional conference be convened. Smith considered such a proposal as the "most irresponsible and evil" suggestion he had ever heard.[35] Such ostrich-like thinking stiffened white resolve not to negotiate and thereby helped to prolong the unwinnable war.

The government's misreading held, at least until 1975, that the blacks could not mount any credible and sustained opposition. Not fearing a serious insurgency, the whites saw no need to reform politically. Salisbury calculated that it could control the internal situation and that the regional and international equations would not significantly change. The government drastically misread the regions' future and its effect on Rhodesia.

Salisbury's military strategy of tracking down the guerrillas rather than addressing economic and political grievances with security protection reflected its political assumption. Rhodesia's major combat units followed mobile—rather than area—defense. Coupled with a paucity of resources, this policy, by not providing permanent presence, precluded serious civic action programs and placed any Africans considering cooperating with the government at greater risk. In the early 1970s Rhodesia probably could have accomplished area defense by training village militias who expressed loyalty to the local chiefs. Beginning in 1978 the government-initiated Security Force auxiliaries initially showed promise but by then the Patriotic Front forces had made substantial inroads, and government training capabilities and finances were too thinly stretched.

Rhodesia's military operations, while tactically sometimes brilliant, suffered several strategic flaws. The government did not heed Special Branch intelligence regarding ZANLA's changeover to protracted war and its buildup in Mozambique's Tete province in 1970 and 1971. Reid-Daly, among others, considers this a major flaw of the conflict. Not only did the government largely permit the buildup inside Tete until 1972, but it failed to comprehend the severity of increased ZANLA infiltration from 1970 to 1972. Attempting to stem guerrilla infiltration following their entry into the Tribal Trust Lands was like closing the door after the horse's departure.

The government exacerbated its manpower weakness by maintaining its chiefs interventionist units—RLI and SAS—as all-white. These groups obtained massive-killed ratios, but their small numbers (the SAS had only one hundred combat soldiers as of 1975) limited their effectiveness. This racially exclusive policy appears questionable when considering the effectiveness and loyalty of blacks in the Selous Scouts and the Rhodesian African Rifles.

Rhodesia's lack of a strong central commander permitted personal and group rivalries. Following the 1976 Nyadzona raid, the SAS felt that the Scouts had gone beyond its mandate of intelligence gathering and had assumed the SAS' primary role of external interdiction. Reid-Daly writes of "petty jealousies between Special Forces . . . severely limit[ing] effective external operations."[36] Reid-Daly himself received an official reprimand for accusations he made against Rhodesia's army commander. The guerrillas also faced internal problems, notably 1974's Nhari rebellion and Kaunda's expulsion of ZANLA in 1975 following various disputes and Herbert Chitepo's death. Yet the guerrillas' emphasis on political mobilization lessened the military effect of such divisions.

External events turned against Rhodesia in the mid-1970s. Rhodesia's unexpected ability to weather UN sanctions helped prolong white belief in eventual victory. But by the mid-1970s the Rhodesian economy was beginning to demonstrate how chimerical this hope was. The UN's three, increasingly severe, sanctions packages (from voluntary, to selective mandatory, to comprehensive mandatory) did not play the leading role their proponents had predicted but, over time, they had several key effects. Sanctions did not keep Rhodesia from international markets, but they did force Rhodesia to purchase at a higher rate (at least 10 percent) than market price, usually in scarce foreign exchange, and to sell often at below-market prices. Whether out of political conviction, desire to conduct only legal operations, or fear of being exposed, many countries and companies refused to trade with Rhodesia. Only rarely and through fraudulent means (e.g., doctored, end-use certificates) did Rhodesia obtain major military ordnance. Had the prosperous Rhodesia of the early 1970s openly purchased required modern military and economic equipment, it could have, at the least, significantly prolonged and, conceivably, even won the war. Although imperfectly enforced, UN sanctions prevented this possibility. Sanctions and their economic effect probably stiffened white Rhodesian nationalism until the mid-1970s when the longer-run debilitating effects took hold.

Regional assistance turned the war's course around only after 1975 when Angola and Mozambique gained independence.

But starting in the early 1960s, African support gave ZAPU and ZANU the foothold they needed for survival, thus prolonging the war. When the armed struggle began, Tanzania allowed several training bases as well as access to weaponry and training. Its importance as a rear base diminished only around 1975, when ZANU began major infiltration from Mozambique. Zambia also allowed both forces to operate from its borders until the Nhari rebellion and Herbert Chitepo's assassination in 1975.

Mozambique proved most important, since FRELIMO had supported ZANLA even before achieving power in 1975. By 1969 ZANLA was operating out of neighboring Tete province. After Mozambique's independence, ZANU spread its network of guerrilla camps and refugee centers through much of Mozambique, while ZANLA started operations from Manica and Gaza provinces.

The independence of Angola and Mozambique sealed the fate of white minority rule by raising guerrilla capabilities while lowering those of the government. Angola demonstrated significant Soviet and Cuban willingness to intervene militarily in southern Africa, while Mozambique offered sanctuary and bases to ZANLA, the more radical of the two guerrilla groups.

Both Washington and Pretoria realized that the highly capable RDF could not handle direct Soviet involvement or the increasingly real infiltration increases. The longer the war continued, the more radical the guerrilla forces would become. Accordingly, the two nations pressed Smith to compromise, thus weakening his position even further.

Yet the same reasons which dictated initial support for Smith during the Johnson and Nixon administrations eventually changed US policy, under Ford and Carter, to one of opposition. Washington had felt that, whatever Smith's domestic racial shortcomings, he provided a buffer for southern Africa—especially South Africa—against communism. The MPLA and FRELIMO victories in Angola and Mozambique not only signaled to America (and to South Africa) that Smith's days were numbered but also that the longer Smith believed that outside aid was forthcoming, the less interested he would be in compromise, while the more militant the opposition would become.

South African aid proved paradoxically to be Rhodesia's greatest strength and its greatest curse. South Africa furnished crucial economic and military support but increasingly had conflicting goals. Facing growing international pressure and the loss of the buffer states of Angola and Mozambique, South Africa in the mid-1970s strove to increase its limited international legitimacy and to limit the effect of regional insurgencies on South Africa. In 1974 President John Vorster sought "detente" with black Africa and especially the frontline states. Paul Moorcraft describes the resultant December 1974 cease-fire as "a major psychological setback for Salisbury," since it stopped government operations against a faltering insurgency and allowed the release of leading nationalists.[37] Equally, in August and September of 1976, Vorster placed powerful and effective pressure on Smith again. Yet soon thereafter, following Smith's announcement of majority rule within two years, Vorster opened SADF training facilities to the Rhodesians and allowed elite South African troops to fight alongside Rhodesians on cross-border operations. South Africa was willing to prolong the war if such prolongation would force the guerrillas to the bargaining table. South Africa refused to supply Rhodesia with massive weaponry which, when used against the frontline states, could provoke Soviet/Cuban military involvement on South Africa's border.

The war's prolongation, greatly aided by Angolan and Mozambican independence, allowed the guerrillas, especially ZANLA, to recruit increasingly from a large, and increasingly politicized, manpower pool while draining the dwindling pool of white manpower (and by so doing weakening the white-run economy and made the manpower pool even smaller by encouraging emigration).

While the combatants' patrons—the frontline states and South Africa—greatly prolonged the military struggle, they effectively ended the struggle before its military resolution. Zambia and Mozambique pressed for peaceful resolution as the nations in conflict increasingly bit at their own economies. Just as the frontline states could threaten their clients with deprivation, so South Africa could use its transport and military system to cajole Rhodesia to the negotiation table. In

1976 ZANU and the Rhodesian government attended the Geneva conference only because of their patrons' urging. Three years later the military situation increasingly favored ZANU's hopes for a full military victory. Yet the Rhodesian defense force had managed to inflict serious losses on Zambia and Mozambique. A grim future of rising losses for all sides convinced patrons of the combatants to force negotiations. That both the RDF and the guerrillas retained some military advantages presaged the political compromise reached at the Lancaster House (and invoked Mao Tse-tung's dictum that "you cannot win at the conference table what you have not won on the battlefield").

Notes

1. Prime Minister Harold Wilson, quoted in Paul Moorcraft, *African Nemesis* (Brassey's: London, 1990), 124.

2. Until 1976 Prime Minister Ian Smith believed the US and South Africa would rush to Rhodesia's side should communist-backed insurgencies seriously threaten white control. The US condemned the Smith regime but between 1972 and 1977 supported Rhodesia through mineral purchases (the Byrd amendment). South Africa was Rhodesia's primary economic and military supplier throughout the war.

3. *Chimurenga* is a Shona word meaning struggle or revolution. Zimbabwe's first chimurenga was in the 1890s, when blacks fought against white occupation. The struggle of the 1960s and 1970s became Zimbabwe's second chimurenga.

4. Quoted in Martin Meredith, *The Past Is Another Country* (London: Pan Books, Ltd., 1980), 79.

5. D. Cowderoy and R. Nesbit, *War in the Air* (Galago, Alberton, 1987), 47–48. Quoted in Moorcraft, 125.

6. ANC and ZAPU cooperation occurred because both received their military aid primarily from the Soviet Union and its East European allies. Also, ZAPU's Matabeleland geographically lay between friendly Zambia and target South Africa.

7. Ron Reid-Daly, "War In Rhodesia—Cross-Border Operations," in A. J. Vener, *Challenge*, 153.

8. Reid-Daly, 155.

9. Quoted in Barbara Cole, *The Elite: The Story of the Rhodesian Special Air Service* (Amanzimtoti: Three Knights Publishing, 1985), 40.

10. Ibid.

11. J. K. Cilliers, *Counter-Insurgency In Rhodesia* (Croom Helm: London, 1985), 9.

12. David Martin and Phyllis Johnson, *The Struggle for Zimbabwe: The Chimurenga War* (Faber and Faber: London, 1981), 88. This important book provided one of the first substantive examinations of the war from the perspective of the guerrillas.

13. Ibid., 82.
14. Ibid., 77.
15. Ibid., 73.
16. Henrik Ellert, *The Rhodesian Front War* (Gweru [Zimbabwe]: Mambo Press, 1989), 22.
17. Ibid., 29.
18. Paresh Pandya, *Mao Tse-tung and Chimurenga* (Braamfontein [South Africa]: Skotaville Publishers, 1988), 173.
19. Reid-Daly, 175.
20. Quoted in Cilliers, 34.
21. Moorcraft, 130.
22. Ibid., 131.
23. Reid-Daly, 149. The formal name for the Rhodesian police was the British South African police.
24. Quoted in Pandya, 203.
25. Meredith, 171.
26. Quoted in Meredith, 241.
27. Moorcraft, 133.
28. Cilliers, 43.
29. Interview with a former Selous Scout, 1988.
30. During Operation Uric, an RPG-7 killed 11 SAS troops, an action that Cole regards as the "worst single military disaster of the war." See Cole, 335.
31. Ellert, 47.
32. Cilliers, 240.
33. Ibid.
34. Ibid., 56.
35. Quoted in Martin and Johnson, 96.
36. Reid-Daly, 178.
37. Moorcraft, 130.

Civil War in a Fragile State—Mozambique

Christopher Gregory

Mozambique has been wracked by almost uninterrupted internal conflict since 1964, the year in which the Front for the Liberation of Mozambique (FRELIMO) began a protracted guerrilla war against the then 400-year-old Portuguese colonial administration in Lourenco Marques (now Maputo). That conflict endured 11 years, coming to a conclusion in 1975, principally because of the officers' coup in Lisbon on 25 April 1974.

This analysis, however, focuses on the conflict which began only 12 months after FRELIMO formally became the first government of independent Mozambique on 25 June 1975. At the time of this writing, the conflict between FRELIMO and the Resistencia Nacional Mocambicana (RENAMO) had become intense and had acquired a longevity envisaged by few, if any, of its principal protagonists and supporters.[1] This chapter identifies and analyzes the reasons for the prolongation of the war. To that end, the key events of the war are first sketched.

The War in Mozambique Since 1976

RENAMO's activities inside Mozambique are generally accepted as having commenced in 1976, the year in which the organization was first given training and logistical support on a sustained basis by Rhodesia's Central Intelligence Organization (CIO).[2] The Rhodesians were attempting to counter cross-border incursions by ZANLA, the military wing of the Zimbabwe African National Union (ZANU). These incursions had been stepped up in 1972, following a FRELIMO offer to ZANU to use the northwestern Mozambican province of Tete as a springboard. Relations between the two movements were to strengthen steadily during the 1970s.

Initially, RENAMO's actions were largely confined to intelligence gathering within Mozambique for the Rhodesian army and joint sabotage operations with elite Rhodesian units. As the conflict escalated, RENAMO operated deeper inside Mozambique, establishing a base in 1978 at Gogoi in the southern Manica province, and in the Gorongosa mountains of Sofala province. At this time, RENAMO numbered roughly 500 combatants.

Zimbabwean independence in 1980 signaled the end of Rhodesian support for RENAMO. The entire organization was transported to South Africa. Meanwhile, RENAMO activities continued to be confined largely to the central provinces, Manica and Sofala. In 1980 an offensive by the Mozambican army, the Forcas Armadas de Mozambique (FAM), reestablished control over the Gorongosa area and overran the Gogoi base in June. Most observers agree that the 1,200-strong insurgent movement was at the nadir of its fortunes. In 1981, however, a more resurgent and more widespread RENAMO surfaced. By the end of that year, a network of semipermanent bases had been established in Mozambique from which attacks were mounted on small towns. While FAM troops captured RENAMO's main base at Garagua, 20 kilometers from the Zimbabwean border, RENAMO reestablished itself in the Gorongoza mountains, whence it had been evicted a year earlier.

In 1982–83 RENAMO continued to expand both its geographical spread and the scale of its operations, deploying some 5,000 to 8,000 combatants.[3] Notwithstanding a number of successful FAM counteroffensives in 1983, RENAMO fought back, claiming by late 1983 to have a "sphere of operations" comprising all of Mozambique, with the exception of Niassa and Cabo Delgado in the far north, western Tete, and southern Maputo province.

A flurry of diplomatic activity marked 1984. In March the South African and Mozambican governments signed the Nkomati accord which ostensibly provided for the end of the South African government support for RENAMO. Further diplomatic moves, including the first face-to-face meeting between RENAMO and FRELIMO representatives in Frankfurt, West Germany, culminated in the "Pretoria Declaration," thereby defining terms under which to declare a cease-fire.

As war, however, continued, both FRELIMO and RENAMO spokesmen soon repudiated the cease-fire clause. Indeed, by the end of 1984 RENAMO had for the first time established a substantial presence in the northern regions. It now claimed 15,000 men under arms and "consolidated control" over Nampula, Niassa, Zambezia, Tete, Manica, Sofala, and Inhambane provinces. By 1985, RENAMO was carrying out hit-and-run attacks throughout the country.

The next few years saw first one side, then the other, take the military advantage. In 1985–86 FAM again went on the offensive, bolstered this time by increased military support from the Zimbabwe National Army (ZNA) and Tanzanian army units. The RENAMO base at Casa Banana was captured twice in a span of six months. The insurgents responded with a major offensive of their own, concentrating their efforts in the Zambezi valley and threatening by their actions to cut Mozambique in two. Indeed, the October 1986 offensive was the largest launched by RENAMO to date. For the first time since its inception, the movement seized and held several sizeable towns. In the same month, President Samora Machel was killed in an aeroplane crash. The smooth accession to the presidency of his successor, Foreign Minister Joaquim Chissano, quelled any speculation that the war effort would be disrupted.

An FAM/ZNA counteroffensive in February 1987 reversed many RENAMO gains of the previous year, recapturing the five provincial towns held by the insurgents since October 1986. RENAMO countered with its first raid *inside* Zimbabwe itself, having declared war on Zimbabwe in the previous year in response to an increased and more geographically dispersed ZNA presence inside Mozambique. By the end of 1987, there were reports of RENAMO tightening its grip on Maputo's environs and of almost daily attacks by RENAMO inside Zimbabwe. By 1988 a determined joint offensive occurred in central Mozambique by 8,000 FAM, ZNA, and Tanzanian troops, which, however, failed to push back the insurgents.

Talks resumed in 1989, one year later. RENAMO leader Afonso Dhlakama announced his organization's willingness to halt the insurgency and negotiate an end to the fighting as long as FRELIMO agreed to a number of reform measures, including a multiparty constitution and a free market economy. Notwithstanding a determined FAM/ZNA offensive against RENAMO's Gorongosa headquarters in July 1989, a first round of talks was held in Nairobi, Kenya, a month later. FRELIMO adopted a triple-track strategy against RENAMO: continue with negotiations while maintaining the military pressure in the field and accelerate internal political and constitutional reforms designed to undercut RENAMO's political demands. Five rounds of negotiations were conducted in various centers over the next 18 months, resulting in the signing of the Rome agreement, which took effect on 1 December 1990. The parties negotiated a partial cease-fire; ZNA troops in Mozambique were to be confined to the Beira and Limpopo transport corridors; and RENAMO guaranteed not to attack what the agreement termed these corridors of peace. A 10-country Joint Verification Commission was established to monitor the cease-fire.

Much of the continued fighting in 1990 favored FRELIMO and its allies. In March a joint force of ZNA and FAM troops launched yet another offensive against Gorongosa. Unsuccessful in its efforts to dislodge RENAMO, the joint force followed up with a similar attack in May, then another in September. Consequently, Dhlakama was forced to move his headquarters north to a point just 50 kilometers south of the border with Malawi. Another blow was struck against RENAMO in Zambezia province, where, after having yielded control of the populous province to RENAMO for the previous five years, Mozambican armed forces succeeded in regaining control over the thousands of peasant farmers in the province who had acted as a significant base of RENAMO support. FAM forces received assistance from an unexpected quarter: a spear-carrying militia, known as Naprama, that follows a 25-year-old traditional healer, Manuel Antonio, who claims inspiration from Jesus Christ and knowledge of a magic plant that renders his followers bulletproof.[4] Antonio recruited several thousand militia and used them to good effect against RENAMO in Nampula and Zambezia provinces.

In early 1991, RENAMO set about recouping its losses, focusing its offensive on Manica and Sofala provinces. The cease-fire collapsed in March, following RENAMO allegations of ZNA infringements of the Rome agreement. The insurgents increased attacks on the Beira and Limpopo transport corridors and intensified pressure on the areas surrounding Maputo.

At the time of this writing (November 1991), a situation existed perhaps best characterized as one of "violent equilibrium." The conflict inside Mozambique had become a prolonged war conducted in tandem with intermittent negotiations between the two principal protagonists. Little progress with the negotiations had been made during 1991 to the visible irritation of the current co-mediators, the Italian government, and the Catholic Church, neither of which appeared able to demonstrate leverage over the warring parties. On the ground, RENAMO deployed more combatants than ever, some 20,000, yet was unable to take and hold urban centers of any significant size for a sustained period. FRELIMO, on the other hand, was unable to dislodge the countrywide RENAMO presence, much less militarily defeat the organization. The security situation was effectively one of stalemate.

The 15-year-old war continued to exact a terrible price from the inhabitants of Mozambique, at least 100,000 of whom were reported to have died as a direct consequence of the fighting. Of an estimated total population of 16 million in 1990, 1.3 million Mozambican refugees were reckoned to be in neighboring countries, particularly Malawi, while some 3 million were said to be displaced persons or internal refugees. Sixty percent of the population lived in absolute poverty. The rural-based economy had collapsed, with 78 percent of marketed grain supplies having to be imported in 1988. This figure did not even include the considerable volumes of food aid imported under the UN emergency relief program. The national economy was the world's most aid-dependent, with official development assistance accounting for 76 percent of the gross national product (GNP) in 1988–89 against a sub-Saharan average of 11 percent. With an estimated per capita GNP of US $80.00, Mozambique had become the world's poorest country.[5]

The Civil War and Its Prolongation

The disputed origins of the war, the incompletely documented nature of the prime combatants, the characteristics of

Mozambican rural society, the many sources of external support, and a host of imponderables pose a formidable challenge to the analyst attempting to penetrate and explain the conflict in Mozambique. In particular, an adequate understanding of the war on the ground has been hindered for over a decade by a lack of dispassionate writing on the primary protagonists, FRELIMO and RENAMO, a state of affairs which has been perpetuated by the substantial links between the conflict in Mozambique and the regional bête noir, South Africa. Analysis also has been hindered by the difficulty of access to information in a country the size of Mozambique, and where the limited communications infrastructure has since independence been successfully dominated by AIM, the Mozambican government information agency. To cap it all, most observers approach the conflict with tools of analysis drawn largely, even exclusively, from the conventional occidental political science tradition. Analysis of the conflict in Mozambique, however, benefits from a more eclectic approach, drawing inter alia on ethnography, ethnology, and social anthropology.[6] In Derluguian's evocative prose, "The truth of the huts is very different from the truth of the palaces of the capitals."[7]

The prolongation of the civil war in Mozambique owes much to the origins and early development of RENAMO. This first section dwells at some length on this organization before turning its attention to the other principal protagonist, FRELIMO.

Origins and Development of the Resistencia Nacional Mocambicana

RENAMO is an atypical insurgency movement, whose origins complicate attempts to characterize the early period of the Mozambique conflict. RENAMO may be understood as having gone through three phases of development: the first, from the mid-1970s to 1980, was characterized by Rhodesian patronage of the organization; the second, from 1980 to the mid-1980s, was marked by South African government support; the commencement of the third, post–South Africa stage is more difficult to establish. It began during the period after the signing of the Nkomati accord in March 1984 and is characterized by RENAMO's greater military and political self-reliance in the face of dwindling levels of South African support.

Observers differ on the question of the extent to which Rhodesia's Central Intelligence Organization created RENAMO in 1976.[8] Of more direct relevance to this analysis, however, is the role played by the CIO in supporting and directing RENAMO. For it appears beyond dispute that, regardless of the nature and level of Mozambican opposition to FRELIMO which existed prior to 1976, the Rhodesians were instrumental in taking at least part of this opposition and giving it much-needed support and direction. One can argue that the CIO created RENAMO in that the assistance with equipment, training, and logistics provided by the Rhodesians was indispensable to RENAMO's growth and development in the 1970s.

Intensifying the insurgency war in the period after 1972, particularly the offensive northeast of Mozambique by ZANLA forces acting with the military and logistical support of FRELIMO, convinced CIO chief Ken Flower and his colleagues of the necessity for "a pseudo-terrorist operation directed from Rhodesia into Mozambique."[9] The Rhodesians established training camps for RENAMO at Bindura and Odzi, near Umtali (now Mutare) and close to the Mozambican border, and set up a radio station, Voz da Africa Livre, near Gwelo (now Gweru).

The movement acted as the CIO's eyes and ears, gathering intelligence inside Mozambique for the Rhodesian army. RENAMO personnel would also engage in joint sabotage operations with such Rhodesian elite units as the Selous Scouts and the Special Air Services. A Rhodesian former senior instructor subsequently described RENAMO strategy as follows: "To start off with it was sabotage, to disrupt the population and disrupt the economy which really comes under sabotage, to come back with decent recruits at that stage and hit any FRELIMO bases they came across. And if they came across Zanla they were to take them on."[10]

RENAMO, under Rhodesian tutelage, seems, ironically, to have been as much—if not more—a product of the Rhodesian civil war, than of the Mozambican conflict. It is not clear to what extent the movement's leaders and followers enjoyed the latitude to formulate and pursue their own objectives and to conduct their own military operations in Mozambique independently of the Rhodesian CIO.[11]

While the anti-FRELIMO insurgency had its origins in this period, to characterize the activities of RENAMO from 1976 onwards as underpinned by a conscious decision at this early stage to follow a particular insurgency strategy, Maoist or otherwise, would be to ignore the role of RENAMO in assisting with Salisbury's objectives. For Ken Flower and his colleagues, RENAMO was little more than a military tool for the prosecution of the Rhodesian war effort. In the telling words of the late Evo Fernandes, a longtime RENAMO spokesman, "It was not our war, but a civil war in Rhodesia."[12]

Indeed, it is important to consider what might have happened in the early 1980s to the "organization without a name," as Fernandes once called it, had the South Africans not taken over the movement in its entirety. Certainly, the loss of external patronage would have been a severe setback. Moreover, RENAMO suffered several other blows at about the time Rhodesian support was withdrawn. The movement's first leader, Andre Matsangaiza, died in a FRELIMO attack on the movement's Gorongosa base camps in October 1979. The camps themselves were captured by FRELIMO in the assault. Afonso Dhlakama succeeded Matsangaiza—but apparently only after a bloody succession struggle. Moreover, his defeated opponent's supporters surrendered to Mozambican authorities, leaving RENAMO, according to one source, with fewer than 500 members.[13] Dhlakama himself is reported to have written in mid-1980 that "when Andre died the MNR was on the road to destruction."[14] Were it not for South African support, the RENAMO insurgency might well not have become prolonged.

Given the nadir of RENAMO's fortunes in 1980, it appears beyond dispute that South African support was instrumental in perpetuating and redirecting the conflict in Mozambique. Again, to ensure continued external support, RENAMO played its part in the furtherance of objectives which were not necessarily its own. In any event, South Africa took over RENAMO, lock, stock, and barrel, giving the organization a new lease on life and a far more ambitious role than that originally conceived of by the Rhodesian security services and committing far more resources to RENAMO than Salisbury ever did.

Ken Flower's previous attempts to interest South Africans in supporting the RENAMO operation had been without much success.[15] By 1980, however, South Africa's regional security concerns had changed—and those senior officials with the ultimate responsibility for defining and pursuing them had been replaced. It is important to understand the regional context of South Africa's dealings with Mozambique.

The "cordon sanitaire" of white-ruled and moderate black countries in the 1960s had been brought to an end by events following the April 1974 coup d'état. Southern Africa in the 1970s had entered a period of rapidly escalating conflict marked by the active involvement of outside powers. The colonial administrations in both Angola and Mozambique were succeeded by avowedly Marxist-Leninist governments. Soon after his electoral victory in 1980, Zimbabwean Prime Minister Robert Mugabe broke off diplomatic relations with South Africa and set about intensifying the international diplomatic offensive against South Africa. In the same year, nine countries in the region set up the Southern African Development Coordination Conference (SADCC) with the intention of reducing their economic and transport reliance on South Africa. Mozambique, because of its history and position, was assigned the transport portfolio—key to the meaningful reduction of independence on South African trade routes.

The South African government responded to regional developments with a more militaristic defense posture. Pretoria officially adopted the view that a communist total onslaught was being waged against the republic and responded to the perceived threat with a total national strategy for the defense of the republic. The ninth point of the 12-point plan enunciated by Prime Minister P.W. Botha in August 1979 stated, "South Africa [has a] firm determination to defend itself against interference from outside in any possible way."

To be sure, there was no military involvement in Mozambique comparable to that in Angola. However, while Mozambique did not close the border with South Africa as the country did in 1976 with Rhodesia, FRELIMO pledged full support for the African National Congress, South Africa's principal guerrilla opposition, which was enjoying something of a resurgence in activity following the Soweto riots of June 1976. The extent to which FRELIMO was willing to go in supporting insurgent groups in neighboring states had been illustrated since the early 1970s by its support for ZANU and ZAPU in Rhodesia.[16]

The impetus for South African support for RENAMO was aimed at achieving the following objectives: putting an end to Mozambican logistic support for the ANC, undermining in general SADCC member-states' efforts to lessen their dependence on South African transport routes, undermining in particular Robert Mugabe's attempts to intensify diplomatic pressure against South Africa, and demarxificating the region by demonstrating that black Marxist governments could not efficiently run a country. Thus, the South African government's involvement in Mozambique was a product of its *regional* foreign policy objectives, rather than a particular bilateral focus on that country. In the period immediately following Mozambican independence, and prior to the declaration of FRELIMO support for the ANC, relations between the two countries began to have a more peaceful footing.

RENAMO's second stage is characterized by more than just the replacement of Rhodesian patronage by that of South Africa. Pretoria's objectives were such that it needed a larger insurgency movement with more territorial reach, and one which would specifically target Mozambique's communications infrastructure—SADCC's arteries. Pretoria also planned a higher public profile for the movement to present it as a genuine nationalist movement rather than keeping it clandestine, as had been the Rhodesians' preference. Accordingly, the South African government began funding RENAMO from mid-1980 on a scale greatly exceeding that of the Rhodesians, working to build it up substantially from the organization they had been bequeathed. Large numbers of RENAMO personnel were airlifted into Mozambique in late 1980 and early 1981.[17]

It is not apparent that the SADF in 1980 intended RENAMO to adopt a strategy of protracted war against the FRELIMO government, much less that prolonged conflict was in the cards. RENAMO was part of a strategy of limited- or low-intensity conflict designed to reduce what Pretoria saw as a propensity for FRELIMO's leadership to meddle in South Africa's internal affairs. It appears that Pretoria did not anticipate—or underestimated—what RENAMO was to become by the end of the decade.

Significant South African support for RENAMO accounts in large part for the resurgence of RENAMO in 1981 and the expansion of its zone of operations. The movement now focused its attentions on Mozamique's transport infrastructure disrupting the railways from Maputo and Beira to Zimbabwe, the pipeline to Mutare, and the main paved roads in the center and along the coast of the country. Again, observers suggest, RENAMO had to pay for external patronage by engaging in actions the objectives of which were more South African than RENAMO's.[18] RENAMO bands of some 100 to 150 combatants also began to attack and destroy small towns, capturing foreign technicians and aid workers. Some would be killed; others would be held for ransom or political propaganda purposes. RENAMO soon acquired an international reputation for frequent and violent acts against the rural population.[19]

However, the question of RENAMO's post-1980 resurgence raises other issues central to an understanding of the reasons for the rapid spread and prolongation of the insurgency to 1991. Conventional wisdom has it that, in the (1989) words of a longtime FRELIMO supporter, "[T]he conflict in Mozambique . . .is, in fact, an undeclared, low-intensity, covert war waged by the South African Defence Force through surrogate forces."[20] On that view, the principal reason for the prolongation of the war in Mozambique was and is South African support for the insurgents, who enjoy insignificant levels of rural support. Furthermore, as the argument often runs, the conflict would rapidly wind down were (presumed) South African government support to be curtailed.[21]

This characterization appears to be not so much incorrect as incomplete. Field research conducted since the mid-1980s has analyzed the RENAMO phenomenon in the context of the FRELIMO leadership's attempts to impose its ideals on Mozambican society, concluding that the deep-rooted presence of RENAMO throughout Mozambique can only be understood in the context of the fertile ground for insurgency created by FRELIMO in the years after independence.

FRELIMO: Government and Ruling Party

By independence in 1975, FRELIMO's leadership had moved from minimalist nationalism, which had characterized the Front at its formation, to a belief in Marxism-Leninism. The leadership regarded itself as the vanguard of social revolution in Mozambique and sought to centralize control of both the state and the economy, establishing a mechanism for the central planning of the economy, and implementing extensive nationalization measures in the two years following independence. FRELIMO also attempted to transform itself from a broad-based liberation movement into a narrow-based vanguard party, composed of ideologically committed cadres, purging its ranks in the process of individuals whose ideological orthodoxy or personal integrity were deemed doubtful, including many longtime FRELIMO supporters who could not understand why they were suddenly excluded. On the foreign policy front, FRELIMO signed a 20-year treaty of friendship and cooperation with the Soviet Union and a similar treaty with Cuba the next year.

Of particular pertinence to this analysis is the impact of FRELIMO ideology on its agrarian policies in the rural sector whence came the challenge from RENAMO. Recent research indicates that both the rapid spread and the deep-rooted nature of the Mozambican conflict may be explained, in large part, by examining the catalytic impact of RENAMO activities in areas of the country where FRELIMO's increasingly commandist implementation of highly interventionist agrarian policies antagonized the local inhabitants into either active support for RENAMO or passive neutrality.

FRELIMO regarded agriculture as the basis of the Mozambican economy, a sector which had to be developed in a politically and economically correct fashion by means of the "socialization of the countryside." However, FRELIMO thinking was characterized in general by a modernization mentality and a concomitant contempt for (and ignorance of) indigenous practices. During the independence celebrations in 1975, Machel publicly called for an "ideological offensive to wipe out the colonial and racist mentality which is deeply rooted in the urban areas, as well as the feudal traditional mentality which is predominant in the rural areas."[22]

FRELIMO's initial interventions in the agricultural sector largely took the form of crisis management. Ninety percent of the Portuguese settler population left in the 18 months following independence, confronting the new government with a worsening rural situation which included abandoned plantations, farms and farm shops, and a disintegrating distribution system. However, a concrete policy was soon formulated, emphasizing the formation of cooperatives and communal villages and the creation of large state farms. The predominant mode of agricultural production, the so-called family sector, in which peasant families typically worked two to three hectares of land, was virtually ignored by FRELIMO. Within the ministry of agriculture, there was not even a department responsible for the family sector.

FRELIMO's major, albeit unintended, contribution to the prolongation of the war was its attempted villagization of the peasantry, a strategy that accelerated after 1980.[23] In the process of constructing collective villages, many peasants were relocated from their scattered individual land holdings to the site of the new village. This process disrupted the traditional social hierarchies, the reins of authority frequently being taken over by the original inhabitants of the land on which the village was located. As the most fertile and best located land was held by these people, the newcomers also often found themselves economically marginalized. Over time, many of the newcomers were unable to grow enough food to survive and had to labor on the plots of others for food, a dependency which was both resented and difficult to break. Social marginalization was often added to their woes.

In the initial stages of villagization, participants largely volunteered. However, as FRELIMO encountered resistance to its policies and the threat from RENAMO heightened after 1980, coercion was increasingly employed with those who resisted being often arrested and imprisoned. Villagization came to take on a counterinsurgency dimension similar to that of the previous colonial administration—an irony not lost on the Mozambican population who had resented and resisted the Portuguese efforts to combat the FRELIMO insurgency only 10 years previously.

Some peasant groups preferred to live outside the villages altogether, their menfolk guaranteeing the group's existence by means of pillage. This practice hinders a fuller understanding of the Mozambican conflict, as for most of the 1980s RENAMO and armed bandits were regarded as synonymous by most commentators—a conflation encouraged by FRELIMO and many of its sympathizers. Derluguian comments that only after living for some time in the town of Tete did he receive "testimonial proof that local administrators and many field officers deliberately exaggerated the role of 'political terrorism' in current violence."[24]

Elements of the rural youth have been identified as another important ingredient of the conflict. Their disaffection resulted from a sense of alienation from both the traditional and modern sectors. Few of those who migrated to the urban centers in search of education and/or employment were able to join the originally small and now shrinking industrial sector. Many were deported to the rural areas in 1983, the year in which FRELIMO implemented Operation Production, a coercive program intended to clear the cities of undesirables. Deportees were frequently expelled to rural areas foreign to them and consequently were forced to rely on their wits to survive.

Meteorological conditions were not on FRELIMO's side in the early 1980s. During 1981–84 eight of the 10 provinces were severely affected by drought, while in 1984 the southern provinces were affected by floods and cyclones. With its limited institutional capacity, the Mozambican government failed to provide adequately for people hard hit by disaster and in this way demonstrate the benefits of its administration.

The arrival in a rural settlement of RENAMO was often enough to bring into the open what was until that time a largely latent conflict. Frequently, RENAMO would blame the ruling party and the state for local problems: exploiting grievances at the suppression of traditional authority structures, undermining of religion, and the imposition of FRELIMO's economic and agrarian policies. In this respect, RENAMO followed traditional patterns of rural mobilization by insurgents—mirroring tactics employed by FRELIMO in the previous decade. Building on the rural disaffection with FRELIMO, RENAMO expanded both in size and sphere of operations by the mid-1980s.

At the same time, RENAMO demonstrated an increasing self-sufficiency, owing in large part to its practice of taking whatever booty—food, clothing, tools, roofing, vehicle parts, and the like—it found in the settlements it raided or captured. Commentators report that this practice has earned RENAMO the nickname among the Mozambican populace of locust people. In addition to this pillage economy, there are reports of RENAMO involvement in potentially lucrative smuggling of ivory. RENAMO is also reported to be deriving income from the sale of arms. In 1991 the South African press reported a thriving arms trade between Mozambique and South Africa, with AK-47s of Mozambican origin finding their way into Transvaal townships for use in the fighting between their extra-parliamentary groups.

Thus, the fallout in South African government support in the 1980s did not bring about the collapse of RENAMO that many observers had expected. In the third stage of its development, RENAMO moved out from under the wing of its South African government sponsors. In 1979 the frontline states, Mozambique in particular, had been instrumental in bringing ZANU and ZAPU to the Lancaster House conference table at which an end to the protracted Rhodesian guerrilla war was negotiated. In 1989 RENAMO's former principal sponsor, the South African government, was in no position to exert similar leverage over RENAMO.[25] To be sure, it is uncertain that all support from South African sources has been curtailed; rumors of low-level unsanctioned support from elements within the armed forces recently received new impetus with allegations of ongoing collaboration between RENAMO and at least some elements of the SADF's special forces.[26] However, the support commonly alleged is not of a nature or level vital to RENAMO's continued survival given the degree of disaffection with government authority structures engendered by FRELIMO's policies.

The FAM and RENAMO: A Fragile Balance of Forces

By the mid-1980s RENAMO had become an organization whose size, geographic spread, method of operation, and increasing self-sustainability were such that, while able to drive the insurgents from gains they had made, FRELIMO was unable to inflict an overwhelming defeat. To end the conflict, FRELIMO had to capitulate to or defeat the insurgents decisively; RENAMO, by contrast, enjoyed the traditional advantage of the insurgent—merely having to remain militarily active in the bush. The Mozambican conflict had become a prolonged, seesaw war by virtue of the balance of forces inside the country.

One should beware of constructing too universalistic a model of RENAMO, given local and regional dynamics, because in recent years an increasingly coherent picture of the insurgents has emerged. RENAMO does not appear to be a loose collection of bandit gangs but a hierarchical military body with an apparently effective system of command and control founded on efficient radio communications which facilitate regular communication with the provincial base. Efficient radio communications hold the logistical key to RENAMO's ability to organize its offensives over the expanse of rural Mozambique. The efficacy of this organization was indicated in the period immediately following the announcement on 1 December 1990 of a partial cease-fire. This period witnessed a marked decrease in attacks on the cease-fire zones demarcated in the Rome agreement.

The basic operational unit is the company of approximately 100–150 men. Each company is divided into platoons and sections, as well as special operations groups. A battalion comprises two or three companies, each provincial base being able to call on two or more such battalions. Troops carry a variety of ill-assorted and poorly maintained rifles and automatic weaponry. RENAMO's tactics have hindered FAM efforts to counter the insurgency. While several sizeable towns were held for several months during the Zambezia offensive of 1986, insurgents have traditionally preferred hit-and-run operations for the most part rather than try to hold territory.

Notwithstanding its hierarchy and tactics, RENAMO's military capability should not be overestimated. One analyst concluded from first-hand observation that, given RENAMO's total disregard for fundamental infantry practice for posting sentries or for erecting elementary defense perimeters, the insurgents owe much of their tactical success to their opponent's weakness.[27] Notwithstanding its own guerrillas' experience against the Portuguese in the 1960s and the 1970s and the difficulties inherent in prosecuting an effective counterinsurgency strategy, FRELIMO has demonstrated a manifest inability to engage its opponent, exasperating even its closest military ally, Zimbabwe.

Following independence, FRELIMO took the advice of its Soviet supporters and transformed the existing guerrilla army into a conventional defense force, the FAM. Given Mozambique's lack of strategic depth—the Mozambican capital is only 70 kilometers by road from the South African border—FRELIMO feared an attack by Rhodesian or South African conventional forces. Indeed, during the previous decade the Rhodesians had made some 350 hot pursuit raids into Mozambique against the 10,000 ZANLA guerrillas in various camps inside the country. A January 1981 raid by South African commandos on an alleged ANC house in the Maputo suburb of Matola, resulting in the deaths of some 14 people, deepened FRELIMO's anxieties. In August 1982 ANC activist Ruth First was killed in Maputo by a parcel bomb, allegedly the work of South African agents, while the South African military attacked alleged ANC houses in Maputo in May and October 1983, dividing the Mozambicans' attentions at a time when the RENAMO threat was growing.

FRELIMO officials concede that, in the euphoria that accompanied Robert Mugabe's electoral victory in 1980,

FRELIMO lowered its guard against RENAMO. There were several successful offensives against RENAMO in that year, encouraging the view that the threat from RENAMO was on the decrease. Neither a protracted nor a prolonged conflict was envisaged by FRELIMO. Consequently, FRELIMO was ill equipped and ill prepared for RENAMO's resurrection from 1981. By 1983, however, all of Mozambique's provinces had military governors, and defense expenditure had become the largest single budget item, placing a further burden on any already crippled economy. Mozambique's security forces grew from 10,000 at independence to over 30,000 in 1985, most of whom had a counterinsurgency posture. But, for the most part, the FAM consisted of arbitrarily conscripted, ill-equipped, and poorly trained men who—the FRELIMO government itself acknowledged—indulged in the widespread abuse of power, or even deserted.

A larger, better-equipped, and better-trained army than the FAM would have been hard put to control an insurgency in a country with the military geography of Mozambique. The country extends for some 800,000 square kilometers, an area roughly twice the size of the state of California. Its rather elongated shape, straddling 16 degrees of latitude, further complicates the task of the FAM, as it does the Mozambican civil administration. North-south communications lag behind communications from east to west, owing to Mozambique's historical role as entry point to the southern African hinterland. Moreover, Mozambique is bounded on the east by an Indian Ocean coastline almost 2,500 kilometers in length, which has been frequently used to resupply RENAMO units by sea. Mozambique also shares common borders with no fewer than six neighboring states, at least three of which have, voluntarily or otherwise, provided sanctuary for RENAMO units at one time or another.

Mozambique: A Fragile State

The fragilities inherent in the origins and development of the Mozambican state are another factor in the outbreak and prolongation of the conflict. Notwithstanding 400 years of Portuguese colonial rule, the complete pacification of

Mozambique took until the second decade of the twentieth century.[28] Thomas Enriksen comments that "the turbulence of contending groups was to make Mozambique one of the least peaceful regions of Africa and was to give it a legacy of war, slavery, misery, and turmoil that has lived on into the present."[29] Even more than most of its peers, Mozambique is a fragile state without the habit and precedent of peaceful change.

The institutional weakness of the Mozambican state predates the current FRELIMO administration. The country only became a single administrative unit with a national economy in 1941, when the last of a number of company charters granted in the nineteenth century lapsed. For the previous one-half century, Mozambique had effectively comprised a number of colonial districts and company concessions administered from head offices in Lisbon, London, Paris, Monaco, Brussels, and Durban. Different currencies obtained in each region and tariff barriers were erected between them. Economic development, such as it was at that time, tended to take the form of regional projects reflecting the individual interests of the respective companies rather than (albeit unidentified and unarticulated) national priorities. Hence, the emphasis on the development of east-west communication links at the expense of a more spatially complete network linking the northern and southern provinces is quite clear. Consequently, successive Mozambican administrations have enjoyed a limited presence and concomitantly a limited influence in the northern provinces of the country, a situation that RENAMO—and FRELIMO before it—has exploited for politico-military gain.

More recently, the communications imbalance has been further exacerbated because of the civil war. Projects with international funding have been put in place to upgrade the handling facilities of the ports of Maputo, Beira, and Nacala, and the road and/or rail transport corridors linking these ports with the hinterland, the intention being to revive the traditionally large services sector of the Mozambican economy.[30] Troops, particularly the ZNA, have been deployed to protect these rail tracks, and the relative security thus afforded the transport corridors has attracted thousands of displaced Mozambicans, who pose a further challenge to rural reconstruction once the war has been brought to an end.

Mozambique in the 1990s has, in the opinion of some observers, increasingly come to resemble the territory 100 years previously: central government authority is largely confined to the larger urban centers, most of which are to be found on the coast.[31] Much of the rest of the country has effectively been abandoned to RENAMO and to bandits, the latter group frequently of a local and semiautonomous nature in a fragmented rural milieu, where few inhabitants have strong affiliations beyond the regions in which they live. Mozambique in 1991 poses a number of challenges to those seeking to apply theories of the state to a country which faces the prospect of becoming a "permanently weak state" structurally prone to internal conflict and with a long-term dependence on foreign aid.[32]

The Internationalization of the Conflicts

RENAMO's success in capturing and holding several sizeable towns in the central province of Zambezia during the 1986 offensive suggests that, were it not for increased military commitments by FRELIMO's allies, RENAMO might have moved on to major urban centers and pressed home its military advantage. The reversal of the insurgents' gains in 1987 and the further prolongation of the war owe much to the internationalization of the conflict.

While the most significant external intrusion into the Mozambique conflict in the 1980s has been that of South Africa, it has not been the only one. South African support for RENAMO served to kickstart the conflict in 1980; the provision of external support to both the principal protagonists by states both within the region, and beyond, has served to sustain and prolong the conflict. However, that support—military, financial, and diplomatic—has been sufficient to sustain both FRELIMO and RENAMO but insufficient to deliver either protagonist a decisive advantage. Indeed, in FAM's case, the creation of various special forces and the deployment of troop contingents by neighboring states arguably made coordination of operations and logistics even more difficult for the overstretched military high command.

Not surprisingly, Zimbabwe has been FRELIMO's staunchest military ally. In 1981 they signed a defense agreement to regard an attack against one as an attack against the other. By the mid-1980s, some 10,000 Zimbabwe National Army (ZNA) troops were committed to the defense of the Beira Corridor. This number at least doubled in size by 1989 as ZNA became more involved in taking on the insurgents. Regional support for FRELIMO was also forthcoming from Tanzania (3,000 troops) and, after some bullying by its frontline neighbors, Malawi (500 troops). The Zimbabwean contingent in particular was deployed to great effect in 1987, being instrumental in repulsing RENAMO offensives and reversing a number of significant RENAMO gains. However, Zimbabwean President Robert Mugabe came under mounting domestic pressure to pull the ZNA contingent out of Mozambique as casualties mounted and the Zimbabwean economy suffered.[33]

A number of countries outside the subcontinent have also provided support for the FRELIMO government. The Soviet Union remained the Mozambican military's biggest benefactor as recently as 1988, providing nearly all of the army's aircraft, arms, and ammunition and training the "Red Beret" commandos, who reportedly have been the most effective FAM unit. However, to FRELIMO's disappointment, that support had some clear limits. Even in the late 1970s, when adventurism characterized the Soviet Union's third world policy, Moscow refused to become as extensively involved in Mozambique as it had in Angola, forcing FRELIMO to seek other sources of assistance. By 1990 the last Soviet military adviser had left, as Gorbachev scaled down the USSR's third world commitments. In the same year, Mozambique's preferential pricing arrangement with Moscow for the provision of crude oil came to an end, placing considerable additional pressure on Mozambique's balance of payments—and on FRELIMO's security of tenure.

Somewhat more surprising at first sight, given FRELIMO's ideological proclivities at independence, is the support rendered by a number of Western countries. Britain's central role in Zimbabwe's transition to majority rule and its postindependence role in training Zimbabwe's army, was extended after 1980 to the provision of emergency relief aid in Mozambique and the training of several hundred Mozambican army officers. Given South African support for RENAMO in the 1980s, aid to Mozambique could be viewed as a cost-effective expression of opposition to apartheid and solidarity with the frontline states. Though undoubtedly of assistance to FRELIMO, Britain's aid has been criticized, along with that offered by other Western countries, for being too closely tied to the individual sponsor's pet projects rather than being integrated into more centrally driven economic reconstruction and security efforts.[34]

US support for Mozambique from 1984 onwards was provided, according to the assistant secretary of state for African affairs at the time, Chester Crocker, to wean FRELIMO from the Soviet bloc.[35] American assistance largely took the form of development assistance and emergency food aid, of which the country has become the largest supplier. In 1986 the United States along with the Netherlands and Scandinavia began a US $300-million program for the rehabilitation of the Beira transport corridor that links Zimbabwe with the port of Beira. Congress blocked military assistance—lethal or nonlethal—even though some of its members vehemently opposed Crocker's constructive engagement policy towards Mozambique. However, American support was stepped up following FRELIMO's 1989 decision to drop the Marxist-Leninist appellation from its party statutes and following the December 1990 publication in draft form of a liberalized Mozambique constitution. With the advent of the Bush administration, the US has played a more active public role in encouraging negotiations, unsuccessfully putting forward a seven-point peace plan in December 1989 to break the deadlock. However, to date Washington—like Moscow—has resisted deeper involvement in the Mozambican imbroglio, deferring instead to other external players.

Both the International Monetary Fund (IMF) and the World Bank have extended loan assistance since the mid-1980s. This assistance has been invaluable in ameliorating Mozambique's chronic balance of payments problem and foreign currency shortages. However, while assisting the FRELIMO government to remain in power during the 1980s, it has been tied to market-related economic reforms, which have worsened the economic position of the urban population and, in this way, undermined FRELIMO's position among that section of the population which remained least affected by the rural conflict. FRELIMO's economic policy reforms begun in the mid-1980s, together with the IMF/World Bank-supported Economic Rehabilitation Program (ERP) initiated in 1987, have reversed most of the commandist elements of the economy favored by the ruling party in its earlier years. However, without a concomitant improvement in rural security, the structural adjustment program's emphasis on price incentives has impacted negatively on the urban economy at the same time as FRELIMO has been more supportive of the growth of an independent civil society. As this discontent grows and finds expression in the emergent civil society, FRELIMO will almost certainly find its room for maneuver further limited.

Bilateral economic and/or military assistance has also been increasingly forthcoming from the Nordic countries of Canada, Italy, Portugal, Japan, and France. In the late 1980s, even South Africa stepped in, providing a consignment of military supplies and engineers to repair railway tracks and bridges. South African private companies, with their government's blessing, have for several years been heavily involved in the rehabilitation of the port of Maputo, the closest outlet to the sea for the Transvaal province of South Africa—and, until 1975, the route of most of South Africa's bulk exports.

Much of the bilateral aid provided in the preconditionality era of the 1970s and early 1980s provided FRELIMO with the resources it needed to sustain its efforts at ideological experimentation and in this way indirectly served to prolong the conflict in Mozambique. Indeed, some of the leading donors of that era actively supported many FRELIMO goals and were not opposed to some of their methods. By the early 1980s, however, other donors less amenable to centralized, statist, and commandist policies became more supportive of the FRELIMO government but used the leverage of FRELIMO's dependence on such external assistance to encourage political and socioeconomic reform in Mozambique. The United States, for example, tied the increase of its development aid to the condition that the funds be channeled to the private sector. While FRELIMO's reform agenda lengthened, it saw little concomitant improvement in its grassroots support, as few demonstrable benefits of reform could be shown. Aid conditionality, however, is on the increase.

Foreign development assistance has proved crucial in prolonging the war by keeping the economy afloat, accounting for 30 percent of the GNP in 1986 and 76 percent by 1989. Food aid has been essential in preventing mass starvation in a country in which, in times of peace, 80–90 percent of the total working population is involved in agriculture. In 1988, 78 percent of marketed grain supplies were imported, excluding the significant contribution made under the United Nations emergency relief program.[36] Bad as the situation in rural Mozambique undoubtedly is, it would have been a lot worse without the famine assistance that, by 1988, one-third of the population depended on for survival. Foreign aid thus served to ameliorate the lot of the rural peasantry, albeit minimally, while being insufficient or unable to resuscitate an economy plagued by war. This inability can be attributed in large part to donor preference for urban or large showcase projects at the expense of basic rural infrastructure. Nongovernment organizations, of which 128 reportedly operate in Mozambique, have been accused of creating organizational structures parallel to those of the government, leading to a situation where Mozambican civil authority exists in name only and posing problems for future government attempts to reestablish its authority after the war.[37] Whether Mozambique will shake off its increasing aid dependency is another major concern.

While FRELIMO's main strength has been its diplomatic efforts, pioneered by Machel and sustained by Chissano, to increase its international support, RENAMO has been less successful at presenting itself on the international level as a legitimate nationalist movement worthy of support. RENAMO has been dogged by a reputation for widespread atrocities against the Mozambican peasantry, a reputation which the US Department of State-sponsored Gersony Report confirmed for many international observers. The report, issued in April

1988, concluded that RENAMO members had murdered some 100,000 Mozambican civilians in the 1980s.[38]

RENAMO's external representatives also have presented a confused and fractious front to the world community; indeed, the world community has never been quite sure who in Europe and who in the United States was authorized to speak on behalf of RENAMO's leader, Afonso Dhlakama. The untimely death of a number of RENAMO representatives has further set back RENAMO's cause. Nevertheless, RENAMO succeeded in securing financial aid and some lobbying support from a number of sources, principal among them elements in Portugal and the Portuguese diaspora, West Germany, the United States, and the United Kingdom.[39]

Closer to home, Malawi provided an invaluable springboard for RENAMO in the early 1980s, enabling the insurgency movement to strengthen its presence in the northern provinces of Mozambique. On joining forces in late 1982 with Africa Livre, a guerrilla movement of suspected Malawian origins which had been operating in the north of Mozambique for a number of years, RENAMO was able to increase its territorial presence. In the mid-1980s, however, pressure by neighboring states on Malawian President Hastings Banda to crack down on RENAMO bases and domestic support for RENAMO put the insurgency movement under logistical pressure and contributed to its mixed military fortunes in the latter years of the 1980s. In mid-1991 reports surfaced again that showed RENAMO operating out of Malawi.

Since 1984, Kenya has provided internal passports for RENAMO officials who, not being recognized by the United Nations as political refugees, would otherwise have found it impossible to travel. In the late 1980s, Kenyan President Daniel Arap Moi was, along with Zimbabwean President Robert Mugabe, a mediator in the prenegotiations between FRELIMO and RENAMO. However, Kenya's neutrality regarding the negotiations has come under suspicion as in 1990 RENAMO's Afonso Dhlakama was frequently accompanied by Kenya's Foreign Ministry Permanent Secretary Betwell Kiplagat. Some observers regard Kenya as having taken over from South Africa as RENAMO's rearguard, even training Dhlakama's presidential guard.

Other countries that at one time or another have been mentioned as housing elements sympathetic to RENAMO include Morocco, Saudi Arabia, the Comores, Israel, Oman, and Zaire. This diversification of RENAMO's support in the late 1980s robbed the movement's former principal sponsor, South Africa, of much if not all of its previous influence over the movement. This diversification further complicated efforts to bring about a negotiated end to the civil war.

What Road to Reconstruction?

Widespread consensus suggests that an end to the present countrywide conflict is a prerequisite for reconstruction. However, while an end to the war is a necessary condition, it is not a sufficient one, and will mark only the beginning of what will be a drawn-out and difficult process of rebuilding Mozambique.[40] Moreover, extending the war also extends economic recovery, economic growth on which sustainable peace will be dependent.

Several factors key to the prolongation of the present conflict no longer applied in late 1991, suggesting that an end to the prolonged war could be in sight. The South African government, whose support for RENAMO from about 1980 was instrumental in reviving the insurgent movement, now favored an end to the Mozambican conflict. For its part, FRELIMO had largely abandoned those political, economic, and social policies, which had alienated many of the rural populace and which had formed the basis of RENAMO's official political program. Indeed, FRELIMO had from the mid-1980s embarked on a program of fundamental political and economic reform, which in effect undercut most of the demands made by RENAMO throughout the 1980s. A new constitution, which took effect on 1 December 1991, ended 15 years of one-party rule, ensured universal suffrage by secret vote, abolished the death penalty, and guaranteed the right to strike and the freedom of the press, although the television and radio media remained state monopolies. Yet, during 1991, several rounds of negotiations in Rome yielded little progress. The war continued against the background of a declining economy:

increased absolute poverty in the rural areas and growing difficulties in obtaining international aid partly because of Western European concerns about developments closer to home in central Europe and the Soviet Union.

In the course of 1991 a number of developments appeared to militate against an early end to the conflict. The seventh round of talks, in August 1991, centered on the fundamental issue of state power and legitimacy: who would rule Mozambique following the signing of a cease-fire and who would control the transition to multiparty free elections. RENAMO continued to resist what it regarded as an unacceptable continuation of FRELIMO unilateral control of the process of reform and transition in Mozambique. FRELIMO, for its part, held on to its internationally recognized position as the government of a sovereign state.

RENAMO appeared willing and able to continue exercising the military option while apparently lacking confidence that it could pose a significant electoral challenge—a challenge which, judging from the law on political parties passed by the Mozambican Parliament the previous December stipulating that political parties must prove electoral support in all 11 provinces, FRELIMO was determined to undermine. Continued factionalism within the organization hindered efforts to build a unified and consistent RENAMO negotiating position. While reports of factionalism developing within the insurgency movement, and of further military gains by the Naprama militia tempered the military option, such developments had the potential to encourage RENAMO intransigence rather than increase third-party leverage over the organization. RENAMO was having difficulty making the transition from a military organization to a political organization but also was attempting to gain the time to do so. In October RENAMO embarked on a renewed diplomatic offensive: in a single week it sent Afonso Dhlakama on a tour of Europe, started a radio station broadcasting daily into Mozambique, and stepped up a funding campaign in London.

For its part, in 1991 FRELIMO's traditional collegial leadership—virtually unchanged since 1977—and its integral cohesion were being sorely tested at the same time as it was coming under considerable pressure in the cities. There were increasing rumors of factionalism and resurgent ethnicity within the party, along with divisions of opinion over the correct strategy to deal with RENAMO. Rumors of centripetal developments were given new force with the allegations of an attempted military coup in June. These were developments whose significance could not be underrated for a ruling party which, in marked contradistinction to most of its African peers, had enjoyed a reputation for collective and collegial leadership. There were other signs of strain within the FAM: members of the elite Red Beret unit went on strike because of the lack of pay and rations, while reports increased of soldiers holding up civilians at roadblocks and stealing relief assistance.

Meanwhile the by-products of political reform, chiefly the emergence of a growing civil society, together with increasing urban economic hardship because of inflation, sharp increases in rents, repeated devaluations of the national currency and widespread unemployment, severely tested FRELIMO's position in the cities. Already in 1990 there had been a wave of strike action throughout the country to improve wages and working conditions. In 1991 there were reports of a marked increase in urban crime, thought to be the consequence of demobbing thousands of soldiers—in accordance with IMF structural adjustment program requirements—in a rapidly deteriorating economic environment. Corruption in government and ruling party circles also was reportedly on the increase. The move to political pluralism made it all the more urgent that FRELIMO demonstrate the benefits of its reform administration. However, that objective was difficult to achieve. Indeed, owing in large part to the lack of progress on the negotiations front, President Chissano had to postpone the general elections scheduled for 1991.

Looking beyond the question of a negotiated end to the war, observers differed on the impact a national development would have on the security situation in the rural areas. The renewed attacks on the Beira and Maputo transport corridors, which closely followed RENAMO's renunciation of the December 1990 partial cease-fire, seemed to indicate that RENAMO was a directed organization under central control. The corollary was that a permanent cease-fire observed by both FRELIMO and RENAMO forces became a practical proposition that would bring peace to the rural areas. However, it was still not known whether or how many autonomous local RENAMO groups might resist Dhlakama's directives should he sign a final agreement with FRELIMO. Moreover, activity by social banditry and splinter groups, if anything, increased in both rural and urban areas.

Hitherto, Naprama's successes against RENAMO units had redounded to FRELIMO's advantage. However, the movement's leader, a member of the Macua tribe, the largest in Mozambique yet the least represented in government, might not be amenable to stepping down following an end to the civil war. In 1991 reports also surfaced of a new guerrilla movement in Zambezia province, known as the Mukuepas, which, unlike Naprama, was more likely to attack government forces than to attack RENAMO. This phenomenon was further evidence of the dominance of local and regional affiliations over national ones.

The question remained one of where the balance or critical mass lay, with some Western diplomats suggesting that as much as one-third of the attacks on civilian and government targets was caused by groups other than RENAMO. Assuming RENAMO came out of the bush, would the rural situation they left behind have stability to rebuild itself? If so, over what period of time and with what inputs from central government and outside agencies?[41] One seasoned observer suggested that, under the most propitious of circumstances, donors would need to continue pumping $1 billion each year into Mozambique for another 15 years.[42]

Other problems remained. The repatriation of the estimated 1.3-million refugees in South Africa, Malawi, Tanzania, and Zimbabwe was, ironically, being deferred by the ongoing conflict. Within Mozambique, many of the displaced persons who had moved to the urban and periurban areas had become absorbed into the urban cash economy, shaky as it was, arousing government concerns that they would not wish to return to the rural areas after the war. Similarly, the dense communities which grew up within the transport corridors might well reflect irreversible urbanization. Quite apart from the logistical nightmare of holding universal franchise elections in a country with no tradition of national elections, weak local government, and little rural infrastructure, the question of integrating the FAM and RENAMO remained. Fueled and sustained in the 1980s by international involvement, the civil war in Mozambique would need more substantial and more effective international involvement to end it.

Notes

1. RENAMO has been known by a number of names since its formation. This chapter uses the current name.

2. A number of false starts occurred in the period prior to Mozambican independence. According to Ken Flower, the biographer of the former CIO head, "The Mozambique National Resistance (MNR) [was] developed as 'eyes and ears' for CIO in Mozambique" in 1974. However, in a discussion of the 1977 Operation Dingo, Flower states that "elements of the MNR had been our 'eyes and ears' in these areas for more that five years," that is, since 1972. Closer examination of the book reveals certain apparent inconsistences with respect to dates, suggesting typographical or author error. See Ken Flower, *Serving Secretly* (Alberton: Galago, 1987), xvii, 192.

3. Annette Seegers cites a member of the Machel government to the effect that "between 5,000 to 10,000 MNR soldiers had the countryside 'to themselves. They were not confronted by anyone'." Annette Seegers, "Revolutionary Armies of Africa: Mozambique and Zimbabwe," in Simon Baynham, ed., *Military Power and Politics in Black Africa* (London: Croom Helm, 1986), 147.

4. This phenomenon finds an echo in colonial Mozambican history. At the turn of the century, spirit mediums of the Barue tribe in Zambezia province claimed to have a secret medicine that would turn European bullets to water. It took the Portuguese some 30 years to conquer the region. See Allen Isaacman and Barbara Isaacman, *Mozambique: From Colonialism to Revolution, 1900–1982* (Boulder, Colo.: Westview Press, 1983), 23–25.

5. A UN development program official, cited in Virginia Curtin Knight, "Mozambique's Search for Stability," *Current History,* May 1991, 217; "Financial Times Survey: Mozambique," *Financial Times,* 15 January 1991; and *World Development Report 1991* (Oxford: Oxford University Press for the World Bank, 1991).

6. In this respect, probably the most insightful analysis to date is Christian Geffray, *La Cause Des Armes Au Mozambique: Anthropologie d'une Guerre Civile* (Paris: CREDU, 1990). Gueorgui Derluguian also brings an unusual and provocative perspective to bear on the problem. See Gueorgui Derluguian, "Social Decomposition and Armed Violence in Postcolonial Mozambique," *Review* 13, 4 (Fall 1990): 439–62.

7. Derluguian, 441.

8. For an analysis which depicts RENAMO as having its origins in legitimate opposition to FRELIMO stretching back to the early 1960s, see Andre Tomashausen, "The National Resistance of Mozambique," *Africa Insight* 13, 2 (1983): 125–29. Van Aswegan documents internal and external

Mozambican opposition to FRELIMO in the immediate postindependence period. See Van Aswegan, "Restructuring Society: The Case of Mozambique," in Calvin Woodward, ed., *On The Razor's Edge: Prospects For Political Stability in Southern Africa* (Pretoria: Africa Institute, 1986), 167–68.

9. Flower, 301.

10. Quoted by David Martin and Phyllis Johnson, "Mozambique: To Nkomati and Beyond," in Martin and Johnson, eds., *Destructive Engagement: Southern Africa at War* (Harare: Zimbabwe Publishing House, 1986), 7.

11. One could, of course, argue that in 1976 RENAMO was in no position to pursue short- and medium-term objectives distinct from those of its sponsors. Not, that is, if it wanted to retain such support. Barbara Cole argues, without making it clear *whose* plan it was, that "the longtime plan" was the replacement of the FRELIMO government "by a democratic Western-orientated government, one sympathetic towards Rhodesia . . . and one inclined to boot ZANLA out of Mozambique." Further on she reports that, contrary to the wishes of the Rhodesian military, RENAMO had no interest in engaging ZANLA as its main objective was the overthrow of FRELIMO. See Barbara Cole, *The Elite: The Story of the Rhodesian Special Air Service* (Amanzimtoti: Three Knights, 1984), 243–45. An observer with ties to RENAMO asserts that there were "limits to the Rhodesian willingness to support the resistance. The supply of material, the group's movements and its numbers were strictly limited by the Rhodesian authorities." See Tomashausen, 126.

12. Martin and Johnson, 7.

13. Steven Metz, "The Mozambican National Resistance and South African Foreign Policy," *African Affairs* 85, 341 (October 1986): 494.

14. Alex Vines, *RENAMO: Terrorism in Mozambique* (London: James Currey, 1991), 17.

15. Flower, 300–302; and Tom Young, "The MNR/RENAMO: External and Internal Dynamics," *African Affairs*, October 1990, 495. Paul Moorcraft asserts that P.W. Botha's interest in RENAMO dated back to 1978, and that financial and materiel support had been forthcoming in that year. Paul Moorcraft, *African Nemesis: War and Revolution in Southern Africa 1945–2010* (London: Brassey's, 1990), 261.

16. See J. K. Cilliers, *Counter-Insurgency in Rhodesia* (Kent: Croom Helm, 1985), 175–85.

17. Estimates of RENAMO's strength in this early period vary considerably. Martin and Johnson quote a Zimbabwe military intelligence report to the effect that "by December 1980 MNR had between 6,000 and 7,000 fully armed men with an estimated 2,000 troops in the pipeline . . . by February 1981 total strength was in the region of 10,000." See Martin and Johnson, 19. Young quotes various other sources, whose figures range from as low as 3,000 to 8,000 for the period to mid-1982. All estimates, however, suggest that RENAMO had grown substantially in the period immediately following its adoption by South Africa.

18. Young, 497.

19. The question of the extent to which RENAMO employs violence against the rural population and uses coercion to gain recruits is the subject

of much controversy which for reasons of length cannot be fully explored here. We can say that, on the one hand, guerrilla movements have frequently used tactics which were more violent than they were prepared to admit. On the other hand, guerrilla movements find it difficult to survive without a minimum level of acceptability by the local populace—though that acceptance can be more resigned than voluntary. Robert Mackenzie, a Vietnam veteran who has trained and fought in Southern Africa for many years and who was involved with the RENAMO, reports having witnessed "a constant stream of youngsters coming in out of the bush to join the (RENAMO) ranks . . . [while] every patrol is constantly provided the latest information by what we called 'mujiba' during the bush war." Private correspondence with the author, dated 3 July 1991. For an opposing viewpoint, see William Minter, "The Mozambican National Resistance (Renamo) As Described by Ex-participants," *Development Dialogue* 1 (1989): 96–106.

20. Rob Davies, "The SADF's Covert War against Mozambique," in Jacklyn Cock and Laurie Nathan, eds., *War and Society* (Cape Town: David Philip, 1989), 103.

21. A somewhat optimistic scenario was recently provided by a prominent journal of African business and development. "The absence of a (South African) sponsor will inevitably result in RENAMO's disappearance; that will, in effect, be the end of the civil war. The battle for the development of Mozambique can then commence." See Mario Sampiao, "Mozambique: la bataille du developpement," *Marchés Tropicaux,* 2 Aout 1991, 1954 (my translation).

22. "Mozambique: How Frelimo Will Govern," *Africa Confidential,* 1 August 1975, 6.

23. For more details of this process as well as its impact on the Mozambican conflict, see Christian Geffray and Mögens Pedersen, "Nampula en Guerre," *Politique Africaine* 29 (Mars 1988): 28–40; Gervase Clarence-Smith, "The Roots of the Mozambican Counter-revolution," *Southern African Review of Books* (hereafter *SAROB*), April/May 1989, and various articles and letters responding to his contribution in the June/July 1989, August/September 1989, and December 1989/January 1990 issues of *SAROB*; Derluguian, 439–62; and Otto Roesch, "RENAMO and the Peasantry: A View from Gaza," *Southern Africa Report,* December 1990, 21–25.

24. Derluguian, 444.

25. Vines reports that in mid-October 1989, Neil van Heerden, director general of the South African ministry of foreign affairs traveled to Nairobi with a mandate to tell Afonso Dhlakama to become more cooperative in the negotiations then in progress. Van Heerden was reportedly kept waiting for some six hours before Dhlakama consented to see him. Vines, 125–26.

26. "Mozambique Government Accuses South Africa of Continued Support for Renamo," *Mozambique Information Office,* 2 August 1991, 1–2.

27. Moorcraft, 1314.

28. David Abshire, "From the Scramble for Africa to the New State," in David Abshire and Michael Samuels, eds., *Portuguese Africa: A Handbook* (London: Pall Mall Press, 1969), 76.

29. Thomas Henriksen, *Mozambique: A History* (London: Rex Collings, 1978), 1.

30. The nature and importance of these "transport corridors" to the Mozambican economy is discussed at greater length in Denis Fair, "The Beira, Maputo and Nacala Corridors," *Africa Insight* 19, no. 1 (1989): 21–27.

31. J. Christopher Alden, "Mozambique: An Abiding Dependency," in Christopher Gregory and Larry Benjamin, eds., "Southern Africa at the Crossroads? Prospects for the 1990s" (forthcoming); Karl Magyar and Christopher Gregory, "LM Revisited," *Sunday Times* (Johannesburg), 24 April 1988.

32. The concept is developed by Jeffrey Herbst, "War and the State in Africa," *International Security* 14, no. 4 (Spring 1990): 36–38.

33. Christopher Gregory, "A Marxist's Rise in Southern Africa," *The World and I*, September 1987, 199.

34. Karl Maier, "The Military Mix," *Africa Report*, July–August 1988, 56–57.

35. Christopher Gregory, "Supporting the 'Right Side' in Mozambique," *The World and I*, October 1988, 115–16.

36. Steven Kyle, "Economic Reform and Armed Conflict in Mozambique," *World Development* 19, no. 6 (1991): 644–45.

37. "NGOs Causing Heartburn," *Southern African Economist*, June/July 1991, 10–11; and Joseph Hanlon, *Mozambique: Who Calls the Shots?* (London: James Currey, 1991).

38. *Summary of Mozambican Refugee Accounts of Principally Conflict-Related Experiences in Mozambique*, Department of State, Washington, D.C., 20 April 1988. For a contrary—albeit minority—view on the report, criticizing Gersony's methodology, see "Mr Crocker's Propaganda Blitz," *The Washington Times*, 29 April 1988. For a more comprehensive rebuttal, based on field research in RENAMO-held territory, see Sybil Cline, *RENAMO: Anti-communist Insurgents in Mozambique* (Washington, D.C.: United States Global Strategy Council, 1989), especially 24–30.

39. For more discussion of these countries' links with RENAMO, particularly speculation as to motive, see Vines, chap. 3.

40. Reginald Herbold Green sets out the goals, strategies, and components of the Priority District Program, the Mozambican government's strategy for rural regeneration, in "Towards Rural Reconstruction in Mozambique," a paper delivered at the International Conference on Agricultural Economics, Namibia, 1990.

41. N. Kansi, "War and Children in Mozambique: Is International Aid Strengthening or Eroding Community-based Prices?" *Community Development Journal* 25, no. 2 (April 1990): 102–12.

42. Tony Hawkins, "First Steps on Road to Peace," *Financial Times Survey*, 15 January 1991, 1. A sobering assessment of the socioeconomic tasks which face the Mozambican government is provided in "Hard Road Ahead," *Southern African Economist*, June/July 1991, 8–9.

The War Over Angola and Namibia
Factors of Prolongation

Garth Shelton and Karl P. Magyar

Africa's sub-Saharan countries began to attain their independence, with Ghana becoming the first in 1957. Seventeen countries attained that status in 1960, with the last colony, Namibia, becoming a sovereign nation in 1990. In broad terms, independence started first in northern Africa, then West Africa, Central Africa, East Africa, and finally southern Africa. Although civil disturbances and political agitation in Europe's colonies intensified rapidly after the close of World War II, most colonies attained independence peacefully. However, a distinct pattern became discernible when the transition process encountered large-scale hostilities. Generally, violent independence movements emerged in those colonies which attained independence last. This statement implies that as the independence fervor swept through Africa, it experienced its greatest impact in southern Africa, where most of sub-Saharan Africa's independence related conflicts occurred. The recent histories of Angola, Mozambique, Zimbabwe, and Namibia—and the violence in South Africa, which also reflects a liberationist element—corroborate this conclusion.

Portugal had been the first and most ambitious of explorers of the African continent. As Europe's poorest colonial power in the 1960s, Portugal sought to counter the tide of Africa's colonial history by ignoring it. With its extensive holdings in Africa dating back several centuries to its initial coastal contact in the mid-1400s, Portugal did a poor job of assessing Africa's notorious "winds of change." That inadequacy caused Portugal to lose all her colonies abruptly, while the government in Lisbon itself collapsed.

Of Portugal's colonies in Africa, Angola was the most important. Initially, Portuguese adventurers pursued mythical silver mines in the interior, but their only profits came from a lucrative slave trade. The Congo kingdom to the north of

Angola was a capable trading partner, but the territory which would become Angola was more amenable to European penetration and control. In part, this ease of penetration may have resulted from the relatively late arrival of the Ovimbindu people who, along with the Herero in Namibia, comprised the southernmost extension of the Bantu migration along the western coast of Africa. In a frequently shown symbiotic maneuver, the Portuguese and various African kingdoms had become accustomed to each other for commercial advantage. Luanda was established as a commercial base in the latter part of the sixteenth century and would eventually become the capital of the colony of Angola.

About 170,000 Europeans of Portuguese descent resided in Angola in the late 1960s. Residing in the capital of Luanda, and a few other modest sized towns, they dominated the large estates which were concentrated in Angola's fertile northern and central region. Whites, especially those living in southern Africa, remained oblivious to continental developments, despite the atrocities committed in the infamous Congo affair. It was simply assumed that the settler colonies would receive the backing of the home governments in perpetuity or that the European residents in the colonies could raise sufficient armies from among their members. Portugal's colonies were encouraged by another factor, namely, that these overseas units were considered integral territories of the metropole. There would be no debate about the limitless defense of these overseas lands. The Portuguese had also made a mistake in assessing the loyalties of the black population of the colonies to the white overlords. What the Portuguese perceived as a progressive "assimilado" policy, which allowed Africans to join the privileged ranks of the whites, was the reality realized by only few Africans. For the others, identifying with the continent's black population was a much more realistic prospect. The same history of self-delusion also afflicted whites elsewhere, most notably in Zimbabwe, and elements of it are manifest in South Africa today.

Another historical prerequisite to understanding Angola's conflict focuses on the cold war. Angola's colonial war, the country's attainment of independence, and the ensuing civil war were shaped largely by the global reach of the Soviet

Union and the US. The Soviets had taken an early interest in Africa's independence movements and had envisioned a predominantly socialist continent which would deny its political, economic, and strategic resources in the West. The Soviets had assumed naively that the third world would be easily attracted to socialist ideology, that the Soviets would have the resources that would make them attractive as developmental partners, and that there would be good prospects for mutual profit in this relationship. The gross miscalculation of this assessment is evident today as we are left to wonder what the Soviet objectives were in Angola, considering their substantial military largesse until well into the late 1980s.

Portugal, on the other hand, was a member of NATO and held such key insular strategic assets as refueling facilities in the Azores, which the European nation thought would guarantee the Western world's tacit and material support for the retention of its colonies. In fact, arms intended for the defense of Europe found their way to Angola, resulting in a general African perception that the US had deliberately traded its goodwill in Africa for an alliance with the forces seeking to retain white control over southern Africa. This predicament posed a problem for the US, which was caught in a tough dilemma.

Although the independence leaders in Portugal's African colonies became active alongside Africa's other leaders in the 1950s, shooting battles did not actually break out until after numerous countries had gained their independence in the early 1960s. Each Portuguese colony had its own independence movements, but as each colony shared the same colonial master, the fate of each became linked. By the early 1970s, half of Lisbon's government budget went to the war effort in Africa. However, the wars in Africa could not be won by a poor colonial power, and, consequently, Portugal's Salazar regime fell in 1974. This development set the pattern for European colonial armies in Africa, which would win numerous battles but lose the prolonged wars. The US in Vietnam and the Soviets in Afghanistan experienced the same fate.

One must view Angola's war in this greater conflict context, which drew inspiration from the US's experience in Vietnam, Fidel Castro's successes in Cuba, and the struggle in Algeria.

In Portugal's African colonies, Amilcar Cabral from Cape Verde best articulated the role of the guerrilla struggle in the African context. The liberation forces were informed at the outset by conceptions of protracted conflict, which accepted as a premise Mao's formulation of a long, drawn-out struggle when facing a superior conventionally equipped enemy. In Angola, after the Portuguese had abandoned the prolonged struggle, the Cuban-backed Movimento Popular de Libertacao de Angola (MPLA) forces, which took power, ironically became embroiled in a prolonged war with Jonas Savimbi's Uniao Nacional para a Independencia Total de Angola (UNITA). Savimbi and UNITA in turn adopted a new version of a protracted struggle. This typified the general tendency for government forces to resort to conventional short-war strategies while insurgents adopt the protracted struggle. Barring an early, unambiguous resolution, the conflict becomes prolonged. Angola's classic example certainly paralleled this phenomenon.

The Evolution of Conflict

The authors view the conflict in Angola in terms of three phases:

1. The phase between the Angolan liberation movements and the Portuguese colonial authorities, 1961 to 1974.
2. At the start of the Angolan civil war, MPLA received Cuban and Soviet support while the FNLA and the UNITA received aid from South Africa and the US from 1976 to 1977.
3. The broadening of the Angolan war in which two alliances expanded their objectives to include Namibia: South-West African People's Organization (SWAPO) joined the MPLA/Cuba/USSR alliance and was opposed by the UNITA/South Africa coalition, which received limited US support from 1978 to 1990.

Angola's Anticolonial Struggle

On 15 March 1961 Angola's anticolonial struggle began. Holden Roberto's Union of Angolan Populations (UPA) movement initiated an insurrection at the Primavera plantation in northern Angola. The insurrection signaled the failure of Portugal to assimilate Angolans into the Portuguese nation.[1] In 1962 Holden Roberto united a number of other nationalist movements to form the Frente Nacional de Libertacao (FNLA) and established the Gouvernment Revolucionaria de 'Angola en Exil (GRAE), a government in exile, with himself as president. By 1964 another liberation movement, the MPLA, under the leadership of Agostinho Neto, began to receive weapons and financial support from the USSR and the People's Republic of China.[2]

Jonas Savimbi, a top leader, left GRAE in 1964, and after unsuccessful attempts to join the MPLA, he established UNITA in March 1966. Savimbi based his organization in the south among his own people, the Ovimbundu. However, lack of external backing severely limited UNITA's ability to compete with the MPLA and the FNLA for popular support.

The liberation movements assumed that limited violence combined with international pressure at the United Nations would end Portuguese domination in Angola. Their long-term goal envisioned a settlement similar to the one that ended French control in Algeria. But insurrection merely strengthened Portugal's resolve to stay in the territory. The liberation movements underestimated Lisbon's determination to maintain control. By 1972 Portugal had sent over 130,000 troops to defend their colonies, the majority of which were based in Angola.

The United States based its support for Portugal on the Nixon administration's famous National Security Studies Memorandum (NSSM) 39 document. NSSM 39 concluded that the African liberation movements were ineffectual and not "realistic or supportable" alternatives to continued Portuguese colonial rule. The authors of the policy document questioned "the depth and permanence" of black resolve and "ruled out a black victory at any stage."[3] However, they did not question Portugal's determination to hold on to Angola. This omission later proved decisive.

Gradually, Portugal's ability to continue the colonial wars began to erode, and as John Marcum points out:

> By the early 1970s there were ample signs—economic disarray, political restiveness, military demoralization—that Portugal's days as a Eurafrican power were numbered. These indicators were visible to those with eyes to see.[4]

Nevertheless, when in April 1974 Portugal's armed forces overthrew the government of Marcelo Caetano, the US seemed surprised. The US responded to developments in Angola after that coup in terms of global geostrategic concerns—it must oppose the Soviet link to the MPLA. The MPLA's Soviet connection also awakened the region's dominant power, the Republic of South Africa.

The first 10 years of war in Angola laid the foundation for further conflict. The liberation movements were divided. Since the three main groups—FNLA, MPLA, and UNITA—had never united against Portuguese rule, they fought their own wars. Then the superpowers became involved, with USSR supporting the MPLA and the US aiding Portugal, a fellow NATO ally.

The Start of Angola's Civil War

In Guinea Bissan and Mozambuzue, Portugal handed over power to the dominant liberation movements with little difficulty. However, in Angola the FNLA, MPLA, and UNITA claimed to represent the Angolan people. The new government in Lisbon attempted to promote cooperation among the liberation movements. It held meetings in Kenya and later at Alvor in Portugal. At Alvor the liberation movements agreed to create a transitional government to prepare the way for independence on 11 November 1975.

However, rivalry among the liberation movements sparked the collapse of the transitional government and marked a new wave of conflict in the territory. Holden Roberto's FNLA attempted to seize control of the country through military action.[5] The US decided to provide covert aid to the FNLA, while the MPLA began to receive additional support from the USSR. In addition, Cuba dispatched military instructors to assist the MPLA.[6] The MPLA expelled FNLA troops from Luanda and launched a military campaign to take control of all provincial capitals before independence day. In response, the FNLA and UNITA formed a coalition to stop the MPLA advance.

Initial successes suggest that by 11 November the MPLA would control most of the key cities and towns in Angola.

Consequently, the FNLA-UNITA coalition urgently called for increased international assistance. The resulting external intervention transformed the low-intensity civil war into a multidimensional conflict. Participants fought the second phase of the Angola war on three levels: (1) a civil war characterized by competition between three rival liberation movement; (2) a regional war involving South Africa and Cuba, and (3) a global power struggle in which the superpowers promoted their own national interests in the region.

Anxious to prevent a Soviet-Cuban backed MPLA victory in Angola, the US authorized a massive expansion of military assistance to the FNLA and UNITA. The CIA initiated its covert operation, "Operation Feature," to recruit mercenaries and to direct operations from Zaire.[7] With increased US support, the FNLA, assisted by units from South Africa, launched an offensive to expel the MPLA from Luanda. Soon thereafter, South African troops entered southern Angola and deployed around the Cunene hydroelectric plant, which served Namibia's electricity needs in accordance with a prior arrangement.

During September 1975 the MPLA, with Cuban support, held off attacks by the FNLA and South African and Zairian troops in northern Angola. Fearing failure in the north, South Africa decided to advance into Angola. On 23 October, under the code name Operation Zulu, approximately 5,000 South African troops launched an assault in support of the FNLA and UNITA. South Africa did not need much provocation to send forces into Angola. Pretoria had hoped that the incursion would end SWAPO's attacks against Namibia, most of which emanated from Angola. Moreover, "Pretoria's response was to a perceived domino effect with its old cordon sanitaire being replaced by a Marxist garrotte that would strangle the Republic."[8] In addition, the internal struggle within Angola presented South Africa with an excellent opportunity to establish its credentials as an ally of the West. South Africa's decision makers saw the conflict as a contest between the Soviet-backed MPLA and the pro-West forces of UNITA and the FNLA. South Africa's official reasons for intervention were given as "hot pursuit" of SWAPO insurgents, protection of the Cunene hydroelectric plant to counter Cuban involvement, and the UNITA/FNLA requests for support "against communist infiltration in Angola."[9]

The South African advance through southern Angola prompted desperate MPLA requests for additional Cuban support. The MPLA made no major effort to solve the country's problems through negotiations. Instead, it chose to escalate the civil war. On 5 November 1975 a Cuban battalion was airlifted to Luanda, and massive reinforcements left Cuba by sea.[10] Cuba's determined commitment to Angola slowed, then halted, the South African advance. Between November 1975 and March 1976, more than 20,000 Cubans arrived in Angola. Their forces, backed by Soviet military equipment and transportation, denied the FNLA a victory in the north and stalled the South African-UNITA in the south.

In February 1976 the US passed the Clark amendment, which ended all US aid to UNITA and FNLA.[11] The ghost of Vietnam had undermined any serious US effort to directly involve itself in the conflict. Washington's failure to deliver the military hardware required for victory against the MPLA-Cuban-Soviet alliance forced South Africa to withdraw from Angola.

Consequently, the balance of power within Angola shifted overwhelmingly toward the MPLA. Without South African support, UNITA rapidly lost control of southern Angola. A major Cuban-directed offensive against the FNLA gave the MPLA control over northern Angola also. On 22 January 1976 the Organization of African Unity (OAU) condemned South Africa's intervention in the civil war but refused any condemnation of Cuba or the USSR. On 11 February 1976 the OAU admitted the People's Republic of Angola (PRA) to full membership, thus recognizing the legitimacy of the MPLA government.[12]

The Broadening of the Conflict

MPLA-Cuban control of Angola did not signal the end of the civil war. The MPLA's decision to extend support to SWAPO immediately ensured that there would be no peace in the region. Anticipating MPLA hostility, the South African Defense

Force (SADF) established a number of new military bases along the Namibian-Angolan border. In addition, the remnants of the FNLA and UNITA were recognized and trained by the SADF. In this context Deon Geldenhuys outlined South Africa's objectives:

> Pretoria's primary objective would be to force the MPLA into a fundamental shift on Namibia. Ideally, the Luanda regime should deny SWAPO bases and protection on Angolan soil. Alternatively, South Africa would want Angola to exert pressure on SWAPO to support an international settlement in Namibia on terms which South Africa would regard as favorable to its own interests. The way to achieve either objective is not to punish Angola militarily in the same way as Israel reacts against Arab hosts of the PLO.[13]

The MPLA could not bring themselves to end support for SWAPO and consequently launched their country into a protracted civil war, which often resembled a regional conflict. MPLA leaders fully realized that allowing SWAPO bases in Angola would compromise the country's own security. South African support for UNITA would be inevitable.

South African trained UNITA insurgents began the third phase of the Angolan war through the sabotage of transport links, especially the Benguela railway. The SADF joined UNITA insurgents, but the targets were different—the SADF attacked SWAPO training camps and control centers. On 4 May 1978 the SADF launched its first major offensive since the 1975 invasion against SWAPO positions in Kassinga. Numerous small attacks and hot pursuit operations followed.

South Africa maintained that its attacks in Angola were directed only against SWAPO targets and that UNITA was operating with complete independence. However, it soon became clear that the MPLA had no intention of ending, or even moderating, their support for SWAPO. Consequently, SADF became more supportive of UNITA's objective to end MPLA control of Angola.

The election of Ronald Reagan as US president opened new possibilities for Pretoria and for Jonas Savimbi's UNITA forces. The Reagan administration supported the objective of securing a UNITA presence in the Angolan government. Consequently, numerous discussions between US and South African officials took place with the objective of strengthening UNITA. Jonas

Savimbi was invited to Washington and moves to repeal the Clark amendment were initiated.[14] (The Clark amendment was finally repealed in July 1985.)

With implicit US support, the SADF launched new offensives into southern Angola to destroy SWAPO bases and undermine the MPLA. Early in 1981, the SADF campaign, code named "Operation Protea," sought to establish a cordon sanitaire in southern Angola to prevent SWAPO attacks against Namibia. The key element of the SADF strategy was the concept of forward defense (i.e., the elimination of SWAPO insurgents before they could launch attacks inside Namibia).

The MPLA objected to SADF operations in southern Angola and thus ordered their armed forces, FAPLA, to aid SWAPO. Luanda also backed its decision on a desire to prevent the consolidation of UNITA control in southern Angola, the area traditionally supportive of Jonas Savimbi.

Throughout the early 1980s, the SADF and UNITA operated against targets in southern Angola. SADF's primary objective remained the consolidation of its forward defense strategy to prevent SWAPO attacks against Namibia. UNITA received South African support for their long-term objective, a role in the governing of Angola. The SADF launched numerous hot pursuit and counterinsurgency operations into southern Angola. The major attacks included:

- December 1981: bombing raids in Moxico province
- May 1982: attack at Kassinga
- August 1983: the occupation of Cangamba
- December 1983-January 1984: "Operation Askari," operation 300 kilometers inside Angola[15]

The success of "Operation Askari" provoked the USSR to warn Pretoria that the overthrow of the MPLA government would not be permitted. In addition, Moscow stepped up the supply of weapons to the Cubans in Angola, thus beginning a process which would later shift the regional balance of power.

Following the signing of the RSA-Mozambique Nkomati Accord, the US stepped up diplomatic pressure to end the Angolan conflict. US negotiators, led by Chester Crocker, were rewarded when on 16 February 1984 South Africa's Foreign Minister Pik Botha and Angolan Interior Minister Alexandre

Rodrigues signed the Lusaka Accord. In terms of the accord, a joint monitoring commission (JMC), consisting of both Angolan and SADF officers, would oversee a South African withdrawal from occupied areas in southern Angola. However, the MPLA government continued to link their own security to SWAPO and the Namibian issue. An official Angolan statement stressed that peace could only be based on "the conditional and immediate withdrawal of the South African army from the part of Angolan territory it occupies and the implementation without delay of United Nations Security Council Resolution 435 on Namibia's independence."[16]

Thus, the MPLA apparently concluded that they should maintain their long-term security only if someone forced South Africa out of Namibia. With continued SWAPO attacks in Namibia and the continued operation of UNITA forces in Angola, the Lusaka Accord was doomed to failure. In June 1985 the SADF once again launched attacks against SWAPO insurgents, who had begun to assemble in southern Angola. Late in 1985 newly equipped and trained FAPLA forces attacked the UNITA headquarters in Jamba in southern Angola. The FAPLA attack was predictable. SADF strategists had made a monumental error in assisting UNITA to establish permanent headquarters in Jamba. Ignoring the writings of Mao Tse-tung, who repeatedly emphasized the need to avoid establishing permanent bases which become easy targets for the enemy, Jonas Savimbi built Jamba with SADF assistance.[17] Thereafter, Jamba obviously became FAPLA's main target, which forced SADF to come to the assistance of UNITA repeatedly. In Clausewitzian terminology, Jamba became UNITA's center of gravity.[18]

Late in 1985 and again in 1986 the SADF was forced to defend Jamba from a Cuban-FAPLA offensive. By late 1986 the USSR's supply of modern weapons to the battlefront was beginning to shift the balance of power. Throughout the early 1980s South Africa attacked any targets in southern Angola with little serious opposition because of the SADF's air superiority. New Soviet radar and ground-to-air missile systems, supported by interceptors with Cuban pilots, dramatically changed the Angolan theatre of operations.

By 1987 the warring factions and their supporters had set the stage for a final showdown in southern Angola. Savimbi remained in Jamba, giving his opponents an obvious and easy target, and the SADF had lost air superiority. In August 1987 FAPLA launched its spring offensive against UNITA forces occupying Mavinga, the first step towards the objective of reaching Jamba. Once again the SADF committed ground forces to defend Jamba. The SADF counterattack led by G5 and G6 cannons proved to be a major success. However, opposing forces halted the SADF advance at Cuito Cuanavale, a town which the SADF will never forget. As Fidel Castro said: "From now on the history of Africa will have to be written before and after Cuito Cuanavale."[19]

Both Cuba and South Africa soon realized that the costs of attempting to achieve victory in the battle of Cuito Cuanavale outweighed the potential strategic gains. The concentrated deployment of forces equipped with modern weaponry suggested that any major offensive by either side would result in losses. The balance of military power in southern Angola opened the way for serious negotiations, which began in London in May 1988. South African negotiators realized that continued commitment to a military solution could spell disaster. Moreover, the cost of the annual emergency counteroffensive to save Jamba had become unacceptable. Mao Tse-tung taught guerrilla armies how to survive with minimum cost, but UNITA and their SADF advisors ignored this wisdom. The Cubans and the MPLA were kept at the negotiating table by their Soviet allies who, because of serious international economic problems, had no wish to continue their involvement in Angola.[20]

After protracted negotiations throughout 1988, the parties to the Angolan dispute signed an agreement that provided for the implementation of UN Security Council Resolution 435. This resolution paved the way for Namibian independence and for the withdrawal of Cuban forces within a stipulated time period. It excluded UNITA from the peace talks at the insistence of the MPLA. Once again the leaders of the MPLA confirmed that they had no wish to seek reconciliation with their Angolan brothers.

The MPLA obviously hoped that the independence of Namibia would remove the SADF from their southern border, and, in time, UNITA would simply wither away. Cuban forces would no longer be needed to ensure Angolan security. However, this expectation proved to be incorrect.

Namibian independence ended South Africa's help for UNITA, but the US promised Savimbi "all appropriate and effective assistance" until the MPLA made "some positive movement toward national reconciliation."[21] UNITA continued to attack economic targets to encourage the MPLA to negotiate. UNITA could no longer call on the SADF to defend Jamba, but the Angolan government was significantly weakened by the withdrawal of Cuban forces—31,000 of them left the country in 1989 alone. The remaining 22,000 left in stages and by 1 July 1991 were gone completely.

Intense diplomatic pressure by the US and a number of African states brought Savimbi and Angolan President José Eduardo dos Santos together at Gbadolite in Zaire on 22 June 1989. The two leaders apparently agreed to a cease-fire and the integration of UNITA into the Angolan government and armed forces. However, differing interpretations of the agreement led to the return of a state of war.[22]

Further diplomatic activity failed to produce significant results. Then in December 1989 the MPLA launched a major offensive against UNITA to bring matters to an end through a military solution. The offensive ended with a draw. The first major FAPLA-UNITA confrontation without Cuban or SADF support from their respective allies proved that the Angolan civil war would not easily be resolved through military means. By mid-1990 the MPLA had come to realize that UNITA was not simply the creation of South Africa—it was an effective political organization with mass popular support. Consequently, continued efforts to destroy UNITA were doomed to failure.

Protraction and Prolongation, The Liberation Process

Leaders of the Angolan liberation movements expected that an armed uprising would soon bring independence to Angola. However, for the reasons discussed below, this outcome proved to be incorrect. Once these leaders realized that Portuguese authorities had no intention of leaving Angola, they adopted a variation of the classical Maoist protracted guerrilla strategy and accepted the realization that the struggle would be long and difficult. The guerrilla war against the Portuguese in Angola was entirely rural. The war was fought over vast distances on extremely difficult terrain. Much of the FNLA's activities were conducted along the 2,100 km frontier of mountain, swamp, jungle, and elephant grass that separated Angola from Zaire.

MPLA operations in the Moxico region ranged over a forest area covering 391 square kms. Angola's difficult terrain clearly worked to the advantage of the guerrilla forces. The Portuguese found it impossible to control the whole of Angola all the time. The guerrilla movements were well aware of this and used this geography to their own advantage. Zaire provided bases for the FNLA, especially after Joseph Mobutu, Holden Roberto's brother-in-law, became president. The MPLA established bases in the Congo Republic and in Zambia after that country's independence in 1964. Bases in neighboring territories made the task of the Portuguese forces much more difficult.

The original nucleus of 300 MPLA guerrillas received their training in Algeria. MPLA guerrillas also received training in Bulgaria, Czechoslovakia, and the USSR. The FNLA accepted support in the form of weapons and guerrilla training from the PRC. The liberation movements benefited from the moral and political support offered by the United Nations and the OAU. The UN General Assembly strongly condemned Portuguese colonialism, while the OAU granted political recognition to the FNLA until 1968 and thereafter to the MPLA.[23]

The leaders of these movements attempted to establish liberated zones inside Angola. They designed these zones to grow food for the MPLA, but these zones were vulnerable to air attack. The FNLA neglected to strengthen its political support through control over land inside Angola. Instead, they concentrated on armed attacks. The failure to promote political support prompted the breakaway of UNITA under Jonas Savimbi. For Savimbi, political education held the key to success. UNITA attracted the support of the PRC, but it failed to establish external bases. Zambian officials expected UNITA after UNITA destroyed the railway line that linked Zambia to the port of Benguela in Angola.

None of the guerrilla movements seriously challenged Portuguese control. In fact, a concerted counterinsurgency neutralized campaign the initial guerrilla offensive in the Bakongo area of northwestern Angola. The FNLA survived the Portuguese counterattack and grew to approximately 6,000 trained guerrillas, but thereafter FNLA tactics concentrated on limited attacks from Zaire and from the Dembos Mountains inside Angola.[24] UNITA operated in the south among the Ovimbundu and Chokwe peoples, but its guerrilla forces were limited and ineffective. The MPLA, numbering approximately 5,000 trained guerrillas, were the most active liberation movement. The MPLA's initial offensive was designed to take control of the Cabinda enclave. However, MPLA abandoned this strategy due to a lack of popular support in the region.

Thereafter, the MPLA operated from Zambia in the Moxico and Bie provinces. Vast distances prevented the MPLA from launching a concerted guerrilla campaign in the Mubundu region around Luanda. Moreover, the Portuguese met their operations in the east with fierce and effective counterattacks. Thus, by 1974 the MPLA had retreated to their original base in Congo-Brazzaville, and the Angolan liberation war had become a low-intensity stalemate.

The strength of the Angolan guerrilla armies lay in their ability to fight a Maoist campaign. As one observer pointed out:

> The war in Angola is a harsh campaign. Black guerrillas dictate the way it is fought. . . . They only engage in battle when they are confident that they have a material advantage. It is for this reason that a Portuguese army of almost 60,000 men is tied down to counter a guerrilla threat of barely a sixth of that number.[25]

Thus, the application of Mao's guerrilla strategy enabled the guerrillas to fight a far superior enemy over an extended period of time.

Portugal's Counterinsurgency Strategy

At the start of the political revolt in 1961, the Portuguese army had little difficulty in overcoming the rebels. They employed air power to attack rebel strongholds. The initial Portuguese response to guerrilla attacks was slow due mainly to a lack of military resources. Troops withdrew into defended outposts and relied on air attacks, but they sent out occasional patrols to contain guerrilla activities. But between 1964 and 1968 the initiative lay with the guerrillas.

In 1968 the Portuguese adopted a more comprehensive antiguerrilla strategy. The key elements of the strategy included:

- The concentration of the population in protected villages.
- A hearts and minds campaign in which houses, schools, and hospitals were built. All blacks were granted automatic Portuguese citizenship, the distinction between indigenes and neoindigenes being abolished. The contract labor system was ended and social services were greatly extended.
- Air power was used to seal off guerrilla supply routes and as an immediate response to guerrilla attacks. (Napalm was used from an early stage, and from 1970 herbicides and defoliants were used against guerrilla villages where crops were being grown.)[26] Coordination of light bombers, helicopters, and ground patrols was used extensively during dry season operations.[27]
- Elite units of the Portuguese paratroops, commandos, and marines undertook most of the fighting. Africans comprised a large percentage of the soldiers in these units.
- Captured guerrillas assisted Portuguese forces. Strong leadership—generals such as Spinola and Kaulza de Arriaga—provided the leadership needed to boost morale. Membership in NATO gave Portugal access to military equipment, including German G3 carbine (the main infantry weapon), the French Alouette helicopter, and the Italian Fiat G.91 jet fighter.

Insurrection merely strengthened Portugal's resolve to stay in the territory. Lisbon continued to deny Africans the right to political self-determination and emphasized instead the links between Portugal and the African "provinces." John Marcum observed: "Without its colonies, Portugal, it was felt, would shrink into a country of little political consequence and limited economic potential."[28]

In response to insurgency in Angola and the other colonies, Portugal increased the size of its armed forces. Before long, Portugal's armed forces had doubled in size and by 1974 had consumed 45 to 50 percent of the government's budget. In terms of the official budget, military expenditure rose from 35.6 percent in 1961 to 40.7 percent in 1969, while expenditure on socioeconomic development declined from 22 percent in 1960 to 14 percent in 1968.[29] In response to the colonial wars and a weakening economy, emigration from Portugal increased dramatically, reaching an annual rate of 170,000 in 1971. By 1972 one observer estimated that 1.5 million Portuguese had found employment abroad, compared with a labor force of only 3.1 million in Portugal itself.[30]

Thus, the colonial conflicts brought major economic problems to Portugal. Portugal's ability to maintain control of the overseas "provinces" required sufficient finances to pay for the military campaign and the education programs which had been introduced. To boost income, Portugal agreed to permit foreign investment in the colonies. Between 1964 and 1970 US, German, and South African investments strengthened the Angolan economy. The Portuguese in Angola benefited from the new economic prosperity, but the demands of the African population grew louder. By 1972 Portugal had sent over 130,000 troops to defend the colonies; this number consisted of more than one half of their total armed forces.

The liberation movements responded to Lisbon's determination by adopting a war of attrition. However, as some observers pointed out, economic obstacles prevented the liberation groups from achieving victory. In may 1970, one analyst suggested:

> One optimistic supposition has been that, in time, the poorest of the west European countries would tire of the cost—in money and manpower—of supporting these rearguard colonial wars. But that thesis is tenable only if the price of holding on to the territories is greater than the return.[31]

In other words, oil revenues from Cabinda's wells, which totaled $61 million in 1972, would dramatically alter the cost-benefit equation. Cash from Angolan oil would enable Portugal to maintain control.

The liberation movements responded to Portugal's improving economic condition by stepping up the casualty rate. The insurgents used mines and booby traps to maximize enemy casualties. Consequently, Portuguese losses from 1967 to 1974 totaled 11,000 dead and approximately four times as many wounded. These rates paralleled those of the United States in Vietnam.

By 1974 the liberation movements were active in the northern and eastern provinces of Angola. Guerrillas ambushed Portuguese convoys, destroyed bridges, and attacked military outposts. However, they failed to launch any meaningful offensive against the settler population.[32] The Portuguese farmers continued their normal daily routine. Significantly, the liberation movement did not mobilize support from the densely populated Ovimbindu region west of the Cuanza River.

By the 1970s Portugal began to reconsider its commitment to the overseas provinces. Domestic opposition to the wars had become a major issue.[33] Portugal restricted opposition at home, but an ever-increasing number of army defections and draft dodgings confirmed a lack of domestic support for continued involvement in Africa. Portugal was divided on its involvement in Africa. Some leaders demanded a military solution and a return to full control by Lisbon; others saw Portugal's future in a united Europe, arguing that the African colonies had become a burden. Portugal's business leaders gradually gave more and more support to the Europe-first argument.

However, Portugal's political leaders refused to make a choice between Africa and Europe. Instead, they adopted the US model of "Vietnamization"—an approach which later failed both the United States and Portugal. "Vietnamization," or "Africanization" as it was applied by Portugal, had three main objectives: First, the provincial governments would increasingly bear the costs of counterinsurgency.[34] Second, local recruits would replace metropolitan troops. The government offered education and technical training to recruits for the Angolan army.[35] Thirdly, increase external support. As a quid pro quo for a two-year extension of US base rights in the Azores (important for US communications with NATO allies), the US extended loans and aid worth more than $400 million to Portugal.

The Early Civil War

In July 1975 South Africa began serious military operations inside Angola. Following guerrilla harassment of engineers working at the Calueque hydroelectric dam, Pretoria authorized the SADF to undertake hot pursuit operations against SWAPO guerrillas in Angola. Soon thereafter, the SADF established a permanent presence in southern Angola. During August 1975 the SADF launched a series of strikes against the MPLA. By October the SADF was involved in a full-scale war with the MPLA. However, South Africa's long-term objective did not seek control of Angola; rather, it sought to assist UNITA and the FNLA under the direction of the CIA station in Kinshasa.[36] The SADF's advance through southern Angola was rapid and decisive. South African armored cars easily overwhelmed MPLA and Cuban defenses.

In November 1975 Pretoria had to decide on a final assault against Luanda. Apparently this decision depended on a promise of more US support. The official SADF account published later referred to the SADF as awaiting orders to withdraw while "mediation by go-betweens" went on during this period.[37] South Africa's advance on Luanda was delayed while the politicians talked, giving the Cubans an opportunity to fly in reinforcements. By December a highly trained Cuban army equipped with Soviet weapons would have made any further SADF advance an extremely costly exercise.

South Africa was obviously not enthusiastic about sending their entire army into Angola; nevertheless the units deployed were extremely effective. Without a doubt, the SADF, with UNITA and FNLA support could capture Luanda with sufficient political and material support from the USA. By November 1975 the SADF stood poised for victory, but the promised US military assistance needed to finish the job did not arrive.

With sufficient US commitment, the Angolan civil war would have been a limited affair that lasted only a few months. An FNLA-UNITA alliance could have taken power in Angola with OAU and full international recognition. US failure to act decisively gave the Cubans an opportunity to prepare a defense of Luanda and to strengthen the MPLA government. This Cuban involvement was the initiative of Fidel Castro but not of Moscow. The USSR was induced to support the war by Castro's enthusiasm to back the MPLA. The Cubans had provided training for the MPLA since 1966, and Neto had established a good relationship with Castro.[38] Between

October 1975 and February 1976, approximately 12,000 Cuban troops arrived in Angola.[39] By September 1977 the figure rose to almost 20,000.[40] The clear Cuban commitment, backed reluctantly at first by the USSR, made the outcome obvious. Without massive US support, the SADF would have been unable to carry out conventional war deep inside Angola.

On 22 January 1976 the SADF began to retreat towards the Namibian border, while Pretoria made it clear that without a more open Western commitment, South African forces would have no option but to establish a defensive line in southern Angola.[41] On 27 January 1976, the US House of Representatives voted to ban aid to Angolan combatants by a vote of 323 to 99. In February the OAU recognized the MPLA as the sole, legitimate government in Angola.

The SADF did well in Angola but were undermined by a lack of courage on the part of South African and American political decision makers. What could have been a short decisive engagement turned into a civil war that lasted for more than a decade.

Fighting the Expanded Conflict

During the final phase and after the withdrawal of the SADF from Angola, the FNLA and UNITA continued to fight on their own. The FNLA failed to launch a coherent guerrilla campaign and consequently ceased major operations altogether. However, UNITA regrouped in southern Angola and launched a guerrilla campaign by using the same approach that they had used against Portuguese rule.[42] The de facto government of Angola, the MPLA, began to provide extensive support for SWAPO including the provision of bases and camps from which attacks could be launched against Namibia. Thus, remarkably, the MPLA chose to provoke South Africa through its support for SWAPO, rather than concentrate their efforts on consolidating control of Angola by neutralizing UNITA. This strategic error proved to be an expensive mistake for the MPLA.

Once it was obvious that the MPLA was in full support of SWAPO, South Africa had no option but to devise a counter strategy. That strategy consisted of the following main elements:

- Attacks on SWAPO facilities in southern Angola;
- Avoidance of attacks on MPLA or Cuban forces;
- Support for UNITA to prevent SWAPO incursions into Namibia through the southeast of Angola; and
- UNITA was encouraged to launch attacks into central Angola, thus putting the MPLA on the defensive.[43]

UNITA was not strong enough to launch a serious conventional campaign, consequently the chosen strategy was guerrilla warfare. Had the MPLA concentrated on eliminating UNITA during the late 1970s, they would have consolidated their control in Angola. Luanda's decision to support an offensive war against the SADF in Namibia guaranteed the prolongation of the conflict in the region.

The MPLA underestimated the popular support for UNITA, and by 1980 Jonas Savimbi had established a fairly secure base area in southeast Angola with significant influence further north. In the early 1980s UNITA designed its offensive operations to extend its control into central Angola. These attacks had limited objectives, and were fairly successful. By the middle of 1983 UNITA dominated most of Angola south of the Benguela railway line and had gained control of Cangombe, Cangonga, and Munhango.[44] This success led to others further north.[45]

The USSR and Cuba responded to the UNITA advance by providing greater assistance to the MPLA. Once they realized that UNITA was a serious threat to the MPKLA, the USSR began to provide more advanced military equipment, and Cubans took control of military operations against UNITA. In 1984 and 1985, FAPLA with extensive Cuban and Soviet support, began a series of counteroffensives designed to recapture areas controlled by UNITA. The key features of these campaigns included:

- extensive use of aircraft: MiG-23s and MiG-24s;
- major ground offensives against towns held by UNITA;
- efforts to infiltrate and neutralize UNITA (the MPLA received assistance from East Germany in this regard);[46]
- a counterinsurgency operation against forward units of UNITA;

- increased employment of SWAPO personnel against UNITA; and
- the direct involvement of Soviet officers to direct operations.

A major offensive against UNITA base areas in 1985 provoked a decisive response from the SADF. Late in 1985 UNITA faced a major defeat at the hands of the Soviet-directed and Cuban-led FAPLA offensive. The MPLA realized that if UNITA were defeated, it could provide bases for SWAPO throughout southern Angola, thus making the defense of Namibia far more difficult for the SADF. Consequently, the SADF came to the assistance of UNITA, stopping the MPLA offensive and forcing its retreat.[47] The SADF counterstrike was not sufficient to destroy the FAPLA army, but it was enough to end the 1985 offensive and give UNITA a chance to regroup.

The following year Soviet Gen Yuri Petrov took control of operations against UNITA, thus signaling the USSR's determination to hold on to Angola. A new Cuban commander arrived with additional Cuban reinforcements, bringing the total deployment of Cuban troops to approximately 45,000.

With SADF support, a new guerrilla campaign in 1986 by UNITA included attacks in northern and central Angola. The SADF-UNITA strategy sought to undermine the MPLA's objective of launching another offensive started in June 1986 and included the use of chemical weapons.[48] With direct SADF support, UNITA countered the offensive and launched new attacks into central Angola. The Soviets and Cubans underestimated UNITA's strength and the determination of the SADF to support UNITA.

The Final Phase of the War

Throughout the early 1980s, the civil war continued in Angola because the balance of power was fairly even. FAPLA forces with Soviet and Cuban support could not overcome the SADF-UNITA alliance. Moreover, neither side was able or willing to commit sufficient forces to achieve a decisive victory. By 1987, however, the USSR appeared determined to lead a new offensive to gain full control of southern Angola and the expulsion of all SADF personnel from Angolan soil. Early that year, the MPLA-USSR-Cuban alliance decided that a new major offensive would be launched against UNITA with the objective of full control of southern Angola.[49] Preparations included:

- providing for new Soviet equipment and supplies;
- improving the existing air defense system;
- receiving T-62 and additional T-55 tanks from the Soviets;
- providing more MiG-23 fighters and Su-22 ground attack aircraft; and
- delivering Mi-25 attack helicopters.[50]

As a backup to the military campaign, Angola opened a new round of talks with the US to promote pressure on South Africa to withdraw support for UNITA.

The MPLA offensive began in June 1987 and once again UNITA faced the possibility of a major defeat. The balance of power clearly favored the MPLA after the introduction of new equipment and the forward deployment of Cuban forces.

Again, the SADF had little choice but to come to UNITA's assistance in the knowledge that a UNITA defeat would mean a new massive SWAPO insurgency campaign in Namibia. SADF units were deployed inside Angola to stop the MPLA offensive. Four major SADF operations—code-named Moduler, Hooper, Packer, and Displace—were launched to stall the advance of FAPLA forces. Numerous attacks and counterattacks, including heavy artillery and tanks, took place during the latter part of 1987 and early 1988.[51]

The 1987 offensive was different from earlier FAPLA attacks. In the 1987 offensive FAPLA was better prepared and enjoyed a comprehensive air protection system over southern Angola. This preparation made SADF's task far more difficult. In addition, in 1987 a time-extension debate occurred within South Africa about whether its air force had lost air control over southern Angola. The government reached no final answer, but it was clear that SADF operations were severely restricted by FAPLA air defenses and the extensive use of fighter aircraft.

The SADF achieved their main objective (i.e., stopping the FAPLA advance on Jonas Savimbi's headquarters at Jamba); however, it came at a far greater cost in lives and monetary terms than any other Angolan engagement. The 1987–88 campaign made it clear that USSR and Cuba were prepared to escalate their commitment to Angola to keep the MPLA in power. At the time, observers predicted that a further FAPLA offensive would be too costly for the SADF-UNITA alliance. An increased deployment of Soviet aircraft and air defense systems would have been decisive.

The middle of 1988 revealed that the SADF and FAPLA were heading for a major confrontation. The 1988 stalemate would set the stage for a further round of conflict in 1989 which would certainly be costly to both sides. Fortunately, by the end of 1988 the USSR was beginning to reconsider their overseas involvements, and Moscow decided that a new round of fighting in Angola would simply be too costly. Soviet pressure on the MPLA and Cuba opened the way for serious negotiations.

South Africa was at first reluctant to negotiate, but once the MPLA agreed that Cuban forces would be withdrawn from Angola in exchange for Namibian independence, attitudes changed. Moreover, there was a realization that it was unlikely that South Africa could bear the costs of another major operation in Angola to defend UNITA base areas. The shift in the balance of power in southern Angola made the next round of conflict a most unwelcome thought for both the SADF-UNITA alliance and the MPLA-USSR-Cuban alliance. Consequently, the success of peace negotiations linked to Namibian independence was inevitable.

Conclusion

When discoursing a prolonged war, one should not confront the conclusion of the analysis with the conclusion of the war. The intense armed confrontation may abate, but the conflict may well continue. This is certainly the case in Angola. The Cubans left slightly ahead of schedule after the peace process brought the two warring sides to the negotiating table.

Elections were held in 1992—under the watchful eyes of the United Nations. Not surprisingly, the loser—in this case Savimbi—charged fraud and postured his return to the bush. He had kept a sizable, well-armed contingent precisely for such a contingency. But even if he were to accept his electoral loss as a legitimate expression of his country's broad sentiment, that same loss might be used to solidify his ethnic power base, which he might easily reactivate if and when the new government's management of the economy begins to fall in line with Africa's general pattern of postindependence frustrations. A termination of war may bring peace to some societies through national reconciliation, but in the absence of economic growth, while investing scarce financial resources in reconstruction, the prospects for lasting peace in such devastated third world societies hold little promise. Events in Chad, Uganda, Somalia, and Ethiopia provide excellent examples of this phenomenon.

By most criteria, Angola's war was a classic case of prolongation. The notable exception is that the cast of participants remained unusually consistent, but this consistency reflected a prominent feature of cold war battles. Other characteristics reflected the usual attributes of conflict prolongation.

The war became transformed in that its origins lay in Angola's independence struggle, but then it lapsed into a civil war and then into a true international confrontation. Its objectives evolved from self-determination and sovereignty to a contest for who should rule, reflecting which power base and according to which ideological guidelines. The indigenous opponents of Portugal's continued colonial rule fragmented when that objective was attained, and, inevitably, power organized along regional lines—along with its unavoidable ethnic identities. This result also elevated the contest of personalities to prominence, and Africa's prevalent urban/rural differentiation emerged. Controlling the capital city is the only objective and whoever maintains that control enjoys a decided advantage. This case was amply illustrated by Frente de Libertacao de Mocambique (FRELIMO) in Mozambique and promises to favor the MPLA in Angola. Rarely in Africa or in Latin America have insurgents managed to unambiguously oust a government in power as Castro had done in Cuba, or more recently forces did when they toppled Ethiopia's government. In Africa the incumbent government had usually had the decided advantage.

Angola's war was also typical in that its fate and the war's conduct were determined in large part by external forces. UNITA had almost disappeared until it was resurrected by South Africa to serve its own purposes. The prolongation of the conflict was financed and enabled by external sources for whom Angola's fate was to serve greater global or regional objectives.[52] The cold war adversaries had backed opposing sides. The Reagan doctrine was clear enough in its support for Savimbi, but a detailed articulation of Soviet objectives never emerged. Had the Soviet Union also undertaken to sponsor Mozambique's government, it would have made plausible the regional domination of southern Africa. South Africa and other overseas right-wing supporters argued this case quote forcefully, but it was never accepted by critics of South Africa's regional perceptions.

After Gorbachev's ascendancy to power, the Soviet Union's political commitment to Angola began to erode although arms still poured in. This seeming paradox may perhaps be explained with reference to Cuba, which had made a substantial commitment to Angola with an estimated 2,000 combat deaths but with nothing to show for the effort. An overnight abandonment of the war was simply not acceptable to Cuba, which had by that time instituted its own agenda. South Africa gained much more in that its perception of a communist advance was kept at bay, and the conflict was not waged within its own borders—which could have activated its substantial internal opposition. Still, this advantage accrued to south Africa by default—by virtue of the Soviet Union's internal unraveling. In short, Angola's own players were but pawns in a global game.

Prolonged wars invariably expand into neighboring countries, which soon threaten or alter the regional order. Zaire was featured early on in Angola's conflict, as it often served as the staging area, if not as a perpetual refuge, for Holden Roberto's FNLA movement. Roberto's power base is among the Bakongo people, who straddle Angola's northern border with Zaire.[53] Indeed, Roberto is closely related to Zaire's President

Mobutu. Over the past few decades, each country hosted its neighbors' own insurgents. Zaire also had allowed the emplacement of staging areas for active US support of Savimbi, and at one point the country was rumored to become the new base for UNITA if the Cubans and MPLA had succeeded in overrunning UNITA's headquarters at Jamba.

In the east, Zambia illustrated an important lesson of prolonged wars. Zambia had at one time allowed Savimbi and his followers safe refuge. However, this permission lasted only until the Benguela railway line, which had been exporting Zambia's minerals to Angola's port of Lobito, became a strategic target and was repeatedly disrupted by UNITA. Zambia had similarly offered staging areas to Rhodesia's insurgents and had paid a stiff price when Rhodesia's white government forces launched several armed raids into Zambia. Mozambique likewise served as a sanctuary for Rhodesia's insurgents and paid a price in that the Mozambique National Resistance Movement (RENAMO), a Rhodesian government-sponsored force created to interdict Rhodesian insurgents, subsequently became the active opponents of the Mozambican government.

These excellent examples illustrate the almost ubiquitous tendency of the involvement of neighboring countries in prolonged wars. More often than not, such involvements become counterproductive, and can even oust from power the heads of neighboring regimes. This example has recently happened to Sierra Leone's government as the result of the prolonged war in Liberia. This is not to say that such involvements simply can be avoided. Instead, it underlines the point that prolonged wars inevitably become a regional problem, and it argues for a concentrated massive effort at the outset. In the future this may also make the use of nuclear weapons attractive for those who have them.

The fate of neighboring Namibia formed an integral part of the Angolan war. South Africa's historical "red fear" guided her response to Soviet-allied activities in southern Africa. Retaining Namibia as a buffer was seen as a necessary strategy. In view of the Western world's objections to apartheid, South Africa would not trust a Western response to a projected "onslaught" on her own territory. Critics had argued that the retention of control of Namibia for strategic purposes was but an excuse for continued colonial subjugation and exploitation of Namibia—which South Africa would never release. South Africa countered that the national party government had no permanent interests in Namibia and that retention of controls was costly. That South Africa did in fact yield independence to Namibia, once the Soviets and Cubans reduced their intervention in Angola, gave credibility to South Africa's contention. Since Namibia never had served as a vital source of wealth, but as a geographic bargaining chip and surrogate battleground, it yielded a great return on the investment. SWAPO, as did Namibia's insurgents, played a diminished role in the conflict, and indeed, that organization's greatest contribution was to serve as an excuse for South Africa's aggressive ventures into Angola in line with its forward strategy. Once again, this illustrates the contention that prolonged wars tend to become regional wars.

May we then view Angola's conflict as a "war without results"? The cynical answer to this question in the affirmation may be a bitter pill to swallow. If the ultimate result is the retention of the MPLA in power, this fact alone cannot justify 17 years of civil war. Angola exported a great part of its oil to pay for the war. This revenue, when added to further funds accruing from South Africa and other international sources would, in theory at least, have financed a substantial development program. Angola should not be poor today, nor should that country be labeled a "land of cripples." Another dimension emerged out of the war's prolongation, namely, the shift in the MPLA's ideological position. As a doctrinaire Marxist movement, the MPLA projected its superior socialist system in opposition to UNITA's inclination towards market systems (although UNITA's contention may have been greatly exaggerated in the West). At any rate, when the Soviets abandoned their commitment to the MPLA, the Marxist paradigm was further discredited, which again raises the question: what was the point of the war and its exceedingly heavy price?

South Africa gained time and, quite by accident, a docile Namibian neighbor, whose quasisocialist veneer poses no threat without Cuban or Soviet support. South Africa also gained measurably by developing, equipping, and training in a live war the best armed forces in the Southern hemisphere. Namibia gained independence but at a heavy price if measured in terms of SWAPO casualties and years of dislocation. However, that country is now embarked on the road toward rapid marginalization. The US gained by virtue of the Cubans' plans being frustrated and by helping to neutralize Soviet expansive ambitions in this distant outpost at a modest price. The MPLA elite remains in power. But in such war-ravaged countries, the particular ideological proclivity of one versus that of an opponent group will hardly justify the enormous destruction of men and materiel. And as that conflict may be far from over, for at least the great bulk of Angolans, this has indeed been a "war without results."

Notes

1. Dr Salazar admitted that assimilation was a slow process. In 1961 he said: "A law recognising citizenship takes minutes to draft and can be made right away; a citizen that is a man fully and consciously integrated into a civilised political society takes centuries to achieve." Quoted in J. Marcum, *Angola—Division or Unity?* in G. M. Carter and P. O'Meara, *Southern Africa in Crisis* (Bloomington, Ind.: Indiana University Press, 1977), 138.

2. See G. J. Bender, *Angola under the Portuguese: The Myth and the Reality* (London: Heineman, 1978), 3.

3. Quoted in R. Bissell, *Southern Africa in the World: Autonomy or Interdependence?* (Philadelphia: Foreign Policy Research Institute, 1978), 21.

4. Marcum, 154.

5. The Portuguese high commissioner Cardoso indicated that if he could not hand over power to "two or three" of the Angolan movements, he would "just get on the plane and leave." The MPLA was therefore never officially given control of Angola. D. S. Papp, "Angola, National Liberation and the Soviet Union," *Parameters* 7, no. 1 (March 1978): 33. US aid went primarily to the FNLA through Zaire. See J. Stockwell, *In Search of Enemies: A CIA Story* (New York: Norton, 1978). An account of the US decision to provide aid can be found in N. Davis, "The Angolan Decision of 1975: A Personal Memoir," *Foreign Affairs* (Fall 1978): 108-24. See also C. K. Ebinger, "External Intervention in Internal War: The Politics and Diplomacy of the Angolan Civil War," *Orbis* (Fall 1976): 669-99.

6. W. A. Durch, *The Cuban Military in Africa and the Middle East* (Arlington, Va.: Center for Naval Analysis, 1977), 17. See also C. Legum, "A Study of Foreign Intervention," in C. Legum and T. Hodges, *After Angola: The War Over Southern Africa* (London: Rex Collings, 1976), 11.

7. Stockwell, 102.

8. J. Swift, *Armoured Car Diplomacy: South Africa's Intervention in Angola* (Cape Town: OUP, 1987), 22.

9. Ibid., 25. See also Steenkamper *Adeus Angola* (Cape Town: Howard Timmins, 1976), chapter 2.

10. G. Garcia Marquez, *Cuba in Angola: Operation Carlota* (New York: Cuba Update), 128. Ian Grieg suggests that the USSR's reasons for becoming involved in Angola included the desire to provide bases for SWAPO attacks against Namibia, surveillance of Western shipping in the south Atlantic, and access to the country's raw materials. See Ian Grieg, *The Communist Challenge to Africa* (Cape Town: Cape and Transvaal Printers, 1977), 248.

11. Senator Clark was defeated in the 1978 elections. South Africa assisted his opponent. See R. Leonard, *South Africa at War* (West Point, N.Y.: Lawrence Hill, 1983), 68.

12. South Africa's defence minister, P. W. Botha, argued that the Angolan war had been lost by the politicians and not by the SADF. Many government leaders blamed the United States for failing to provide the promised military support. See G. Cawthra, *Brutal Force* (London: IDAF, 1986), 148.

13. Deon Geldenhuys, *The Diplomacy of Isolation: South Africa's Foreign Policy* (Johannesburg: Southern Books, 1990), 195.

14. See J. Barber and J. Barratt, *South Africa's Foreign Policy* (Johannesburg: Southern Books, 1990), 195.

15. Seaborne commando raids were directed at Lobito oil terminal, 12 August 1980; Luanda oil refinery, 30 November 1981; Cabinda oil pipeline, 12 July 1984; and cargo ships in Luanda harbour, 29 July 1984. For details of the SADF's attacks into Angola, see W. Steenkamp, *Borderstrike: South Africa into Angola* (Durban: Butterworths, 1983).

16. Quoted in Cawthra, 155. See R. S. Jaster, *The Defence of White Power* (London: Macmillan, 1988), chap. 8. The US exerted significant diplomatic pressure on Pretoria to force South African compliance with the accord. However, similar pressure was not applied on the MPLA or SWAPO by the USSR.

17. See J. Ch'en, *Mao and the Chinese Revolution* (London: Oxford University Press, 1967), chap. 11. See also B. Turok, ed., *Revolutionary Thought in the 20th Century* (London: Zed Press, 1981), chap. 6; and P. Pandya, *Mao Tse-tung and Chimurenga* (Johannesburg: Skotaville, 1988), chap. 2.

18. See Carl von Clausewitz, *On War* (London: Penguin Books, 1976), chap. 11.

19. Fidel Castro, addressing the Non-Aligned Movement Ministerial Conference, Havana, May 1988, quoted in R. Davies, "South African Regional Policy Before and After Cuito Cuanavale," in G. Moss and I. Obery, eds., *South African Review* 5 (Johannesburg: Ravan Press, 1990), 166.

20. *Angolan Peace Monitor* 2, no. 9 (16 October 1990).

21. SADCC-NGO Newsletter, ICDA, April 1989, 12.

22. International Institute for Strategic Studies, *Strategic Survey, 1989–1990* (London: Brassey's, 1990), 83.

23. See N. Muchnik and A. Humbaraci, *Portugal's African Wars* (London: Macmillan, 1974), 119 ff.

24. I. Beckett, "Portuguese Africa," in A. Thompson, *War in Peace* (London: Orbis, 1981), 156.

25. R. Corbett, *Guerrilla Warfare* (London: Orbis, 1986), 144.

26. The Fiat G.91 and the American F-84 Thunderjet and Lockheed P-2V were used for offensive air support. The backbone of Portuguese air operations was the American T-6 converted trainer and the Dornier Do 27.

27. Such as the operations in Moxico province in 1968 and 1972.

28. J. Marcum, *The Angolan Revolution: The Anatomy of an Explosion (1950-1962)* (Cambridge, Mass.: MIT Press, 1969), 142.

29. B. Davidson, *In the Eye of the Storm: Angola's People* (New York: Doubleday, 1972), 125.

30. Quoted in Marcum, 145.

31. Ibid.

32. At the same time, the Portuguese government realized that domestic opposition had been the major factor in both the Algerian and Vietnam wars.

33. Between 1967 and 1971 the percentage of defense costs carried by the colonies increased dramatically. See B. McColm and D. Smith, "Angola: Behind the Lines," *Newsweek*, 13 December 1976, 52–55.

34. Recruiting Africans for the armed forces without providing them with a role in the political process was doomed to fail.

35. See Marcum, 146.

36. H. Heitman, *South African Armed Forces* (Buffalo: 1990), 202.

37. Ibid. See also H. Heitman, *South African War Machine* (Johannesburg: CNA, 1985), 166 ff.

38. See A. Kapcia, "Cuba's African Involvement," *Survey B-A Journal of East and West Studies* 24, no. 2 (1979): 152.

39. A. Parker, "Behind Cuba's Involvement in Angola," Southern African Freedom Foundation Position Paper 1, no. 23 (1978): 16.

40. Ibid.

41. Heitman, *South African Armed Forces*, 210.

42. Ibid., 211.

43. The strategy used against the Portuguese was implemented again.

44. In August 1983 UNITA took control of Cangamba—one of the most serious defeats suffered by Fapla.

45. Early in 1984 UNITA overran Cafunfo near the Zairean border.

46. See F. Bridgland, *Jonas Savimbi - A Key to Africa* (Johannesburg: Macmillan, 1986), chap. 27.

47. See Hietman, *South African Armed Forces*, 212.

48. Bulgarian experts were employed in the CW operations.

49. The USSR provided heavy transport aircraft to assist preparations.

50. Heitman, *South African Armed Forces*, 215.

51. Ibid., 223.

52. Karl P. Magyar, "Darkness in a Dark Landscape: Angola and Mozambique," *Defense and Foreign Affairs*, October 1989.

53. Ieuan L. L. Griffiths, *An Atlas of African Affairs* (London: Methuen 1984), 98–99.

Cambodia
Prolonged War, Prolonged Peace?

J. Richard Walsh

On 23 October 1991 the leaders of Cambodia's warring factions and the foreign ministers of 18 nations signed a peace treaty to end more than a decade of war. The Cambodian conflict is the last of a line of wars that has made Indochina one of the most devastated regions in recent history. What began as a communist-led war of liberation for Vietnam, Laos, and Cambodia in the post–World War II period evolved into an East-West conflict. In the course of its evolution, the nationalist roots of the conflict were forgotten. Only with the final withdrawal of the United States (US) in 1975 and the ultimate victory of communist movements throughout Indochina did the forces of nationalism come again into sharper focus.

When analyzing the nature of prolonged war in Cambodia one can identify three levels of conflict: deep-rooted domestic conflict, regional rivalries, and superpower involvement. Each of these levels of conflict has its distinct origins and was sufficient to prolong the war.

These levels of conflict, however, are also interdependent because of the coalitions that emerged during the evolution of the conflict. Within these coalitions, the different combatants became dependent on one another to sustain the war effort, whatever their objective might have been.[1] This interdependence is a key to explaining both the prolongation of the war and its resolution.

The levels of conflict and their interdependence form the framework in this chapter for analyzing the Cambodian conflict. Within this framework several factors emerge. First, the conflict aggravated the nascent geopolitical division in Southeast Asia. Vietnam's invasion of Cambodia in 1978 drove a wedge between communist-dominated Indochina and the more Western-oriented member nations of the Association of Southeast Asian Nations (ASEAN). This division delayed the regional reconciliation that was emerging at the end of the

Second Indochina War in 1975 by facilitating coalitions of nations and groups with varying interests for prolonging the war. To the outside observer, the Cambodian war reflected a conflict culture deeply imbedded in the history of the region.

Second, the search for patrons by each side (i.e., Vietnam and the Cambodian resistance) introduced the necessary diplomatic, economic, and military resources to prolong the war. Soviet support for Vietnam was necessary for the invasion of Cambodia and essential for Vietnam's occupation of its neighbor. The costs that Vietnam incurred for its support of the Phnom Penh regime would eventually weaken its resolve. A key to Vietnam's costs was the ability of the patrons of the Cambodian resistance—China, ASEAN (especially Thailand), and to a lesser extent the United States—to foster a more effective resistance organization that garnered international recognition. The rise of countervailing power to Vietnam's ambitions forced Hanoi and its patron to reconsider original estimates of the expected length of the war.

Third, as the Cambodian conflict progressed, the diplomacy of stalemate became a factor in prolonging the conflict. It becomes apparent to all the parties involved that the Cambodian war became one that must not be lost, if it cannot be won. The several rounds of talks leading up to the final agreement revealed that no single key could unlock the impasse. As diplomatic talks proceeded, the distinction emerged between balance of power interests and conflict termination interests. There was a shared interest in ending the conflict, but it was not a high priority for either coalition. More important was the need to maintain a favorable balance of power in the region through a policy of continuing conflict. As the conflict reached a stalemate, neither side spoke of the benefits of prolonged war; they simply lacked a compelling reason to change their policies. For successful negotiations, both parties must introduce incentives for changing policy.[2] Given the interdependence between the different levels of conflict, incentives could and did arise from several quarters for a variety of reasons.

Before analyzing the reasons for prolonged conflict in Cambodia, this writer offers the reader a short historical overview of events. The overview is followed by an analysis of prolongation that focuses on geopolitical divisions, the levels of conflict, and the differences between balance of power interests and conflict termination interests. The conclusion addresses the effects of prolonged war on the prospects for prolonged peace.

An Overview of the Conflict

The Second Indochina War ended in Cambodia before it ended in Vietnam. Thirteen days before the final withdrawal of US personnel from Saigon on 30 April 1975, the Khmer Rouge had marched into the Cambodian capital of Phnom Penh. By the summer of 1975, Indochina, Laos, Vietnam, and Cambodia were perceived as under the control of Hanoi-dominated movements. The Khmer Rouge had installed the deposed prince Norodom Sihanouk as titular head of state while keeping him under virtual house arrest. Under the leadership of Khieu Samphan and the control of Pol Pot, the Khmer Rouge began a brutal reorganization of society that resulted in an estimated 1 million deaths. The regime was determined to carry out a thorough socialist revolution with an emphasis on the collectivization of agriculture.

By the following year, Prince Sihanouk resigned and returned to exile in Beijing. The shadowy Pol Pot emerged as prime minister, as the Khmer Rouge were increasingly emphasizing their nationalist character. With uncertainty about Chinese support following Mao's death in September 1976, about relations with Thailand following its right-wing coup in October 1976, and about Vietnamese intentions given their long-standing interests in regional hegemony, it is possible that the regime felt the need to complete the revolution as rapidly as possible, regardless of cost. These uncertainties probably intensified the regime's search for "enemies" within Cambodia and the leadership itself.[3] Increasingly to strengthen its nationalist claims, the regime began to consider Vietnam as the enemy.

China, with its own age-old animosity for Vietnam and a concern about Cambodia's tilt toward the Soviet Union, encouraged Cambodia's animosity toward Vietnam. By 1978 anti-Vietnamese sentiment led to attacks against ethnic Vietnamese in Cambodia and on Vietnamese territory claimed by the Cambodian regime. With Soviet military and economic support, Vietnam struck back with a full-scale invasion in December 1978. Vietnam's invasion was rationalized as support for a national liberation movement (composed of former internal enemies of the Khmer Rouge regime) to overthrow a tyrannical regime whose demise would be welcomed internationally. Vietnam's invasion and occupation of Cambodia provided the rationale for China's punitive action in February 1979. Though Chinese troops devastated areas along the Sino-Vietnamese border, they were fought to a standstill by second-level regional Vietnamese forces.

After the establishment of the People's Republic of Kampuchea (PRK) in Phnom Penh and the Vietnamese army's successful push to the Thai border, the fighting was largely contained around the border area. The resistance effort with its tenuous logistics and refugee camp buffers was stretched along the border. The opposition consisted of three groups who formed the Coalition Government of Democratic Kampuchea (CGDK) in 1982: the Royalists led by Prince Sihanouk, the mainly conservative Nationalists led by former prime minister Son Sann, and the Khmer Rouge represented by Khieu Samphan. CGDK was able through allied help to gain the official international recognition that was largely denied to the PRK.

The conflict quickly became a war of attrition. It settled into a pattern of seasonal ebb and flow with dry season offensives by the Vietnamese forces and more limited opposition guerrilla attacks during the rainy season. The pattern was broken by 1986 when the Cambodian resistance did not recover from the blow delivered during the 1984–85 Vietnamese dry-season offensive. Deprived of their depots and rest and training facilities on the Thai border and their supply networks, the resistance was unable to sustain a rainy season guerrilla campaign. A second factor affecting the CGDK forces was the defensive barrier of mines, ditches, barbed wire, and earthen walls that made infiltration into Cambodia much harder. Finally, disagreements within the CGDK and skirmishes between the Khmer Rouge and its noncommunist allies strained the coalition.[4]

While militarily dominant in Cambodia, Vietnam was in no condition to dictate terms for an end to the conflict. Vietnam's diplomatic isolation and devastated economy were exacting its price. There was a growing realization that, in the long run, military force would have to be supplemented or even replaced by a political strategy. The first indication of this was a declaration in August 1985 announcing the gradual withdrawal of Vietnamese troops from Cambodia. The CGDK responded in March 1986 with a plan that did not make the complete withdrawal of Vietnam a precondition for talks. After pursuing initiatives through various diplomatic channels, the Phnom Penh government called for direct and unconditional talks with the resistance groups in the summer of 1987. The combination of the PRK's failure to secure effective domestic or international support, Vietnam's costs, Sino-Soviet rapproachment with mounting pressure from Moscow on Hanoi for a rapid settlement, and Prince Sihanouk's "leave of absence" from the CGDK presidency propelled the conflict into a new phase.

In December 1987 Prince Sihanouk met with the PRK Prime Minister Hun Sen for the first direct talks. Their joint communique emphasized the need for negotiations among the Cambodian parties to the conflict; once they reached an agreement, they would hold an international conference to guarantee it. Four other rounds of negotiations over the next four months proved inconclusive. Even the nearly complete withdrawal of Vietnamese troops in September 1989 did little to break the stalemate among the warring parties.

During the same period, the diplomatic maneuvering of the parties was more closely linked to international diplomacy over the fate of Cambodia. In January 1990 the five permanent members of the United Nations (UN) Security Council drafted a plan that called for UN peacekeeping forces and administrators during a transition period to new elections. The four Cambodian factions would join together in a Supreme National Council (SNC). The factions accepted the plan and a council was established by late 1990. By September 1991 the SNC had agreed to a cease-fire, to stop receiving foreign arms, to disband 70 percent of their forces with the remainder under UN supervision, and to a compromise system for elections. The combination of international and national efforts finally led to the signing of the peace accords in October 1991.

Analyzing the Prolonged War in Cambodia

An old Khmer adage states that "a curved road is not always to be abandoned, a straight one not always to be taken." The road to a Cambodian peace accord has been a curved one due to the many levels of conflict. To capture the serpentine nature of the prolonged war, we will focus on three factors: the geopolitical divisions, external support and uneasy coalitions, and the role of diplomacy.

Geopolitical Divisions in Southeast Asia

In the late 1970s, Southeast Asia appeared to be a region with few prospects for a prolonged peace. A heterogeneous region, it was the site of various nationalistic and ideological animosities, urban-rural differences, economic groupings, ethnic problems, and population pressures. Southeast Asian nations also had to contend with the interests of extraregional powers. In the midst of all these divisions the Cambodian conflict emerged.

Cambodia's prolonged conflict was precipitated by the Vietnamese invasion in December 1978. In making this costly move, Vietnam was reacting to threats to its regional hegemony. The threats were two-fold: the Khmer Rouge at the regional level and great power relations at the global level. Reflecting the long-standing animosity between the Khmers and the Vietnamese, the Khmer Rouge had been pursuing anti-Vietnamese policies since 1971. After their victory in April 1975, the Khmer Rouge leadership did little to acknowledge debts or fraternal ties to other communist movements, Vietnam in particular. The leadership was "ideosyncratically national," making it easier to define enemies of the revolution who were foreigners or those who served them.[5]

The evidence suggests that Vietnam tried to reach a border agreement with Cambodia in early 1978. The Khmer Rouge responded by continuing to shell the Vietnamese side of the border with Chinese-supplied artillery. They also continued to allege Vietnamese involvement in assassination attempts against the Cambodian leadership. During this period, the Vietnamese may have hoped to effect a coup and bring a more pro-Vietnamese regime to Phnom Penh.[6] Such a turn of events would have allowed Vietnam to avoid the difficult option of large-scale invasion.

Recognizing the different levels of conflict, the origins of Cambodia's prolonged war are also linked to the interests and actions of outside powers. Vietnam's belief that it must be treated as the dominant power in Indochina conflicted with China's own historic interests in the region. The conflict was muted during much of the Second Indochina War (1964–75), but China's rapproachment with the United States in 1972 precipitated the Sino-Vietnamese conflict. The United States enlisted China's help to pressure Vietnam to accept a compromise, not unlike the half-loaf that Vietnam was forced to accept after the Geneva Conference in 1954. As China moved to secure its links with the West and concentrate more resources on economic modernization, it reduced its assistance to Vietnam. Combined with the refusal of the United States to honor post war aid agreements, the decision by China forced Vietnam to rely more on the Soviet Union. By November 1978 a treaty of friendship and cooperation formalized the Soviet-Vietnamese alliance. The Soviets committed themselves to $2–$3 billion annually in economic and military aid and doubled the Vietnamese arsenal in the months just prior to the December invasion.

Long-standing territorial disputes and the treatment of ethnic Chinese in Vietnam compounded distrust between China and Vietnam. Hanoi's tilt toward Moscow reinforced Beijing's concerns about possible encirclement and was a factor in China's rapid normalization of relations with the United States in late 1978. An assessment of Vietnamese intentions drove the United States to reaffirm its strategic interests in the region and pursue a policy of isolating Vietnam.

Because of Cambodian, Chinese, and US policy, Vietnam appeared to be driven toward an alliance with the Soviet

Union. This outcome though was not inevitable. There is evidence that Hanoi wanted to avoid taking sides in the Sino-Soviet dispute. Potential benefits could be gained by playing one side against the other. After 1975 Vietnam pursued more moderate policies to reduce Soviet influence in the region. Hanoi hoped for a more positive relationship with the United States and retained membership in Western financial institutions like the World Bank, International Monetary Fund, and Asian Development Bank. The Vietnamese premier visited Western Europe for aid and resisted Soviet pressure to join COMECON. Though Vietnam viewed ASEAN with suspicion, it did court its individual members who sought economic ties to offset dependence on the Soviets and who encouraged a climate for normal relations.[7]

However, finding itself politically isolated, impoverished, and increasingly under military pressure, Vietnam turned to the Soviet Union for help in meeting its regional security challenges. Vietnam's difficult decision to invade Cambodia resulted in tremendous national and regional costs.[8] First, despite attempts to avoid excessive dependence on the Soviet Union, Vietnam was almost totally reliant on the Soviet bloc for economic, military, and diplomatic support. This dependence ran counter to the strong desire for self-determination that ran through Vietnam's wars of liberation. Second, by alienating both China and the United States and allying with the USSR, Vietnam reintroduced great power rivalry into Southeast Asia. The invasion was a catalyst for the uneasy coalitions that would form on each side of the battlefield. The interaction among the United States, China, and the Soviet Union would effect both the prolongation of the war and its resolution.

Third, Vietnam's invasion provoked a tenacious insurgency in western Cambodia that was directly supported by China and tacitly assisted by Thailand. Vietnam's action did not destroy the Khmer Rouge but returned it to its more natural habitat in the isolated countryside. From its remote bases, the Khmer Rouge would attempt to wage Maoist-inspired protracted war. Actors and events would intervene to constrain the Khmer Rouge, but they would not lose their capacity for independent action, a concern of allies and opponents in both war and peace. The need to support the Khmer Rouge would lead China to court a new ally and an old adversary of Vietnam and Thailand and would give China more direct influence over the balance of power in the region.

Fourth, Vietnam alienated ASEAN and catalyzed its development into a more effective regional organization. Prior to December 1978, Vietnam's goal had been to neutralize ASEAN to prevent alignment with China or the United States. Vietnam's alliance with the Soviet Union was perceived by ASEAN as a challenge to its "zone of peace, freedom, and neutrality" and a deliberate effort to repolarize regional politics. Members of ASEAN traditionally more oriented toward Vietnam than China (i.e., Indonesia and Malaysia) were disillusioned by Vietnam's flagrant disregard for the principles of peaceful coexistence which it had proposed as a framework for relations with ASEAN. Vietnam's action provoked a united front that became a core element of the opposition.

Fifth, there were the human costs. Tens of thousands of refugees fled to camps along the Thai-Cambodian border, where they would become pawns in the prolonged conflict. Thousands more would flee to other Southeast Asian countries, receiving increasingly hostile receptions from native populations. The eventual occupation of Cambodia by more than 200,000 Vietnamese troops would also divert scarce resources from the reconstruction and development of Vietnam. The country would become further isolated from the economic development taking place in the rest of East Asia.

Polarization of regional politics and sunken costs resulting from the failure to quickly subdue its enemy would drive Vietnam to prolong the war in Cambodia. Despite being militarily dominant in Southeast Asia, Vietnam faced a coalition that could exploit the geopolitical divisions. The key to the opposition's ability was the support it received from outside powers and the formation of an uneasy coalition.

External Support and Uneasy Coalitions

During his November 1978 visit to Thailand, Chinese leader Deng Xiaoping remarked to Thai leaders that "the hegemonists have stepped up their expansionist activities . . . particularly in Southeast Asia. It is only natural that some Asian and

Southeast Asian statesmen and men of vision should have perceived . . . the attempts of the hegemonists . . . and taken positive measures to counter them."[9] Deng's statement was a not-so-veiled reference to the Soviet-Vietnamese alliance and the need to counter Soviet-backed Vietnamese expansion. The recognition of their mutual interest in resisting Vietnam would lead to the development of a Beijing-Bangkok axis.

The Moscow-Hanoi axis was borne not only of Vietnamese need but also from the belief that "the correlation of forces" had shifted in Moscow's favor. The Vietnamese conflict had little effect on the détente between Washington and Moscow while the Soviets continued to heavily arm Vietnam for victory in 1975. By the mid-1970s, as Soviet-American relations deteriorated, the return of more zero-sum superpower relations meant there was little opportunity for cooperation in Southeast Asia. The Soviet Union sought to expand its political influence in Southeast Asia by supporting Vietnam's objectives and establishing a more direct military presence in Cam Rahn Bay. In doing so, the Soviets could not only better contain China but also directly challenge US support for status quo powers.

The Soviet-Vietnamese alliance and Vietnam's invasion of Cambodia provided a favorable climate for China's punitive action against Vietnam in February 1979. China undertook the action just after Deng Xiaoping's successful visit to the United States, giving the appearance of US approval. International condemnation of China's action was muted. The members of ASEAN privately welcomed China's military action as a means of signaling to Vietnam that it could not act with impunity. China's incursion into Vietnam was inconclusive, however, and did little to slow the Vietnamese push toward the Thai-Cambodian border. In the absence of geographic proximity to Cambodia and its Khmer Rouge allies and at the risk of further loss of prestige, China would need more effective means to pressure Vietnam. Those means lay with more cooperative relations between Thailand and China.

Before the Vietnamese invasion of Cambodia, the protracted conflict with a communist insurgency was Thailand's most pressing security threat. This movement was perceived as an export from Beijing, which participated in its founding, provided arms and advice, and continued to do so after the establishment of diplomatic relations in 1975. The Thai government met the challenge, but the challenge seriously constrained closer Sino-Thai relations. Vietnamese aggression, though, provoked a great change in Sino-Thai relations. Vietnam's occupation of Cambodia was denying Thailand a necessary buffer state and creating new sources of internal security threats with the presence of large numbers of refugees. Thai and Chinese interests quickly converged on the need to contain Vietnamese aggression.

China wanted to see the eclipse of Soviet and Vietnamese influence in Southeast Asia and to establish a more influential regional presence. More cooperative relations with Thailand allowed China to successfully pursue both of these objectives. The PRC could militarily support the Cambodian resistance, particularly the Khmer Rouge, in their sanctuaries along the Thai-Cambodian border. By offering support directly to Thailand, China could strengthen the united front against Vietnam while influencing Thailand's policy toward Cambodia and, indirectly, the policies of other ASEAN nations. Arms transfers through and to Thailand became the principal means by which China's objectives could be met.[10]

The potential for Chinese influence over the Cambodian issues and in Southeast Asia as a whole was not lost on ASEAN. The growing dependence of the resistance groups on China and the dominant position of the Khmer Rouge could limit ASEAN's influence over the outcome of the war. ASEAN military assistance to the Cambodian resistance was a contentious issue within the organization. Indonesia, though wary of China's intentions, was equally concerned that direct arms transfers from ASEAN to the resistance would militarize ASEAN's relationship with Vietnam. ASEAN did not resolve the issue, leaving it up to member states to pursue bilateral links to the resistance groups.[11]

The central problem for ASEAN's coalition efforts was the need to disengage from direct support of the Khmer Rouge while keeping them on the battlefield to goad Vietnam toward negotiations. Beginning in late 1979, ASEAN led the way in an attempt to fashion a more neutral alternative to Khmer Rouge dominance that would also satisfy the interests of China. In successive visits to Beijing by the Thai and Singaporean prime ministers, the Chinese leaders were convinced that Pol Pot and the Khmer Rouge were not viable international leaders of the resistance. By December 1981 ASEAN had fashioned a diplomatic "fig leaf," the coalition government of democratic Kampuchea, that masked their support of the Khmer Rouge. The most conservative group, Son Sann's KPNLF, and the most radical, the Khmer Rouge, were forced to accept the coalition as the price for continued ASEAN international support and Chinese arms.

The CGDK allowed ASEAN to promote the noncommunist members of the coalition, Sihanouk and Son Sann, while keeping the military pressure, largely the Khmer Rouge, on Vietnam. The presence of the CGDK after June 1982 also made it easier for ASEAN to mount the annual defense of the resistance groups' UN seat. Thus, Vietnam could be denied both international recognition of its sponsored regime in Phnom Penh and a decisive victory in the field.[12]

The difficulty with which the CGDK was sustained was due to the mutual suspicions among the members of the resistance movement. The fundamental contradiction was between the noncommunist Sihanouk and Son Sann factions and the militarily superior Khmer Rouge. Mistrust also existed between the noncommunist factions because of Prince Sihanouk's popularity and potential influence. Mutual suspicion, therefore, made the promotion of each resistance group's interests as important a goal as maintaining the coalition. Coordinated military action was undermined by protection of each group's power and territory. The failure of coordinated action—combined with the fear that too rapid an end to the war would bring more intense intracoalition rivalries—helped to prolong the conflict.

The discord within the coalition reinforced the ambivalence of the ASEAN states toward the CGDK. ASEAN member states did not relish the possibility of a return by the Khmer Rouge or the expansion of Chinese influence that could follow. The alternative of a Vietnamese-dominated Indochina was equally undesirable. Cynics could and did argue that ASEAN states were in no hurry to see the conflict end in order to wear down all the communist participants. With the exception of token and discreet support from Singapore and Malaysia, little material support flowed to the resistance groups. While professing its neutrality in the conflict, Thailand also placed constraints on the flow of Chinese military aid. ASEAN support for the CGDK was enough to ensure its viability but not an early or decisive victory until the military power of the Khmer Rouge could be checked.[13]

In the face of such ambivalence, it would appear that the combined power of Vietnam and its Cambodian and Soviet allies was greater than that of the anti-Vietnamese coalition. At the height of the conflict, Vietnam was fielding a battle-tested and well-equipped force of 150,000 and 180,000 supported by 30,000 Cambodian troops. With these forces, Vietnam undertook dry-season offensives in 1984 and 1985 that heavily damaged the resistance groups. The result, though, was not a decisive victory, but only prolongation of the conflict. Tensions between Hanoi and Phnom Penh and between Hanoi and Moscow caused this outcome.

The deeply ingrained enmity between the Khmer people and the Vietnamese characterized relations between the Heng Samrin regime in Phnom Penh and the Vietnamese occupying force. Distrust meant that Vietnam could not expect much real support from the Cambodians. At the same time, the Vietnamese could not afford to expand the Cambodian forces beyond its 30,000-troop size and maintain effective control. The Kampuchean People's Revolutionary Armed Forces (KPRAF) was by independent accounts a weak force lacking in motivation and training and plagued by a high desertion rate. The KPRAF did not even gain an independent identity until 1987. Vietnam's failure to build a strong and reliable KPRAF delayed its withdrawal and further prolonged the war.

From Vietnam's perspective, Soviet aid was desirable only to the extent that it facilitated the initial invasion and enhanced Vietnam's position vis-à-vis China and the ASEAN states. After prolonged struggles against France and the United States and the costs of alienating China as an ally, the Vietnamese were in no mood to be dominated by the Soviets. Therefore, Vietnam preferred only indirect Soviet involvement in the Cambodian conflict. Because of Vietnam's mistrust and limitations, the Soviet Union provided only the amount of aid necessary to sustain the stalemate on the battlefield and preserve its access to bases in Vietnam. The lesson of Vietnam's drastic break with China was not lost on the Soviets (i.e., Vietnam would turn away from the Soviet Union as soon as it could). From the Soviet perspective, Vietnam's turn would most likely come after securing its hegemony in Indochina. An early end to the war, therefore, did not necessarily benefit the Soviet Union.[14]

Larger geopolitical considerations also motivated Soviet policy. The Soviets had to consider carefully the consequences of their support for Vietnam on Sino-Soviet relations. China was the more strategically important country with whom the Soviets shared a long, contested, and heavily armed border. Finding its own support circumscribed but a prolonged war suitable for bleeding Vietnam dry, China also turned its attention to larger geopolitical concerns. Beginning in October 1982, China and the Soviet Union began the first of several rounds of negotiations to normalize relations. China maintained that the Cambodian issue was one of "three obstacles" standing in the way of better relations. By the summer of 1986, the Soviets under the leadership of Mikhail Gorbachev had compromised on two of those obstacles, the Sino-Soviet border issues and Soviet occupation of Afghanistan. Deng Xiaoping noted positive Soviet moves, but in a notable change of emphasis, stated that an end to the Vietnamese occupation of Cambodia was "the main obstacle in Sino-Soviet relations." If Gorbachev could remove this obstacle, Deng then would meet with the Soviet leader.[15]

The possibility of Sino-Soviet compromise on the Cambodian issue indicated the extent to which great power interests could conflict with those of its coalition partners. ASEAN and Vietnam were prompted to take more effective diplomatic steps but found their efforts thwarted by deep geopolitical divisions and a fear of the possible outcome. As the conflict was prolonged, it entered into the final stage of stalemate diplomacy.

Diplomacy: From Stalemate to Compromise

The resumption of Sino-Soviet normalization talks in October 1982 prompted Vietnam to consider concrete diplomatic moves. At the first session of the talks, the Chinese offered a settlement plan that linked Sino-Vietnamese normalization talks to the phased and unconditional withdrawal of Vietnamese troops from Cambodia. Although the Soviets reassured Vietnam that it would not sacrifice Vietnamese interests, it left Hanoi feeling uneasy because a Cambodian settlement was an ongoing agenda item and because of the history of Soviet compromise over Indochina issues in 1954 and 1972.

Vietnam's approach was to push for "five plus two" negotiations between the five members of ASEAN and Laos and Vietnam. Such an approach sidestepped the thorny issues of the role of the CGDK and the PRK regime in Phnom Penh. The proposal was coupled with an announced program of troop withdrawals that amounted to little more than a thinning of the Vietnamese ranks. ASEAN's response reaffirmed the UN-sponsored International Conference on Kampuchea (ICK) formula which called for a cease-fire and a quick withdrawal of all foreign forces and UN supervision of the withdrawal and subsequent free elections. ASEAN's position was intended to demonstrate its flexibility and less confrontational stand while masking confusion within its own ranks.

The prolongation of the war increased the opportunities for other interested parties to propose diplomatic solutions. Paralleling the multilateral ICK and Vietnamese approaches were bilateral talks by Thailand and Indonesia. Indonesian diplomacy was particularly important because Indonesia became the authorized conduit between the ASEAN nations and Vietnam. Indonesia was an effective intermediary because of its shared concern for great power dominance and a desire for a strong Vietnam as a buffer to China.[16]

Attempts by the ASEAN states and Vietnam floundered as a result of Chinese opposition and Vietnamese dry-season offensives in 1984 and 1985 that dealt major blows to the CGDK's armed forces. As a result, ASEAN members closed ranks to support the status quo in the absence of effective leverage over Vietnam. Yet between 1981 and 1985, several issues emerged whose resolution would form the parameters of any settlement. These issues were the withdrawal of Vietnamese troops, recognition of the Phnom Penh regime, an end to external assistance, preventing the return of the Khmer

Rouge to power while including them in any settlement, power sharing during the transition to free elections, and the role of the UN in Cambodia.

ASEAN support for the status quo and the stalemate on the battleground played to the advantage of Vietnam. Following the successful 1985 dry-season offensive, Vietnam announced in August 1985 total troop withdrawal by 1990. The CGDK responded with a plan that no longer made the complete withdrawal of Vietnamese troops a condition for negotiations. Vietnam understood that the progress of the war was decisive in shaping and changing the CGDK's policies. By 1986, however, the Vietnamese were admitting that effective military leverage was becoming difficult to sustain because of disease, malnourishment, and low morale. Economic conditions at home and mounting pressure from the Soviets were also beginning to take their toll. In January 1989 Vietnam announced a complete troop withdrawal by September 1989, if an agreement were in place. In April 1989 Vietnam amended the timetable by announcing the unconditional withdrawal of its troops by September.

The unilateral and unconditional declaration by Vietnam was a result of three factors: (1) the recognition of the Phnom Penh regime; (2) the linkage of the end of external assistance with Vietnamese troop withdrawal; and (3) recognition of the need to prevent the return of the Khmer Rouge. The recognition of the Phnom Penh regime followed from its willingness in the fall of 1987 to discuss issues of national reconciliation with other Cambodian groups, with the exception of Pol Pot and his closest supporters. Prince Sihanouk, who had stepped down from the presidency of the CGDK for one year in May 1987, accepted the offer. The two rounds of talks in December 1987 and January 1988 resulted in little substantive change but did much to reinforce contacts between a Phnom Penh regime in need of international recognition and the one member of the resistance coalition who commanded international support.

Prince Sihanouk's discussion with Prime Minister Hun Sen did not sit well with his coalition partners, who feared a bipartite agreement. But before his contacts could rent the coalition asunder, an Indonesian initiative under ASEAN auspices was gaining momentum. In a joint communique with Vietnam in July 1987, Indonesia offered to host an informal gathering of *all* the parties to the conflict. The Jakarta Informal Meeting (JIM) took place in July 1988. Despite inclusive results, the JIM was a boon to Vietnam and its Phnom Penh ally. The declaration issued by the Indonesian foreign minister in his capacity as conference chairman established a link between Vietnamese troop withdrawal and the phased reduction of external assistance to the resistance forces. ASEAN had resisted the linkage but was now committed to a position that equated Vietnamese withdrawal to safeguards against the restoration of the Khmer Rouge. International reaction following the JIM also appeared to be one of resurgent condemnation of the genocidal Khmer Rouge.[17]

Vietnam and ASEAN nations had found common ground on the issue of the Khmer Rouge. ASEAN's flexibility was as much a product of the battlefield stalemate as the waning influence of China over the Cambodian issue due to changing great power relations. Declining influence was due in part to the initiatives taken by Mikhail Gorbachev. Under Gorbachev, the Soviet Union formulated a more effective Asian security strategy that recognized the importance of mutual security and emphasized the defensive nature of its military capabilities. Gorbachev also increased the pressure on Vietnam for a negotiated settlement. One interpretation of the doubling of Soviet economic aid to Vietnam to $2 billion annually during the 1986–90 period was to increase Hanoi's dependency and thereby Soviet leverage.[18]

By August 1988 China and the Soviet Union had "found common ground" on the Cambodian issue. In a joint communique at the Sino-Soviet summit in May 1989, China and the Soviet Union defined the common ground as the gradual reduction of military aid to all the combatants, the establishment of a provisional government under Prince Sihanouk, and the conduct of free elections under international supervision.

Within three months of the joint communiqué, China appeared to abandon its flexibility for a harder line to regain leverage over the Cambodian issue. Following the Tiananmen incident in June 1989, relations with China were cooler with the more reform-minded Soviets and the more human rights-minded Americans. In a show of independence at the August 1989 Paris Conference on Cambodia, China explicitly rejected a Soviet-American proposal for a temporary end to foreign assistance for the opposing Cambodian factions. The intransigence of the Chinese and their Khmer Rouge allies doomed the international conference to failure.

Following the pattern of other prolonged wars, the diplomatic gridlock resulted in a shift back to the battlefield. The completion of the Vietnamese withdrawal in the fall coincided with the beginning of the dry season. The Khmer Rouge went on the offensive and succeeded in capturing a district capital, Pailin. Although an insignificant population center, Pailin is the gem-mining capital of the country, giving the Khmer Rouge a valuable source of income. Some Vietnamese troops returned to support the Phnom Penh regime. When the dry season ended in the spring of 1990, the stalemate continued. The Khmer Rouge plan to create a large "liberated" zone in Cambodia failed to materialize. The Phnom Penh regime survived the loss of substantial Vietnamese and Soviet assistance and acquired self-confidence in the process.

By the spring of 1990, China was becoming disillusioned with the military capability of the Khmer Rouge. Despite Chinese military assistance and their claims to have liberated a large section of Cambodia, the Khmer Rouge had not effectively challenged the control of the Phnom Penh regime. China's ties with the Khmer Rouge were also becoming a political liability for attempts at post-Tiananmen diplomatic rehabilitation. The dramatic diplomatic action of the United States in July 1990 underlined the liability. Moving closer to the Soviet position, Secretary of State Baker strongly condemned the Khmer Rouge and withdrew recognition of the CGDK. With the renewal of dialogue with Vietnam, the United States clearly signalled its intent to prevent the return of the Khmer Rouge to power.

A second factor was also driving change in both China and Vietnam's policies. The rapid collapse of communism in Eastern Europe beginning in the fall of 1989 left these two countries with few ideological allies. With a stalemate on the ground, China was more eager to mend fences with Vietnam. With a sharp drop in Soviet aid after January 1990, Soviet rapproachment with China gained a new urgency for Vietnam. At an unpublicized summit meeting in September 1990, Chinese and Vietnamese leaders reached a broad accord on a political settlement in Cambodia. The members of the CGDK were reportedly shocked to learn of the meeting and drew the appropriate conclusions about the prospects for peace.[19]

Sino-Vietnamese rapproachment parallelled the multilateral efforts at a Cambodian settlement. Following an Australian proposal, the five permanent members of the UN Security Council drafted a peace plan that dealt with the remaining issues of the role of the Khmer Rouge—a coalition government—and the UN role in administration, peacekeeping, and organization of free elections. The Perm Five plan was finalized in November 1990 and met the approval of the Cambodian Supreme National Council, the coalition government of the resistance forces, and the Phnom Penh regime. Its implementation as the basis of a peace treaty, however, became the final obstacle on the serpentine road to peace.

The role of potential spoiler fell to the Phnom Penh regime. The regime had shown moderation throughout the negotiations over the Perm Five plan. Moderation was repaid with the removal of the Khmer Rouge-dominated coalition from the UN seat, the beginning of direct dialogue with the United States and assurances from the UN that the Phnom Penh government would not be dismantled. The essence of the UN plan, though, was the inclusion of the Khmer Rouge in the political process. The Phnom Penh prime minister, Hun Sen, insisted on language that would condemn the Khmer Rouge's genocidal past and prevent its possibility in the future, effectively excluding the Khmer Rouge from the process.

When these demands were first made, Western observers viewed them as tactical moves to obtain concessions in other areas. Hun Sen's persistence suggested that he saw the benefits of prolonging the war. As the war dragged on, however, these observers expressed a concern that a return of the Khmer Rouge might force some governments to recognize Phnom Penh. By forcing the West in particular to choose between Hun Sen or Pol Pot, the Phnom Penh regime could break the stalemate by suggesting the possibility of a stark future. Some observers also saw another factor behind Hun

Sen's moves, namely the popularity of Prince Sihanouk on whose coattails the Khmer Rouge could return to power.[20]

Hun Sen's position was tenable as long as his principal ally, Vietnam, believed that China was using the Khmer Rouge to bleed Vietnam. Though concerned about the Khmer Rouge's resurgence, the momentum of Sino-Vietnamese normalization convinced Hanoi to pressure Phnom Penh to abandon its opposition to the Perm Five plan. Prince Sihanouk's growing independence and desire to return to Cambodia also convinced Hun Sen that he could be a valuable political ally against the Khmer Rouge. In a series of meetings beginning in June 1991, the members of the SNC agreed to a cease-fire, to end the flow of foreign arms, to disband 70 percent of their forces and place the remainder under UN supervision, and to an electoral system of proportional representation. The peace treaty was signed in October 1991. The war was formally over, but was the basis created for prolonged peace?

Conclusions

Several factors combined to prolong the war in Cambodia. First, the legacy of over 30 years of war created a complex situation with different levels of conflict. The regional and great powers—Vietnam, ASEAN, China, the Soviet Union, and the United States—used Cambodia as a proxy war to play out larger conflicts among them. Maintaining a favorable balance of power was a more important objective than terminating the conflict. In the process of using Cambodia as a means to address other interests, external powers contributed to a prolonged conflict, the roots of which will not necessarily disappear with the formal end of hostilities.

Second, the war aggravated the geopolitical divisions that were another legacy of the First and Second Indochina Wars. The failure of Vietnam to achieve a decisive victory in 1979 heightened these divisions along which coalitions emerged. They were uneasy coalitions, reflecting the concern for balance of power over conflict termination. Mirroring US involvement in Vietnam in the late 1960s, the Cambodian war became one not to lose. Tens of millions of dollars in military aid was poured onto the battlefield to secure the stalemate. The aid was never quite enough to win, because it was not necessarily in the interest of the patrons to terminate the conflict. When reasons were found to negotiate an end to the conflict, they underscored the sad fact that the war did not have much to do with Cambodia itself. Despite the fact that the peace treaty commits the signatories to the reconstruction of Cambodia, the history of conflicting interests, ambivalent support, and the magnitude of the task do not ensure Cambodia's future prosperity.

Third, as a result of prolonged conflict, diplomacy became another means of waging war. It is therefore uncertain whether the October 1991 peace treaty does lay the basis for prolonged peace. Reflecting on the Cambodian accords, Lee Kuan Yew, the long-time former prime minister of Singapore, has argued that the Cambodian pact is "less advantageous to the non-communist groups." He states that the settlement was largely due to a compromise between China and Vietnam. Therefore, it is not as "clear-cut a solution as it would have been if there had been no reconciliation between China and Vietnam." Lee concludes that because of the interim SNC power-sharing arrangement, China and Vietnam will continue to exert influence over Cambodia's fate.[21]

Lee Kuan Yew also may be expressing ASEAN's anxieties over the pact given the organization's ambivalence about supporting the Khmer Rouge. Khmer Rouge officials describe as misplaced any international concern over a military takeover by their forces. In a widely circulated 1988 speech, Pol Pot outlined a strategy in which the Khmer Rouge would be restored to power through political organization. Building on its rural base, the group has begun public works projects in western Cambodia and a public relations campaign to improve its image. Given the recent violent reception and flight of some of its leaders from Phnom Penh in November 1991, the Khmer Rouge remains a cause for alarm.

The reason for this alarm is that the Khmer Rouge maintains the capacity for independent action. Though China announced that it stopped shipping arms to the Khmer Rouge in November 1990, it is believed that the group has stockpiled a two-to-five-year supply of weapons and ammunition in its remote sanctuaries. The Khmer Rouge also has large reserves of cash and controls lucrative gem mining and teak forests, which they exploit on the Thai markets. In violation of the October agreement, the Khmer Rouge has also used their access to the refugee camps to force the migration of Cambodians into their controlled areas to strengthen their electoral base.

Given the long-standing concerns of its former allies and enemy, it is possible that the Khmer Rouge could revert to a strategy of protracted war if rejected at the polls. Consequently, maintaining a balance of power within Cambodia may be the most important objective of the noncommunist groups and the PRK. To that end, the Phnom Penh regime has joined forces with the mercurial Prince Sihanouk to counter the Khmer Rouge. Though the Khmer Rouge continues to participate in the SNC, Hun Sen and Sihanouk have formed a political coalition to exploit Sihanouk's popularity and reclaim the nationalist legacy tarnished by an alliance with Vietnam. It is uncertain, with less than two years before UN-sponsored elections, where this political maneuvering may lead. But since the balance of power continues to be a concern, the opportunity for external powers' involvement remains.

The peace treaty stripped away the levels of conflict and returned Cambodia to its civil war roots. Years of prolonged war created the opportunities for great power involvement and the polarization of the conflict into uneasy coalitions. In their haste to remove Cambodia from the international agenda, the powers left that nation in the uncertain status of neither war nor peace.

Notes

1. Chang Pao-min, "Kampuchean Conflict: The Continuing Stalemate," *Asian Survey* 17, no. 7 (July 1987): 748–52; and Nayan Chanda, "Indochina," in Anthony Lake et al., *After the Wars* (New Brunswick, N.J.: Transaction Publishers, 1990), 79–80.

2. David W. P. Elliot, "Deadlock Diplomacy," in David A. Ablin and Marlowe Hood, eds., *The Cambodian Agony* (Armonk, N.Y.: M. E. Sharpe, 1990), 69.

3. David P. Chandler, "A Revolution in Full Spate: Communist Party Policy in Democratic Kampuchea, December 1976," in Ablin and Hood, 178.

4. Nayan Chanda, "Cambodia in 1986: Beginning to Tire," *Asian Survey* 17, no. 1 (January 1987): 116–17.

5. Chandler, 177.

6. Sheldon Simon, "Vietnam," in Edward A. Kolodziej and Robert E. Harkavy, eds., *Security Policies of Developing Countries* (Lexington, Mass.: Lexington Books, 1982), 147.

7. For an overview of Vietnam's efforts to avoid the polarization of regional politics, see Alexander Woodside, "Nationalism and Poverty in the Breakdown of Sino-Vietnamese Relations," *Pacific Affairs*, Autumn 1979, 381–409.

8. Simon, 145.

9. "Vice Premier Teng Visits Thailand," *Peking Review*, 10 November 1978, 3–4.

10. R. Bates Gill, "China Looks to Thailand: Exporting Arms, Exporting Influence," *Asian Survey* 31, no. 6 (June 1991): 224–25.

11. Donald E. Weatherbee, "The Diplomacy of Stalemate," in Donald E. Weatherbee, ed., *Southeast Asia Divided: The ASEAN-Indochina Crisis* (Boulder, Colo.: Westview Press, 1985), 4.

12. Ibid., 5.

13. Chang Pao-min, 756–57.

14. Ibid., 754.

15. "Deng on Sino-Soviet, Sino-American Relations," *Beijing Review*, 15 September 1986, 5.

16. Weatherbee, 14–22.

17. Khatharya Um, "Cambodia in 1988: The Curved Road to Settlement," *Asian Survey* 29, no. 1 (January 1989): 75–76.

18. Hiroshi Kimura, "Gorbachev's Agenda for Asia," *The Pacific Review* 1, no. 3 (1988): 220.

19. Nayan Chanda, "Cambodian Talks Bear Fruit of Uneasy Peace," *Asian Wall Street Journal Weekly*, 30 September 1991, 24.

20. Nayan Chanda, "Hope for Settlement in Cambodian War Wanes As Hun Sen Government Hardens Its Position," *Asian Wall Street Journal Weekly*, 25 February 1991, 20.

21. Nayan Chanda, "Singapore's Lee Kuan Yew Airs Reservations About 'Fuzzy Situation' After Cambodian Pact," *Asian Wall Street Journal Weekly*, 4 November 1991, 1, 14.

El Salvador's Prolonged Civil War

Steffen W. Schmidt

Most wars are fought to obtain maximum gain, at least expense, in the shortest time. Most wars or conflicts, however, are not short—they continue over long periods of time. These wars can be either prolonged wars or protracted wars. A *prolonged war* is a conflict which the relevant protagonists would like to win promptly and decisively; however, its resolution is delayed unintentionally. Karl P. Magyar argues that *protracted conflict*, on the other hand, is war in which one or more of the contending parties develop a deliberate strategic plan of incremental, gradual, or extended fighting.[1] In modern times, protracted conflict is principally associated with Chinese revolutionary leader Mao Tse-tung and his disciples.

This chapter describes the dynamics of protracted conflicts in El Salvador with special emphasis on the most recent one (i.e., from 1979 to 1991)—a war which killed almost 1 percent of El Salvador's population, displaced from their homes over 20 percent of the population, and cost the United States over $4 billion in assistance.[2]

My description and analysis of El Salvador's conflict begins with a reference to Harry Eckstein, who argues that explanations for the cause of internal war can be found in some combination of five basic clusters of hypotheses. These are as follows:

1. Hypotheses emphasizing intellectual factors such as "the failure of a regime to perform adequately the functions of political socialization or the alienation (desertion, transfer of allegiance) of the intellectuals."
2. Hypotheses emphasizing economic factors: increasing poverty, imbalances between the production and distribution of goods, and so forth.
3. Hypotheses emphasizing aspects of social structure such as the "inadequate circulation of elites," "too much recruitment of members of the non-elite into the elite, breaking down the internal cohesion of the elite," and social stagnation.

4. Hypotheses emphasizing political factors, among them the "estrangement of rulers from the societies they rule," bad and/or oppressive government, and divisions among the governing classes.

5. Hypotheses emphasizing no particular aspect of societies but general characteristics of social process, including rapid or erratic patterns and pace of social changes.[3]

I will not test these assumptions and hypotheses in depth. However, if one disaggregates the Salvadoran conflict into cycles or phases, one finds that many of the five factors apply. The long and arduous war has been extended over time by the division among the governing elites, subelites, and the counterelites (the insurgents). This conflict has produced an unusually complex, shattered pattern of elite political behavior. This fragmentation, in turn, resulted in a failure by the elites to reach sufficient consensus to rule effectively and a lack of cohesion on the part of the insurgents on how to win the war against the government.

I consider it imperative to add to Harry Eckstein's inventory on the importance of externalities (i.e., global affairs) to internal war. The convergence of global events in 1980 hardened both sides in the new cold war between the evil empire as Ronald Reagan called the Soviet Union and the USSR (Union of Soviet Socialist Republics), which projected force into Afghanistan, for example. This force spilled into El Salvador by way of Managua, Nicaragua, and Washington, D.C.

These factors can be conceptualized in an interactive model with three dimensions, each consisting of two dynamics. First, socioeconomic conditions and trends in the aggregate can have either centripetal effects (i.e., pulling social classes and regions of a country together) or they an be centrifugal (i.e., tearing groups apart and pitting them against each other). Second, on the ideology and political dimension, we can broadly identify center-seeking trends (i.e., in which polarization grows as groups abandon moderation and recruit from left and right extremes). Third, we can identify international dimensions—I call them global affairs or externalities. These can have the effect of encouraging consensus, on the one hand, or they can exacerbate conflict, on the other hand.

A review of events that contributed to El Salvador's war is necessary to understand how these events contributed to the protracted nature of that conflict. In reading the material on the following pages, one should note the strategic importance to the guerrillas in conducting a protracted struggle. Tactics within this strategic plan for victory by the insurgents have included a vast and complex arsenal of terrorism, conventional warfare, kidnappings, economic warfare, provocation of institutional elites and other techniques. The overall objective sought to exploit real, structural grievances and to provoke a host of atrocities and calamities over time. In the end, the objective was to demoralize, to cause collapse in the domestic structure, and to erode outside support.

The Salvadoran military and government on the other hand initially sought a quick victory. However, for them and their US allies, the conflict soon became a prolonged war, one which continued longer than they wanted. Circumstances beyond their control dragged the war out. These included guerrilla determination, their own interminable internal quarreling, and a lack of cohesion as rulers. On the other hand, the conflict endured because of strategic decisions on how to pursue the war. A low-intensity conflict policy—a protracted conflict strategy—became an integral part of the government's strategy to check and block the insurgents.

Background to the War

Inside El Salvador a complex brew had been cooking for decades.[4] The military, the 14 leading families, and the revolutionary communist party had been locked in a no-holds-barred struggle for power long before the United States discovered the Salvadoran issue in the 1970s. In addition, there were labor groups, small reformist parties, farmers' organizations, clergy, student groups of diverse coloration, and a host of other factions, all seeking their rights in a fertile soil of chaos and injustice.

Some of El Salvador's political forces indeed shared the American liberal vision of reform. They called for judicial change, fair and open elections, the end of violence, land reform, and an expansion of health, housing, and educational programs for El Salvador's people. But by 1970 the opportunity for reformist change had long passed for many Salvadorans. The left had grown weary of repression and violence. Reformism smacked of a tactic by the forces of the status quo to avoid change. Moreover, the left perceived reformists largely as agents of the right rather than as viable alternatives. The hard left had come more and more to despise any of the bourgeois forces. To a sizeable right-wing faction, reform was nothing more than an open door to Marxism. To others, reform meant a radical change not too far removed from outright revolution. The antireform faction in El Salvador did not consist solely of the extreme right, but it could count on a sizeable portion of the middle and professional classes. This paranoia about reform had its roots in the 1930s.

The Legacy of the Left

The seeds of the Salvadoran insurgency are found in the 1932 massacre of peasants, the same insurgency in which the young Trotskyite, Agustin Farabundo Marti, the son of Indian peasants and one of the founders of the Salvadoran Communist party in 1930, was killed. The Salvadoran guerrilla forces, the FMLN, bears his name.

After crushing the communist rebellion, President (general) Maximiliano Hernandez Martinez ruled until a 1944 coup removed him from power. This marked the emergence of the military as the dominant force in Salvadoran politics.

The Right Attempts Top-Down Reform

Following the overthrow of Martinez, Gen Castañeda Castro gave Salvadorans socioeconomic reform through martial law and strong repressive measures which ended in the

"Revolution of 1948." In this so called radical leaning, colonels and majors, backed by reformist civilians, removed the incumbent government and set up a five-man military-civilian junta. The 1948 revolution sought to make a transition in the face of an increasingly polarized political foundation, in particular the acceleration of center-fleeing political parties and structure.

A 1950 constitution provided new and more liberal political ground rules. The armed forces were to be nonpolitical and obedient to the government in power. Nonetheless, reformist factions in the military saw an activist role for themselves. A new political party, the Partido Revolucionario de Unificación Democratica, was founded by Col J. M. Lemus, who won the presidency in 1950 and again in 1956. Still, the fragmentation of the political elites around a series of alternative and often deeply conflicting alternatives continued and set the stage for political breakdown and finally all-out war. A fragile economic infrastructure is often the weak link in the process of social change and political reform. Compounding the interelite quarrels and fragmentation was the economic vulnerability of El Salvador, a tiny, overpopulated, resource-poor country.

Political unrest grew. A recession led to conspiracies by powerful economic groups who feared that the modest socioeconomic reforms would undermine their wealth and power. Both the left and the right yearned for an opportunity to reverse the modest reforms of the past 10 years.

Prelude to War

In 1960 and 1961 the military overthrew incumbent governments, as leftist violence escalated and groups allegedly controlled by admirers of Fidel Castro and by communists were blamed for the chaos. Another military man, Col Fidel Sanchez Hernandez, was elected president in 1967. His move into the presidency was a commitment to continue the top-down reforms initiated by his predecessors. However, he inherited a nation plagued, as never before, by land shortage and poverty. It is estimated that by the late 1960s, over 300,000 desperate, landless Salvadorans had illegally migrated across the border to Honduras. This serious problem was to erupt into perhaps the major crisis of contemporary Salvadoran political history—the 1969 war with Honduras. This, the so-called Soccer War, seriously disrupted trade and created a massive flood of Salvadoran refugees who poured back into their homeland.

The 1972 election was to be the culmination of a gradual process through which electoral blocks and moderate reform policies gained ground. However, the extreme left repudiated the election and launched a series of attacks against the national guard. Arson broke out in the capital city. Government agents raided the university, turning up caches of arms and communist pamphlets. The atmosphere, particularly in San Salvador, was electric.

The Start of War

On 19 July 1972 the army occupied the university's grounds and buildings. The government claimed that a university under communist influence and control was not in the national interest. The crisis and the war that were to take shape and grow over the next 19 years began with these events. On 16 February 1973 the government reported the discovery of an "international terrorist plot" led by the Salvadoran Communist party. More than 100 left-wing labor leaders and politicians were arrested.

Once again, the fragmentation of the military and political elite precipitated a crisis which played into the hands of the protracted war strategy that was now being crafted by the opposition. The left began to refine and intensify its use of guerrilla attacks on the government. On 27 February 1973 a bomb exploded outside the police station in Chachualpa. On 3 March, less than a week later, members of the People's Revolutionary Army (ERP) seized two radio stations in San Salvador and began to broadcast revolutionary messages. Three days later, guerrillas attacked the National Elections Council, where two soldiers defending the offices were killed. These acts, preceding the 10 March elections, underscored the contempt opposition groups, including the extreme left, felt toward elections in El Salvador. Usually elections lend credibility to a government. In El Salvador hostility had developed toward elections because of fraud.

When peasants organized and began to voice demands, they were terrorized and allegedly even massacred by government and paramilitary forces. As a consequence of this violence, new insurgent groups were formed, including the ERP. ERP kidnapped wealthy industrialist Francisco Sola, collecting $2-million ransom, then issued a pamphlet calling the ransom "a war tax for the Salvadoran revolution."[5] The technique of raising funds through abduction and bank robbery became a powerful tool in financing revolutionary activity and was widely used by leftist groups. The abductions as well as death threats precipitated the creation of "self-defense" organizations by business and agricultural leaders. These groups were the initial seeds for the "death squad" that later became one of the instruments of war against the left.

Clearly, this period represents a distinctive phase in the evolutionary development of the war in El Salvador. In protracted war unconventional combat techniques play an important role. In the case of El Salvador, terrorism and unconventional war were at a developmental, low-intensity conflict stage in the early 1970s. They produced the structures and defined the battle lines for more intensive protracted war later on.

The armed forces then entered a period of soul-searching and of political alignment—left, right, center, or apolitical. Caught in the tide of forming an active part of the process of political action and of political change, many of the armed forces leaders felt increasingly uncomfortable with the role they were playing. However, the treasury and cantonal police, the national guard, the federal police, and the paramilitary ORDEN were heavily involved in these political battles. By 1976 the regular armed forces were also engulfed in the violence.

In the degenerating environment, the opportunities for personal and institutional enrichments were growing. Robbery, rape, theft, extortion, and appropriation of land and property were increasing daily. These variables are not elements of political value but simply opportunistic "benefits" of war. Unless these factors are structured as a permanent backdrop to the war in El Salvador, the analysis will be flawed because this war would appear as a "high road" of ideology, reform, high politics, and principle. A substantial explanation for the protracted nature of the conflict is personal greed and opportunism.[6]

The 1977 Elections—The Eve of War

The next phase in Salvadoran politics centered on the new president, Col Carlos Humberto Romero, elected in March 1977. His election was widely reputed to have been fixed. President Romero unsuccessfully attempted to organize a forum to explore ways to end violence.

One reason for this failure was that aside from the official policy emanating from the executive office of the president, there were innumerable minipolicies undertaken by various elements in the five or more security forces. Local commanders and even noncommissioned officers often made local decisions. This factor is important because, as we shall see later, the US Congress increasingly attempted to hold central commanders and national leaders responsible for events in their country (especially human rights violations). It is clear that central command in a country such as El Salvador has ineffective command and control over military and police units.

Externalities 1979—The War Escalates

On 17 July 1979 the Sandinistas overthrew Anastacio Somoza in neighboring Nicaragua. On 15 October 1979 another military coup d'état was carried out in El Salvador. Former US ambassador to El Salvador, Frank Devine, noted, "We were told that the point of no return had been passed, that President Romero's statements had lost credibility, and that he was incapable of providing the moral, effective leadership to carry the nation to honest elections."[7] According to Devine, young, reform-minded officers were responsible for the coup. He denied rumors that the US State Department was involved. But after evaluating the five political choices available in El Salvador (a leftist revolution, an election, a continuation of the status quo, a hard-line right-wing coup, a reformist coup) the United States sent out signals that it favored the last alternative.

One of the first actions of the new junta was to abolish the paramilitary organization ORDEN. The junta also promised to investigate incidents of torture, murder, and abduction. The junta planned to establish diplomatic relations with Cuba and to strengthen ties with the new Sandinista government in Nicaragua; offered amnesty to political prisoners and exiles; reshuffled the military command, retiring officers closely implicated with the Romero regime; and proposed presidential elections as soon as possible.[8] The junta also issued a decree prohibiting owners of more than 100 hectares of land from transferring or encumbering their property to forestall plans for a new land-reform program. These actions triggered a strong resistance from the right, especially landowners. At the same time, these moves were not sufficient to convince the left and the insurgents that there was now reason to cease hostilities.

In Washington, the Carter administration believed that the junta was the best alternative to the Salvadoran crisis. Devine argues that the left, after assessing the situation created by a reform junta, "came to recognize displacement of the Romero administration as a serious blow to their revolutionary interests."[9] American conservatives, however, were clearly distressed fearing the political slide leftward could seriously reduce the strength of conservative, anticommunist groups in El Salvador. The leftists simply dismissed the junta as a device that would temporarily delay a future and an inevitable revolutionary victory.

Ronald Reagan's election as president of the United States on 4 November 1979 added another clear-cut externality, which hardened both sides. Reagan already had projected a strong stance against the Soviet's "evil empire" in his campaign. The Salvadoran right liked this position. The left stiffened. In late 1979 the armed, insurgent left worked on strategic, tactical plans to obtain either a political or a militarly victory. The insurgents also sought help from socialist countries and other international revolutionary forces. They also stepped up their military activities.

In 1980 El Salvador's national guard chief of investigations was gunned down, and two major newspaper offices were bombed. The Popular Revolutionary Bloc emerged as the most aggressive insurgent group, claiming responsibility for political shootings, bombings, kidnappings, and seizing buildings. Most damaging to a possible rapprochement between the political extremes was the internal fragmentation of the left. Just as the Popular Revolutionary Bloc negotiated a truce with the junta, the United Action Front announced an all-out war against the government. Eventually the cleavages in the guerrilla movements even led to internal purges (one of these cost radical Salvadoran Cayetano Carpio his life).

One point is clear from the events that unfolded in El Salvador in 1979 and 1980—reconciliation, reform, and peace became unacceptable to both the extreme right and the extreme left. In either case, politics was considered an all-or-nothing proposition—a zero-sum game. The rightists were justifiably anxious about the friendly gestures the government made toward Cuba and Nicaragua. Land redistribution and other reforms perceived as anticapitalist were distasteful to both Salvadoran and US conservatives.

In turn, the left believed that it could eventually win an outright victory. The strategy of protracted war had set a precedent in Nicaragua. Moreover, leftists did not trust a government which had vast connections with existing rightist institutions, particularly the military. It is difficult to second-guess the intentions of the junta, but it appears that, given half a chance, it might have tried to restore stability and a measure of civil rights and push for economic reforms. To do this, however, moderates needed the cooperation of armed groups from both the left and the right. Neither group was prepared to yield to an uncertain compromise.

After the Sandinista victory and the election of Ronald Reagan, the role of the polarizing, conflict-based externalities became extremely important. In fact, the strategy of protracted conflict was given renewed viability by the external political, financial, and military resources which suddenly became available to both sides in El Salvador from the concatenations of the 1980s.

In 1982 the US magazine *The New Republic*, reflecting on US options, argued that roughly four options were available for solving the problem of El Salvador: The first was to simply concede the country to the guerrillas, end all US support to El Salvador (which was now growing), and then contain the leftist regime, which had gained power.[10] A second alternative was to push for negotiations and arrange a coalition between the revolutionaries and the government. The third option was to place priority on winning the war militarily, even if it meant using US forces in El Salvador. As a fourth option, the United States could pressure the existing Salvadoran government to hold elections and expand land reform and human rights.

El Salvador held an election in 1984. Officials decided to hold this election after months of consultation, both within El Salvador and with the United States. Christian Democrat José Napoléon Duarte, himself a victim of corrupt elections, said, "It would be the first free election in the history of this country." The United States hoped that Duarte and his Christian Democrats would win and then forge a peace settlement to end the growing war.

The Salvadoran guerrillas ridiculed the elections by labeling them "By the right, for the right; by a minority, for a minority." The guerrillas promptly called for a boycott, threatening to cut off the fingers of, or kill those who voted (in El Salvador after voting, one's finger is dipped in indelible ink).

The scenario, which both Washington and the Duarte government hoped for, was risky. Salvadorans, it was said, were tired of violence, war, terrorism, and repression (i.e., prolonged war). The efforts by moderate and conservative Christian Democrats to implement reforms would be an incentive for voters to pack the political middle. Since no leftists presented themselves as candidates, the choice would be between extreme rightist candidates and the moderate center-right Christian Democrats.

As the campaign began to unfold, the vigor and determination of rightist groups surprised many. Especially disturbing was the prominent role assumed by Roberto d'Aubuisson of the newly created political party, Alianza Republicana Nacionalista (ARENA). Youthful and surrounded by a mystique of violence, d'Aubuisson would be precisely the wrong person to win the contest from the US perspective. A former army major, he was linked to the death squads. The 38-year-old d'Aubuisson had been drummed out of the army after he had plotted a coup against the government in 1978. The United States and Salvadorans also accused him of excessive brutality in the war against leftist groups. D'Aubuisson said on several occasions that if elected, he would exterminate the guerrillas within three months (i.e., that he would end the prolonged war with a quick and decisive—and also terribly bloody—victory).

Despite the pre-election violence, the turnout at the polls was large. However, the distribution of the votes was far from promising. The elections had pitted six political groups against one another. The Christian Democrats received roughly 41 percent of the vote, giving them 24 seats in the 60-member constituent assembly (seven short of a majority). D'Aubuisson's ARENA with 29 percent got 19 seats. The results of the election disturbed Washington. ARENA and d'Aubuisson had formed a coalition which would, in effect, give them the power to influence, or even dictate, postelection policies. Alvaro Magaña, an independent, was elected provisional president. He saw himself more as an administrator than as a political leader.

The new and tenuous Salvadoran government's cabinet included ARENA on the far right, the centrist National Conciliation Party (PCN), and the Christian Democrats to the left. As a concessionary and bargaining gesture, d'Aubuisson was given the presidency of the constituent assembly. D'Aubuisson, no novice to hardball politics, had a powerful tool with which to rule: "Decree No. 3," which gave the assembly the authority to name top officials in the provisional government, the authority to ratify all cabinet appointments, the authority to make all legislative decisions, the power to write a new constitution, and the initiative to organize new national elections.

This election, reinforced the center-fleeing trend already under way and therefore injected further energy into the confrontational environment between the government and its opposition.

The worst fears of the United States came true when a Salvadoran deputy acknowledged that the next step might be the breakup of the farm cooperatives which had been formed during phase one of the agrarian reform movement. It seemed as if the clock was turning backward (assuming that the "reforms" pushed by Washington were a viable means to ending the war). The weakening of distributive policies, such as land reform which can (at least temporarily) reduce socioeconomic divisions, contributed to the centrifugal socioeconomic trends.

The election came at a time when the junta was faltering. It injected new life into the Salvadoran government (albeit shifting it to the civilian right) and forced the guerrillas to temporarily rethink their tactics. But late in the summer of 1983 nothing had really changed in San Salvador. The election seemed more like an expedient excuse to delay the implementation of a real solution.

Major Factors in the War

First, a growing number of politically mobilized Salvadorans came to view the crisis in their nation as not only domestic but international as well.

Second, as Mexican journalist Mario Menendez Rodriguez wrote in 1981, El Salvador's elite were to be maintained by the United States; thus, the fate of that nation was controlled by the interests defined and manipulated from Washington. Menendez concluded with the warning, "El Salvador is the strategic piece of the United States in its game of Central American dominoes. If it falls, it will drag along with it Guatemala and Honduras."[11]

Third, the Marxists in El Salvador saw the United States as their enemy and El Salvador as America's next Vietnam. Commander Isabel, one of the leaders of the Farabundo Marti Popular Liberation Forces, the most powerful of the guerrilla groups, said it in no uncertain terms: "The fundamental point in the strategy of confronting imperialism is the Central

Americanization of the revolutionary struggle because the United States considers Latin America as its exclusive property in which it can directly intervene when its interests and those of its associates are threatened."[12] Vietnam, in fact, became a metaphor for El Salvador. Even Salvadoran revolutionaries believed they were the cutting edge of a battle against imperialism in all of Central America.[13]

In the eyes of the left, events in El Salvador were a function of global forces manifesting themselves with increasing ferocity in Central America. Opposed to the freedom fighters was the international imperialist alliance headed by the United States. In this way Salvadoran Marxists saw their struggle as one part of a global anticapitalist and antiimperialist war and at the same time as a convenient vehicle for greatly increasing the support for their cause by the Soviets, Cubans, and Nicaraguans, as well as supporters in the United States and Western Europe.

Fourth, there was the conservative view of El Salvador in the United States. This was expressed by Brig Gen Albion Knight, director of the National Security Task Force of the Conservative Caucus, retired, in his testimony before the US House Committee on Foreign Affairs: "The main source of trouble in El Salvador is not a relatively small group of left-wing guerilla fighters," he said, "but the Soviet Union and its American agent provocateur, Castro's Cuba." He continued by tying the fate of El Salvador to the rest of Central America, and indeed, to the third world.[14] Again I note the importance of this perspective to my thesis that conflict-based global affairs were instrumental in adding vigor and resilience to the protracted conflict approach.

Fifth, from the moment the Republican convention picked its national ticket in Detroit (with Ronald Reagan at the head), the situation in El Salvador sought another metaphor in the United States shift to the right. When Ronald Reagan won the 1980 presidential election, Secretary of State Alexander Haig defined it as the natural place to draw the line on communism. In El Salvador several corpses were found with signs around their necks saying, "Now that Reagan is President, we will kill all communists." The election of Reagan (and of the Thatcher government in Britain) served as further externalities in global affairs, which breathed legitimacy and self-confidence into the right in El Salvador. This was extremely important and has been generally overlooked in the literature.

Sixth, overcoming the post-Vietnam isolationism was said to be a necessary catharsis to bring maturity and renewed self-confidence to the United States. The necessity of such a cleansing was shared by the right in El Salvador, Chile, and many other countries, where anticommunism played an important domestic role. Perceived US policy needs to "take a strong stand against communism," and the perceived domestic policy preferences by the center and center-right in El Salvador both converged seamlessly on a policy of winning the war. This policy was especially salient when the time came to apply the "lessons learned" in Vietnam about how to fight a protracted guerrilla conflict with a counterstrategy of protracted and nonconventional warfare.

Finally, the Carter administration was seen by conservatives in El Salvador as the best proof that an emphasis on human rights was not rational. The Carter presidency witnessed sharp increase in global violence and strife, reversal of women's rights, liberal use of capital punishment, and repression of regional, ethnic, and religious minorities. Iran fell to radical fundamentalist Moslems and the USSR invaded Afghanistan. These global developments indeed helped to shape the battle lines inside El Salvador and were the subject of intense debate in that country.

As this analysis strongly suggests, the struggle in El Salvador was seen by many by the early 1980s as an international and regional crisis, not just a domestic conflict. In September 1981 the US Senate attached provisions to the foreign aid bill which required a review of human rights conditions in El Salvador twice a year. US aid to that country would be contingent on "progress" in several areas.

While admirable, the emphasis on human rights produced a warning flag against all-out war, which might exact a quick victory for the government but with a heavy toll in civilian casualties. The extremely important point we must make here is that this reinforced the option of a prolonged, low-intensity war.

In sum, political and economic reforms in El Salvador may have been doomed from the start, because, as former US ambassador to El Salvador Robert White once stated, the

United States was anxious to support a "contrived" political center in El Salvador even though most analysts would agree that there was no such political center. Moreover, the "reforms" (i.e., land distribution) were seen as coming from the outside and therefore were cynical moves to avert more profound change (i.e., a victory by the insurgents). Also, there was a mistaken notion, especially among US development agencies, that the initiation of reforms automatically and concurrently generated political support for the regime. Since the land reforms envisioned would eventually affect most farmers in the country, the first effect was to drive all existing landowners from supporting these reforms. Another result of the land reform was the false assumption that newly propertied peasants would become visible, articulate, and passionate supporters of the center, and thus help build this as a base on which future stability might rest. On closer inspection it becomes clear that there was never an effort to mobilize the supporters and beneficiaries of these reforms. Without conscious political mobilization such support could never materialize, and indeed, it did not.

Attacked by the right and the left, and not fully in control of its own security forces, having cut itself off from the last vestige of support of the conservative center, the government of El Salvador lurched from crisis to crisis. Each week that passed exacerbated the government's problems with human rights. President Duarte soon found his base of support limited to a small segment of the US government and a portion of the armed forces of El Salvador.

Some observers argue that even the US government never really believed that the center could maintain its power. President Reagan and Secretary of State Alexander Haig really hoped for a tough rightist solution. Liberals in the United States hoped Duarte would leave gracefully, making room for only people who, in their eyes, could give the Salvadoran people a better life, namely the coalition of leftist antigovernment forces.

The United States began to transfer millions of dollars in economic and military assistance to the Salvadoran government. Looking back at most of the 1980s, one must conclude that it may indeed have been the United States-backed reforms which made it possible to fight a protracted war. Land reform, food marketing to low-income groups, massive spending on public works to rebuild damaged bridges and other infrastructures carried out regularly by the guerrillas, and the trickle-down effect even of military assistance certainly had an effect on the country's socioeconomic conditions. The rich fled El Salvador to safer havens, and some of the most ostentatious examples of wealth and power were eroded because of the war and because of distributive policies. These may in the long run, say by 1990, have reduced the centrifugal socioeconomic forces which were such important contributing factors to the war.

Talks with the rebels never got far, even after moderate president José Napoléon Duarte was elected in 1984. These peace talks took place in 1984 and again in 1987 as part of a Central American peace plan to control irregular forces (i.e., El Salvador's guerrillas, the Contras fighting in Nicaragua and El Salvador, and Guatemalan death squads). In 1987 Rubén Zamora and Guillermo Ungo, two prominent leftist leaders connected to the insurgents, returned to El Salvador from exile as part of a tenuous political opening (apertura).

In March of 1989 the far right ARENA party, the party of Colonel d'Aubuisson, won the presidency with Alfredo Christiani at the head of the ticket. The insurgents launched the largest offensive of the war in November of that year. Killing more than 2,000 people, the insurgents controlled substantial parts of the capital city of San Salvador for several days. The evidence suggests that the insurgents had prepared for this offensive for a long time, supplying and hiding ordnance and personnel in various parts of the city (again typical of prolonged war, where the belligerents prepare and wait for opportunities to engage the adversary).

One of the important components of this prolonged war is the role of the church and religious persons. In El Salvador, as in the rest of Latin America, the Roman Catholic Church has always played a significant role in politics. Father Jon Sobrino, a Jesuit priest affiliated with the Jesuit University in El Salvador, wrote a fascinating piece called "Death and Hope for Life."[15] In it he indirectly argues for the legitimation of insurgency, prolonged struggle, and war of liberation by elements of the Catholic church. This was an important factor in the guerrilla strategy of prolonged conflict in that it gave moral sanction to the process. Moreover, the world paid little attention to El Salvador until three momentous events occurred—the killing of San Salvador's Archbishop Oscar A. Romero on 24 March 1980, the murder of four American church women on 3 December 1980, and the killing of several Jesuit priests and two lay women at Jesuit University in November 1989.

Rafael Minjivar, in his excellent analysis of Salvadoran politics, sees the church as a tactical instrument in the Central American war. "The CIA," he observed, "counsels not to attack the church as an institution, but instead to establish a division between progressives and those who are not."[16] In his view, having identified communist priests who are betraying the evangelical message of Christ, the CIA would then be able to kill or persecute the clergy in the name of God. Interestingly, during the mid- and late-1980s, conservative evangelical groups and even representatives of the Unification Church founded by the Reverend Moon became active in El Salvador with the intent of winning the war against communism. Thus, religion came to be a moral sanction for the justification for prolonged struggle for both sides, the government as well as the guerrillas.

Modalities of Violence and Protracted War

El Salvador's death and injury rate climbed to over 36,000 people by 1982, and by 1991 almost 100,000 people had lost their lives in this war. The war precipitated a variety of forms of violence which are important in understanding the effects of protracted conflict.

One type of violence, the direct-revolutionary and counter-insurgency-operations types, pits government troops against guerrillas. The purpose of these operations has been to win ground, control the populace, and eventually win the war. When guerrillas clashed with troops, many people died both intentionally and accidentally. Many of the troops and guerrillas in the early years were inexperienced youngsters who had a tendency to shoot indiscriminately. The enemy was everywhere and could be anyone.

Terrorism is a more widespread type of violence used by both left- and right-wing extremists. The existence of terrorism on one side is given as justification for its use by the other side. Terrorism is also a tactic for killing the trained personnel of the opposition and for forcing an adversary into submission. Left-wing extremists often use terrorism to force excesses on incumbent governments to erode their public support. In a turbulent situation such as El Salvador's, a death threat from the left or right is enough to make people flee. This type of violence also paralyzes the economy; its practitioners, primarily the guerrillas, blew up electrical power grids and bridges, destroyed crops, burned buses, blocked highways, ambushed railroad cars, and bombed buildings.

Salvadoran revolutionary groups have used terrorism to frighten away foreign capital and business, to encourage flight of money from the nation's economy, to cast an aura of fear over soldiers and security troops, to persuade villagers into supporting the revolution, and to encourage defection from the government's armed forces.

According to James Berry, the terrorist justifies his actions by arguing that "society is sick and cannot be cured by half-measures of reform. The state is violent and can be overcome only by violence, and the truth of the terrorist cause justifies any action that supports it."[17]

National armies find it hard to deal effectively with prolonged terrorist violence. Police intelligence services and paramilitary forces are often even less capable. The secret police has been used for decades as an instrument of political control by the government; therefore, they have no long-term perspective on security, making them little more than "hired guns" ruling politicians. Of course, the average citizen is the least prepared to cope with terrorist brutality.

A third type of violence is the crime of opportunity, which accompanies all unstable situations. Just as looting inevitably follows natural disasters; theft, kidnapping, and holdups have followed political chaos in El Salvador. Personal and impulsive crimes are easily committed when large quantities of guns are accessible, and law enforcement is almost nonexistent.

Fourth, many societies use violence as a tool of justice. Chain gangs, public beatings, solitary confinement, stocks, and capital punishment are common forms of official violence. In traditional societies, personal vendettas are often accepted as justice. Feuding and crimes of passion are often personally vindicated.

Fifth, a more sadistic type of violence took place in El Salvador. It was caused by the deep scars and traumas of war and brutality. This violence produced horrendous cases of torture and mutilation, which became a powerful personal reason for revenge on all sides. These acts transform a significant part of politically motivated war into communal-type violence and deepen legacies, which prolong violence (even after the strategy of deliberate protracted war ends).

The annual human rights report to the US Congress explained the historical background of Salvadoran politics thusly: Killings and terrorist acts are the work of both leftist "Democratic Front" forces who often claim responsibility for them, and of rightist elements with whom some members of the official security organizations are associated.

There is little to quarrel with because a substantial explanation for killings and tortures is the same strategy and tactics of protracted, nonconventional, diffuse, permeable war in which there are no clear battle lines and in which the line between civilian and combatant is deliberately and tactically blurred. The report also indicated that summary executions were common. Both the government and the guerrillas went into villages carrying lists of suspected opponents and executed those whose names appeared on the list.

A further feature of political violence is the "join us or die" tactic used to intimidate people and to trigger violence. Lists of victims published by the human rights organizations clearly show that such groups as labor organizations, peasant leagues, and clergy were targets of violence. These groups were often created outside the traditional structures of authority. Violence was directed at some government-sponsored farmer groups by both right and left forces. The remaining reform-oriented segments of Salvadoran government were also being attacked by the same right and left forces.

In sum, the prolonged conflict of El Salvador grew out of (1) an environment of totally fragmented political elites and counterelites fleeing the political center and creating ideological polarization; (2) a fragile economy, several economic crises (recession, the Soccer War, and the economic devastation of war and fighting), and socioeconomic conditions; (3) the self-feeding, multiplex dynamics of the extensive violence itself; and (4) the convergence of international forces and externalities which contributed to the conflict. Moreover, violence and protracted conflict also allowed many to enrich themselves financially, and they became important sources of institutional identity (a new raison d'étre) and resources (US aid) for the El Salvador military.

Low-Intensity Conflict and the War in El Salvador

In completing the analysis of the El Salvadoran struggle, we must now look at the strategic elements that have made it a protracted conflict. For 12 years both political and military strategies unique to El Salvador turned a small-scale internal conflict into a long-term protracted war. We have already analyzed the political elements of weak reform governments pitted against a powerful oligarchy and a radical left guerrilla movement that have drawn out the war and still today prevent any sort of real compromise and peace in El Salvador. Observers of the 12-year strife have also identified specific military tactics to help explain the protracted conflict. In particular Kate Doyle and Peter S. Duklis, Jr., have written an excellent research article on this factor, and it warrants extensive review.[18] They argue that the United States's involvement in the war in El Salvador contributed, directly and indirectly, to the extended nature of the conflict.

US monetary support and advice to the armed forces in El Salvador, according to their analysis, was based on a security policy developed by the United States military during the course of the Vietnam War. This policy was devised to counter the insurgency movements that were replacing regimes previously agreeable to the United States. In light of the cold war, the United States saw many opposition governments spring from these small conflicts. The policy was revised to fit El Salvador. A new doctrine was written to address the new concept of a low-intensity conflict (LIC) in a third world country that was considered to be a major foreign policy and national security concern of the United States.

Low-intensity conflict is a "limited politico-military struggle to political, social, economic and psycho-social pressures through terrorism and insurgency." LIC is usually limited to a specific geographic area and is distinguished by its "constraints on weaponry, tactics, and level of violence."[19]

Doyle and Duklis argue that in its strategy the United States began to turn its attention from strictly military issues and pulled together the resources of such government agencies as the Central Intelligence Agency (CIA), Agency for International Development, and the United States Information Agency to reach the core of the armed conflict. LIC coordinators joined guerrilla-style counterinsurgency strategies with the formulation of civilian juntas and militias for local defense and relief programs to secure the support of the people. Rather than placing emphasis on a military victory, the strategy emphasized reform of the entire El Salvadoran system.[20]

LIC allowed the level of violence to continue while promoting humanitarian relief and political activities to convince the people that the government was on the right track. The idea, according to Doyle and Duklis, was to convince the Salvadoran people that despite all the pain and suffering, conditions were actually improving, and there was justification for the war. Also, an added benefit to the LIC plan enabled the United States to influence and contain a complicated Latin American conflict without having to send any American troops, as was deemed necessary in the Vietnam War.[21]

As a result of this low-intensity conflict strategy, the United States has pumped more than $5 billion into El Salvador. This assistance has helped to increase the Salvadoran military almost fivefold; paid for six elections; helped redistribute land to thousands of peasants; helped to create a more modern judiciary system; established a food management and distribution system for low-income people; and rebuilt bridges, roads, railroads, power grids, and other infrastructures.[22]

Among the groups formed by way of this new policy were the Agencia Nacional de Servicios Especiales de El Salvador (ANSESAL), the elite presidential service that monitored Salvadoran dissidents, and also the Organización Democratica Nacionalista (ORDEN), the rural paramilitary and intelligence network. It is unknown whether the United States actually established these groups. However, CIA operatives appear to have played an important role in structuring an intelligence-gathering system for Salvadoran security organizations.[23]

Presidential candidate Colonel (later general) Medrano revealed that ANSESAL and ORDEN evolved from the US State Department, the CIA, and the Green Berets stationed in El Salvador going back to the Kennedy administration. Most of the credit went to Green Beret Col Arthur Simmons and 10 American advisors who devised a plan to "indoctrinate" peasants. According to Krauss, by 1970 one in every 50 civilians was an *oreja* (ear) for ANSESAL. Being a member of ORDEN was a means for gaining status as a peasant. Medrano claimed that the organizations combined grew to become a 30,000-man military informant network. By the late 1970s, ORDEN had expanded to an 80,000-member political force operating in the countryside for the extreme right.[24]

As indicated earlier, in 1979 and 1980 the new junta of military moderates and civilians attempted to dismantle the US-created ORDEN and ANSESAL. However, a junta member allegedly allowed national guard intelligence chief Maj Robert d'Aubuisson to remove ANSESAL's files. These files were created with the help of the CIA and were allegedly later used to assassinate thousands and to destroy a junta that originally had the backing of the Carter administration.

D'Aubuisson was trained in intelligence in the United States. He later became the deputy director of ANSESAL. According to Krauss, General Medrano put d'Aubuisson in charge of secret assassin missions to identify targets for hit teams. He was later connected to the murder of Archbishop Romero by information in a notebook from a death squad member, air force Capt Alvaro Rafael Saravia, that contained the details of a Romero assassination plot known as Operation Pineapple. Years later Saravia testified that he drove the assassin to the church and later overheard the assassin confirm to d'Aubuisson that the mission was complete.[25]

According to Krauss, ORDEN derived its strength and ability to paralyze an entire nation from two seemingly opposing forces: the rich and the poor of El Salvador. Because both classes benefitted in some way from the activities of ORDEN, both were willing to support the brutal tactics of the national security organizations. The poor, by becoming informants or assassins for ORDEN, could achieve the status and respect that could never be obtained from working someone else's land as peasants. The oligarchies kept the status quo of their wealth and ultimate power by providing the finance for weapons and equipment. Those who did not belong to or who did not financially support terror groups were simply too fearful of these brutal tactics to rebel; they complied, notes Krauss, to survive, not for any sort of social or monetary gain.[26] If Krauss is correct, ORDEN and other fundamentally repressive policies may have actually (and in a perverse logic) helped reduce the socioeconomic cleavages which contributed to protracted war and helped set the stage for a slightly more level social and economic playing field in the early 1990s.

Doyle and Duklis argue that in addition to the clandestine security forces, the structure of the military itself served to prolong the conflict. The armed forces have been ideologically divided and unevenly dispersed throughout the country, thus making it easier for rebel forces to control territory. There has never been a clear-cut strategy to fight a war. Instead, the armed forces relied on paramilitary tactics. This was enough to restrain the peasants from supporting the guerrillas but not enough to keep the guerrillas from roaming the countryside and attacking economic and military targets.

When Ronald Reagan assumed the presidency, he sent some $35 million in military aid along with military advisors to assess the El Salvadoran Armed Forces (ESAF). The Woerner Report sketched a plan to retrain and reequip the ESAF and also noted the analyst's perception of a "complete lack of vision on the army's part."[27] The author of the report, Gen Fred F. Woerner, noted that

> El Salvador indicated somewhere in this process that they had a national strategy. In fact they did not. I read it, and it was a couple of

> pages of political platitudes but nothing substantive that would provide an operative plan in order to establish a democratic, pluralistic society. It was more a statement of grand idealistic, philosophic, ideological objectives - ideological objectives, not even political objectives.[28]

This means that, practically speaking, the externalities (American concern with security and the concomitant growth of radicalism in Central America, especially Nicaragua) became the sources of policy by default. LIC and guerrilla insurgency (i.e., armed struggle) became the leitmotif discussed earlier that ran through (and behind) the domestic political scene of El Salvador. Both of these are protracted conflict strategies.

In reassessing the situation, the United States found the El Salvadoran army unable to overcome the insurgency. The United States set a new goal; to produce a military led by skillful soldiers. This goal not only encompassed the defeat of the guerrillas, but it also revised the internal structures of the country.

> First, the Salvadoran officer corps had to be persuaded to subordinate itself to civilian authority. Second, the armed forces needed to evince a respect for human rights. . . . And third, the military needed to rationalize its own internal methods of governance so that talent was nurtured, success was rewarded, incompetents weeded out, and the officer corps in general became operationally effective.[29]

However, the armed forces proved to be a difficult institution to reform. As one of the strongest in Central America, it has been the most powerful force in the country for half a century. Its main objective over two decades, however, was not primarily military but rather to promote its own institutional interests and privileges. The armed forces used the rich and powerful oligarchy, terrorist tactics, and a tough anticommunist stance to maintain its position.[30]

The United States has spent billions of dollars in economic and military aid on El Salvador. But, as Doyle and Duklis argue, only the military knows the whereabouts of many of these funds. Much of it (over $1 billion since 1979) has probably been taken out of the country, funneled into private endeavors, spent on luxuries, or used as insurance in case the insurgents won, as they did in Nicaragua or Cuba.[31]

Minister of defense, Gen Eugenio Vides Casanova, strongly supported the counterinsurgency strategy. It was General

Casanova who recognized that El Salvador was engaged in a long-term, protracted war and that a quick and decisive victory by conventional means was not possible. Thus, the strategy was to initiate rural pacification programs and to deploy small *cazador* (hunter) battalions that would stamp or drive out guerrillas. Once the rebels were removed from an area, authorities established civil militias for local protection. Economic and social assistance followed. This change in the Salvadoran forces transformed them into mobile units that imitated the tactics of the guerrillas.[32]

According to Doyle and Duklis, another problem area that exasperated Americans was the El Salvador arm's dependence on expensive, high-tech weapons that were impractical for guerrilla-type warfare. However, even though it was a concern for US advisors, American security aid allowed the ESAF to acquire the ineffective equipment.

Another low-intensity strategy was established with the use of the intelligence network devised by the CIA. This strategy was intended to infiltrate the left with vast numbers of informants at the lowest local level. In this way, most Salvadoran citizens were under surveillance at all times, even by their own neighbors and families. Control was achieved over the population through fear and intimidation from the knowledge that one's every move and exercise of political expression was under the watchful eye of the government and carried with it the threat of death. The strategy stressed constant engagement everywhere in the country, with emphasis placed on the civilian aspect of the struggle.[33]

Programs were set up to generate support and loyalty of the Salvadoran people. The United States encouraged humanitarian aid and a reduction of the most visible human rights violations. These strategies enhanced the ESAF reputation through civic programs designed to improve the life of the average Salvadoran.[34]

New security tactics, social assistance programs, and a reorganization of the army to imitate guerrilla warfare were the US and the El Salvadoran governments' new policies, according to Doyle and Duklis. But the US aid, without direct American military involvement, created an increasing El

Salvadoran dependence on economic assistance to keep the war going rather than negotiate an end.

Other economic factors operated to continue the war. Reforms in the 1980s, nationalization attempts, promotion of industry, and the increasing military investment in agriculture and industry increased the government's influence over the economy and made it feasible to continue the war indefinitely. Moreover, the concentration of the armed forces in protecting economic resources has led most analysts to argue that the military itself profited from the war and thus developed a vested interest in seeing it continue.

According to Enrique A. Baloyra, the LIC-protracted war strategy of the government forced the FMLN to adjust their strategies, because the new counterinsurgency strategy by the military was gaining on the rebels little by little. The rebels also had to assume a strategy of indefinite low-intensity conflict. Instead of pressing their national liberation goals, the rebels felt compelled to turn to acts of terrorism and increased violence. Their new terror tactics instilled fear and compliance into the people just as the government's tactics did. However, this change in battle plan caused a loss of ideological support and cost the rebels their status as "freedom fighters." Their acts of sabotage against the government induced the armed forces to support and step up their low-intensity efforts.[35]

In sum, the strategies of low-intensity warfare practiced by both sides in the conflict greatly reinforced the protracted struggle. That's because LIC is not so much focused on the *end* of war as it is based on *managing* war. As can be seen by the Salvadoran example, this is achieved by abandoning conventional methods and then by clustering resources and agencies of the government to fight the battle at all levels of society.

Peace Initiatives 1990–92

This writer argued earlier that the dynamics of protracted war in El Salvador were shaped by the convergence of domestic socioeconomic events and trends, politically and ideologically extreme (center-fleeing politics) events inside El Salvador, global events (most importantly, the Reagan election) and the fall of Somoza and Nicaragua outside the country. At the end of the decade of the 1980s, another set of global circumstances intersected to produce a significant impact on the strategy of war in El Salvador. I am referring, of course, to Soviet President Mikhail Gorbachev's policy of *glasnost*, to the rapid decline of Soviet aid to insurgent groups, and to the sharp reduction of East European and Soviet support for allies, specifically Cuba. The fall of the Berlin Wall and the dramatic end of the cold war on 21 November 1990 by way of the Charter of Paris signed by the United States and the crumbling USSR and the leaders of Europe forced a sharp change in the strategies of both the insurgents and the government of El Salvador. This was because both relied heavily on the cold war model and thus on externalities for political, ideological, and material support.

Already in 1990 a series of intensified developments took place to resolve the 11-year-old civil war. Both sides approached these developments reluctantly because they feared giving in to any reforms that might mean a surrender of power.[36] On 4 April 1990 the FMLN guerrillas and the government of El Salvador signed an accord in Geneva which endorsed United Nations mediation in peace talks between the two warring sides. Peace talks also took place in Mexico City between the leaders of the main political parties. The negotiations centered around the reduction of the role of the armed forces, a strengthening of the judiciary, a serious crackdown on human rights abuses, and an expanded oversight of the electoral process. These were all seen as eventually leading to the terms of a cease-fire.

In March of 1991 the left agreed to participate for the first time in decades in legislative and municipal elections. With substantial international monitoring, these elections were relatively honest, and the leftist parties obtained seats in the national assembly and local government.

An accord reached in the April 1991 talks in Mexico City called for significant curbs on the power of the armed forces by placing them under presidential control, by rescinding their public security functions, and also by allowing the National Assembly to prohibit, by majority vote, any presidential use of the military in a state of emergency.[37]

Human rights issues and the electoral process were given special consideration in the negotiations. An independent supreme electoral tribunal was established to regulate the often-corrupt voting process. A permanent government human rights prosecutor and an independent "truth commission" were delegated the duties of investigating and making public the facts surrounding important human rights crimes over the last decade. The Salvadoran judiciary was also reworked to eliminate excessive political domination.[38]

Two crucial problems slowed the negotiation process. The first was the army's reluctance to surrender its power and system of payoffs and corruption that begin with every graduating military class. Leonel Gomez, an upper-class Salvadoran who promotes social reform, explained the phenomenon thusly: "To reach power in the army you have to reassure those waiting for power that you aren't ruining the system. You have to link up with the *tandas* (military classes) leaving power and cover their killings and gross corruption."[39]

The second problem was the guerrillas' request to hold on to several chunks of their territory with an "unarmed" zone between them and the army. The armed forces have been understandably suspicious of these demands. A notebook discovered on the body of Antonio Cardenal, a known insurgent killed by the army, revealed the rebel's past use of talks to further their objectives while "conserving and improving our military force." Thus, suspicion clearly remains high on both sides. The leader of the Christian Democratic party in 1991, Fidel Chavez stated simply what an end to the civil war would require: "Peace will entail reorganization and purification of the army and the political incorporation of the Farabundo Marti Front—supposedly simultaneously."[40]

On 25 September 1991 the FMLN and the government signed a broad set of accords at the United Nations headquarters in New York. These called for land reform, a civilian police force into which components of the rebel army would be incorporated, and a reduction as well as a purge of the armed forces (the emphasis being on elements in the military which are suspected of human rights violations).

On 1 January 1992 the two sides signed a permanent cease-fire. The accord was cobbled together by United Nations

Secretary General Javier Pérez de Cuellar and facilitated by the so-called Four Friends of the Salvadoran Peace Process—Spain, Colombia, Venezuela, and Mexico. These countries pledged to help implement and supervise implementation of the accord. The four are also to help secure funding for the reconstruction of El Salvador.

What seems to have precipitated the start of reconciliation between the two sides is the conclusion that a military victory by either side would not be achieved and that prolonged conflict based on a virtual stalemate would no longer be viable either. The guerrillas realized that the original Castroite revolutionary, insurgency-based model may be outmoded. The government acknowledged that it would have difficulties eliminating the military threat of the rebels and their ties with segments of the impoverished of El Salvador without unacceptably high costs in lives and with sharply declining US support in the post–cold war era.

One of the serious problems which arises out of a tenuous peace accords is how past human rights violations will be punished. A US military adviser to El Salvador, Maj Eric Buckland, revealed in January of 1990 that the head of the Salvadoran military academy, Col Guillermo Alfredo Benavides, had organized the killing of the six Jesuit priests at the Jesuit University. Other former American advisers echoed similar stories of killings and cover-ups and of the American military being manipulated in such plots.[41] Punishment for crimes on both sides but especially in the highly visible military will, as it did in Chile and Argentina, require great political will and finesse.

Conclusion

These elements of the Salvadoran conflict should be kept in the perspective of larger structural issues which provided a suitable basic environment for conflict. Returning briefly to Eckstein's 20 hypotheses about internal war enumerated at the beginning of this chapter, we can conclude the following regarding the protracted war in El Salvador:

First, in regard to the hypotheses emphasizing intellectual factors, the El Salvadoran conflict was exacerbated by what Harry Eckstein calls "the failure of regimes to perform adequately the functions of political socialization."[42] Indeed, the second hypothesis, which argues that "internal wars are due to the coexistence in a society of conflicting social myths," is clearly borne out by the ideological vacuum in the political center discussed earlier in this analysis. This also reinforces the third hypothesis, namely that "internal wars result from the existence in a society of unrealizable values or corrosive social philosophies."[43]

Second, insofar as the hypotheses emphasizing economic factors are concerned, El Salvador meets the criteria of growing poverty, severe imbalances between the production and distribution of goods, and the phenomenon of some economic improvements interspersed with short-term, crisis-oriented setbacks.

Third, the hypotheses emphasizing social structure clearly demonstrate that El Salvador is characterized by the "inadequate circulation of elites coupled with too much recruitment of members of the nonelite [in particular elements in the military] into the elite, breaking down the internal cohesion of the elite."[44]

Fourth, Eckstein's hypotheses emphasizing political factors are also of significance, in particular the estrangement of many of El Salvador's rulers from the societies they rule and "bad government (government which performs inadequately the function of goal attainment)." Moreover, given the atrocious human rights environment, the war in El Salvador was also "a response to oppressive government," reflecting another of Eckstein's indicators.

Fifth, the convergence of global affairs in a new tension between East and West injected externalities into the process which increased its intensity and also its prolongation.

Thus, the socioeconomic structural conditions of El Salvador were the underlying factors which created an environment suitable for protracted conflict. This was in turn exploited by various elites and counterelites who, for political and ideological but also for purely instrumental reasons, extended the war. Finally, the renewed cold war resulted in external support for the insurgents, and the US strategy (LIC) provided infusions of resources and tactics to the government, which allowed for a "managed war." The objectives of LIC were not to terminate the fighting but instead to "remake El Salvador." In theory this would bring an end to war by eliminating the underlying factors, which were seen as the fundamental causal factor of war.

Indeed, one can argue that LIC may not have allowed for a fast and decisive military victory over the insurgents, but on the other hand, it did ultimately make it impossible for the guerrillas to win the war. In the long analysis, it may even be demonstrated that the social and economic policies aimed at the population and the "mobilization" of people to help fight the LIC did affect El Salvador enough to legitimize genuine sociopolitical reforms in the future.

I would argue as well that the tremendous power of the armed forces during the past 12 years was itself a levelling factor. Most of the officers and soldiers in the various services of the armed forces are at best lower middle class and include substantial lower-class peasants and urban lower-class recruits. The rich don't fight in the military in countries such as El Salvador. The "trickle down" to the middle class produced by this phenomenon should be studied in more detail.

In spite of the formal end to hostilities by way of the 1992 peace agreement, the technique of grinding down and defeating one's adversary will in all probability shift (hopefully more than temporarily) to the political arena. In the classic reasoning behind protracted conflict, as Karl P. Magyar and other scholars have so cogently argued, it is not clear if 1992 marks the permanent end of El Salvador's protracted war or merely a strategic shift in tactics dictated by internal stalemate and changes in the external (global) environment. That final verdict requires the passage of more time.

Notes

1. Karl P. Magyar distinguishes between the use of the terms *war* and *conflict*. "Prolonged Wars," unpublished draft discussion piece, 1990.
2. *New York Times*, 31 December 1991, A3.
3. Harry Eckstein, "On the Etiology of Internal Wars," *History and Theory* 8, no. 3 (1965): 143–44.

4. Substantial portions of this narrative and analysis are taken from Steffen W. Schmidt, *El Salvador: America's Next Vietnam?* (Salisbury, N.C.: Documentary Publications, 1983).

5. Ibid.

6. The case resembles the political violence in Colombia between 1949 and 1965.

7. Frank Devine, *El Salvador: Embassy under Attack* (New York: Vantage Press, 1981).

8. The victory by the Sandinistas in Nicaragua nourished the Salvadoran war with a whole new element of ideological, political, as well as logistical (i.e., hardware) inputs.

9. Schmidt, 91.

10. I review this here because US media coverage was important in the decision by Washington to try and keep a relatively low profile, to avoid using troops, and to design a low-intensity conflict strategy for El Salvador. Cited in Schmidt, 59.

11. Ibid., 68.

12. Ibid., 59.

13. I chose the title *El Salvador: America's Next Vietnam?* for my 1984 book.

14. Schmidt, 102.

15. Jon Sobrino, "Death and the Hope for Life," *Catholic Worker* 11, no. 4 (September 1980).

16. V. Autores, *El Salvador: El Eslabón Mas Pequeño* (San José: EDUCA, 1980).

17. "International Terrorism: A New Mode of Warfare," *International Security Review* 6, no. 1 (1981): 93–111.

18. Kate Doyle and Peter S. Duklis, Jr., "The Long Twilight Struggle: Low-Intensity Warfare and the Salvadoran Military," *Journal of International Affairs* 43 (Winter 1990): 434.

19. Ibid., 435.

20. Ibid.

21. Clifford Krauss, *Inside Central America* (New York: Summit Books, 1991), 14.

22. Ibid., 55.

23. Doyle and Duklis, 433–34.

24. Krauss, 64, 66.

25. Ibid., 72, 75.

26. Ibid., 73.

27. Doyle and Duklis, 436.

28. Ibid., 438–39.

29. Ibid., 438.

30. Ibid., 440.

31. The numbers on funding are quite unreliable. See Doyle and Duklis, 447.

32. Ibid., 448.

33. Ibid., 449.

34. Ibid., 450.

35. Enrique A. Baloyra, "Negotiating War in El Salvador: The Politics of Endgame," *Journal of Inter-American Studies and World Affairs* 28 (Spring 1986): 123–24.

36. *New York Times*, 30 April 1991.

37. Ibid., 29 April 1991.

38. Ibid.

39. Ibid., 4 May 1991.

40. Ibid.

41. *Miami Herald*, 6 August 1991.

42. Eckstein, 143.

43. Ibid., 144.

44. Ibid., 147.

Nicaragua's Prolonged Contra War

Charles L. Stansifer

Grounded in the historically bitter Nicaraguan political rivalry, the conflict between the Nicaraguan government and its most determined opponents in the 1980s evolved into the Contra War, one of the decade's most destructive wars. The war involved, on one side, the Sandinista government and, on the other, its opponents known as "contras," who were backed by the United States. The Frente Sandinista de Liberación Nacional (Sandinista Front of National Liberation or FSLN) came into power in Nicaragua in July 1979 with overwhelming popular support as victors in a revolutionary effort against the dictatorship of Anastasio Somoza Debayle. Within a year the contras (short for contrarrevolucionarios or counterrevolutionaries) began to organize an army of resistance. The war of the contras against the Sandinistas, which began in earnest in late 1981 with the organization of the Nicaraguan Democratic Front (FDN), continued throughout the rest of the decade. By 1989 it was winding down and could be said to have ended with the electoral victory of President Violeta Barrios de Chamorro over the Sandinistas in February 1990.*

*Vast quantities of documentation of the Contra War are available in the *Report of the Congressional Committees Investigating at the Iran-Contra Affair*, 23 vols. (Washington, D.C.: Government Printing Office, 1988) and more recently in publications of the National Security Archive, a nonprofit research institute. The two microfiche publications of the National Security Archive, which successfully persuaded the United States government to declassify Contra War documentation, massively illuminate the details of the war: *Nicaragua: The Making of U.S. Policy, 1978–1990* and *The Iran-Contra Affair: The Making of a Scandal, 1983–1988* (Washington, D.C.: National Security Archive, 1991). The first contains 13,000 pages of documents and the second 20,000. Although these publications have been consulted, the present narrative rests primarily on the extensive secondary literature of the Contra War.

The Three Phases of the War

The war had three phases. The first phase, lasting from 1981 to 1984, involved a buildup of men and munitions on both sides. By the end of 1984, contra forces numbered approximately 15,000 soldiers, with supply bases and other facilities established in neighboring Honduras with the help of the United States. A smaller contra force under the leadership of former Sandinista commander Eden Pastora operated from less well-established bases along the Nicaraguan-Costa Rican border. In 1984 the Popular Sandinista Army of the Nicaraguan government consisted of 60,000 troops backed by a militia force of over 100,000. Military advice and supplies came to Nicaragua from various countries of the Soviet bloc. Cuban military officers, experienced in guerrilla warfare in Africa, were especially influential in Nicaragua's war planning. Fighting during this phase consisted primarily of contra raids into Nicaragua from Honduran bases, although occasionally contra columns operated temporarily deep in Nicaragua. Mining of Nicaraguan harbors and a few light air attacks on Nicaraguan facilities in 1984 seemed to signal an escalation of the war, but this scale of warfare could not be sustained. Invasion scares pulsed through Nicaragua for a year following the United States invasion of Grenada in October 1983, but by the end of 1984, these fears had begun to diminish. Contra raids had damaged the public's morale in Nicaragua, but in the November 1984 presidential elections, the Sandinistas still harvested nearly 67 percent of the votes.

The complementary buildup resulted in the second phase, a stalemate, during the 1985–86 period. By early 1985 the forces facing each other along the Honduran-Nicaraguan border constituted the largest concentration of military forces ever seen in the history of Central America. However, with far fewer soldiers than the Nicaraguan government and with little prospect of parity, the contras could not and did not threaten the Nicaraguan capital or any other major city. Financial support for the contras from the United States waxed and waned with the domestic political concerns of the United States Congress during this period, leaving the contras dependent on private aid projects of questionable legality and with little prospect of further buildup and dictating a war of attrition rather than full-scale assault. Dreams of a quick contra victory now necessarily yielded to reliance on a long and multifaceted campaign of lower intensity. Contra raids continued steadily, and the contras targeted civilians more frequently than before; but the Sandinista defense was adequate to deter threats of a broad-front invasion. The arrival of several Hind helicopter gunships from the Soviet Union gave the Sandinistas added confidence that they could control any substantial contra invasion. Knowing the dangers of antagonizing the United States by attacking contra base camps in Honduras, the Sandinistas remained essentially on the defensive; they were not able to eliminate bands of contras operating inside Nicaraguan territory nor to eliminate bases on foreign soil. Good offices extended by various nations and international agencies during this period did not succeed in bringing the opposing sides to the bargaining table.

Continuing casualties and economic exhaustion on both sides, plus a realization that neither side could eliminate the other, led to fewer military confrontations and a greater emphasis on diplomacy in the 1986–90 period. The contras had to consider negotiations because they recognized that without substantially increased support from the United States, which was clearly not a prospect, they had no chance to increase the size of their forces. Capture of the North American soldier of fortune Eugene Hasenfus, who had been on a mission to supply the contras in southern Nicaragua in October 1986, began the unraveling of the Iran-contra scandal and completely eliminated the possibility of significant United States military assistance to the contras. Managua's leaders, on the other hand, although they now spoke more optimistically about the prospect of final defeat of the contras, faced declining public support due to continuing political uncertainty and the steadily deteriorating economy. The Sandinistas were therefore increasingly ready to compromise. Leadership of peace negotiations, which had been proceeding desultorily under the sponsorship of the Contadora nations (i.e., Panama, Mexico, Colombia, and Venezuela) now shifted to Central America with the presidential victory of Oscar Arias in Costa Rica in January 1987.[1]

Diplomacy began to show promise in August 1987, with the agreement of the five Central American republics on the

Central American Peace Plan (often called the Arias Plan), and therefore diplomacy took the spotlight from the fighting. The Arias Plan won support from people throughout Central America, including Nicaragua, and ultimately forced the Sandinistas to do what they had said they would never do: negotiate directly with the contras. No doubt the award of the Nobel Peace Prize to President Arias in 1987 gave momentum to the peace movement. The end of the Reagan administration in January 1989 removed another obstacle to peace negotiations. After several cease-fires unilaterally declared by the Sandinistas, the two opposing forces finally met in the Dominican Republic in November 1988 to talk peace. When the talks were moved to the town of Sapoa, Nicaragua (near the Costa Rican border), four months later, the two sides clearly stated they intended to stop the war. On 23 March 1988 the two combatants signed a cease-fire agreement. In the succeeding months, violations of the cease-fire occurred but peace talks continued. Shortly after taking office in 1989, President Bush and the United States Congress arranged for the last significant aid package, consisting of humanitarian aid but no weapons, to be sent to the contras in a forlorn gesture to keep the forces intact. By the time the campaign for the Nicaraguan presidential election of 25 February 1990 got under way, the contras had largely dispersed and were no longer a threat to the existence of the Nicaraguan government.[2] Sporadic fighting between contra and FSLN bands still occurred, but the war was over.

The Issue of Victory or Defeat

Who won? The question is not so easily answered, for the answer, as in many wars, depends on one's perspective. Clearly, from the point of view of the Sandinistas, they had won because they had steadfastly resisted all counter-revolutionary efforts to overthrow them for 10 years. As of January 1990, as the fighting subsided and the contras drifted back to Nicaragua, the Sandinistas were battered, but they were still in power.

Contra leaders also claimed victory. At the end of the prolonged war, the contras had to accept reentry into Nicaragua with the Sandinista government still intact, yet they maintained that they had altered Nicaraguan political realities to the extent that the Sandinistas could not win a free election. The decisive victory of Violeta Chamorro in the presidential election of February 1990 over the FSLN proved them correct, although, it must be added, the Chamorro victory did not mean that contra leaders were in power.[3] Following the election many contras were enraged by Chamorro's perceived softness on the Sandinistas and demanded their elimination from government.

As for the United States, an ally of the contras, victory was also indistinct. Washington had failed to remove the Sandinistas by force, but contra pressure had undoubtedly helped to prepare the way for the Sandinistas' electoral defeat. The goal of removing the Soviet Union as a presence in Central America was accomplished by 1990, although it was debatable whether United States action had more to do with this result than the collapse of the Soviet Union as a world power. The United States, in its zeal to destroy Sandinismo, not only had helped to enfeeble the Nicaraguan economy but also had strengthened militarism in Honduras and Costa Rica. None of these were long-term goals of the United States. Ironically, even though Washington had frequently stated its support of democratization and even contributed substantial funds to Violeta Chamorro's campaign, United States authorities were not completely satisfied with her victory. The problem was that Chamorro, once in office, did not move rapidly enough to de-Sandinize the government. She decided that to rule, she had to compromise with the Sandinistas as well as with the contras. in short, neither the contras nor the United States government could claim a clear victory.

Was It Really a War?

Does the conflict between the Sandinistas and their opponents deserve to be considered a war? This also is a debatable question. No nation officially declared war, although relations between Nicaragua and Honduras, Nicaragua and Costa Rica, and Nicaragua and the United States were severely strained for a prolonged period. The United States declared an economic embargo on Nicaragua in 1985, and both the United States and Costa Rica broke diplomatic relations with Nicaragua. More importantly, spokesmen for the United States government, including President Ronald Reagan, although not calling for a declaration of war and not calling for direct United States involvement, stressed the danger to United States security and repeatedly called for the removal of the Sandinista government even after it had been formally elected to office in November 1984.[4]

To the Nicaraguan government, it was unmistakably a war—a prolonged war for survival, leading to massive mobilization, rationing scarce commodities, including food, and suppressing dissent. Casualties including civilians, numbering in the tens of thousands, intensified the feeling inside Nicaragua that the country was unquestionably at war.

It was, in essence, a civil conflict between two opposing Nicaraguan forces, but the imbalance was such that the contras, unaided, could not hope to win. Massive United States assistance to the contras raised the stakes, moving the conflict from the civil to the international level and turning it into a war, whether officially declared or not. Given the magnitude and duration of the United States commitment to eliminate the Sandinista government, it was a war. From the perspective of the United States, the most appropriate descriptive modifier is not "total," or "popular," but "surrogate"; it was a war fought by contra surrogates in defense of the national interests of the United States as perceived by Washington. The role of the United States was crucial; the war would not have been fought without the commitment of the United States.* To some observers, the war might better be called a low-intensity conflict from the

---

*Similarly, the Filibuster War in Nicaragua in the 1856–57 period would not have been fought without the involvement of the filibusterer William Walker, a United States citizen. Walker's band came to Nicaragua at the invitation of Nicaraguan liberals and ended up taking control of the country. After he took the presidency, liberals and conservatives united to drive him out of the country. Walker received unofficial support from the United States government and from many United States citizens.

point of view of the United States, since few United States troops were directly involved. Economic destabilization measures climaxed by total embargo, the propaganda campaign against the Sandinistas, clandestine support of Sandinista opponents, intimidation tactics such as the unprecedented scale of United States military maneuvers in Honduras, and diplomatic efforts to isolate Nicaragua internationally all support the categorization of low-intensity conflict from the United States perspective. Nevertheless, the crucial measure of casualties forces the observer to tip the scale toward war rather than low-intensity conflict.

Lives lost on both sides reached approximately 40,000. The Sandinistas claimed that 30,000 Nicaraguans, including civilians, died by the violence of the war. The physical and economic devastation of Nicaragua, while difficult to measure, was so great that any impartial visitor to Managua or any other city in Nicaragua in 1990 would have concluded that a war had been fought. If one embraces the figures agreed to by the World Court decision of 1986, Nicaraguan losses as a result of the war amounted to approximately $3 billion.[5] Financial losses to the United States were not so easily quantifiable due to a partial clandestinity, creative use of military assistance to Honduras, diversion of National Guard detachments from various states and regular Defense Department expenditures, and congressional appropriations for the contras, but by any financial measure, they were substantial.

Considering the length of the war, the degree of mobilization, the level of military expenditures, the number of casualties, the physical destruction, the intensity of the debate in the United States about United States indirect military involvement, the degree of international participation, and the devastation inflicted on Nicaragua, the conflict deserves to be designated a prolonged war. Writers who deal with the history of Nicaragua during the 1980s generally agree that the Contra War was indeed a prolonged war.[6] Nevertheless, it should be added that in official communiques and press releases throughout the decade, Washington policy was never to use the "w" word.

The Issue of Prolongation

Neither the Sandinistas nor the contras expected a long war. It is arguable that the Sandinistas were aware that leftist governments in Guatemala in the 1950s and in Chile in the 1970s both failed because of the inadequate attention given to military defense. The Sandinistas were also aware that the survival of the Fidel Castro revolution in the 1960s was due in part to military preparedness, and they began to form alliances and to build military strength in the expectation of a counterrevolution. It is doubtful, however, that they anticipated a 10-year-long struggle for survival.[7]

The contras must have expected major resistance in view of their initial efforts to build up an invasion force, but few contra leaders could have imagined a decade of fighting. They made two critical mistakes: they exaggerated the weaknesses of the Sandinista government, and they overestimated the commitment of the United States to victory. Deprived of even a good opportunity to invade in force, contra leaders had to settle for a war of attrition. A number of historical circumstances, particularly the larger-than-expected support of the Sandinistas by the Soviet bloc and uncertainty of continued support by the United States, brought about a prolongation of the war, which left the contras exhausted, their United States ally divided, and the Sandinistas still in power but ruling over a prostrate nation. The following pages analyze the factors responsible for the prolongation of the war.

Historical Perspectives

This author has attempted to explain the reasons for the prolongation of the Contra War, but I must first deal with Nicaragua's history. Nicaragua's political history since attaining independence in the early nineteenth century has been marked by partisan strife. The strength of the two traditional political parties, conservative and liberal, remained approximately equal well into the twentieth century, when regional and family rivalries centering on the liberal capital of

León and the conservative headquarters of Granada had become deeply entrenched. Even the establishment of Managua as a compromise capital in the middle of the nineteenth century failed to blunt the partisanship. Unfortunately, for the establishment of a stable political system and peaceful political exchange, it had become customary for the party in power not only to repress the opposition but also to make the out-of-power party pay economically in various ways for its political misfortunes.

Outside forces contributed mightily to political partisanship. Since Nicaragua had originally been a part of the Central American Confederation (1821–38), it was common for the losing political party to call for assistance from ideological compatriots beyond the national boundaries. In a period of growing nationalism, as was the nineteenth century, one can see that this behavior opened the parties to charges of treason, thereby aggravating the political debate. In 1856, when the Liberal party found itself in an unusually desperate plight and resorted to calling on United States citizen William Walker and his band of filibusterers for military assistance, the breach between liberals and conservatives widened seriously. In 1909 conservatives compounded the problems of political divisiveness when they accepted official support from the United States government to overthrow a liberal dictator. In effect both political groupings in Nicaragua could accuse the other of treasonous complicity with the colossus of the North. United States military occupation from 1912 to 1928, at a time of conservative ascendancy, further embittered Nicaraguan politics.

The United States agreed to supervise the Nicaraguan elections of 1928 and 1932 from registration to vote counting. Both elections resulted in liberal victories. By 1933 the United States troops had been withdrawn, and a Liberal party president was left in charge. But in the meantime the Nicaraguan National Guard had been trained by United States Marines, who had placed control of the guard in the hands of Anastasio Somoza Garcìa, a Liberal party member. Somoza proceeded to throw out the duly elected president in 1936 and thereafter ruled as a dictator until he was assassinated in

1956. Luis Somoza Debayle, then president of Congress, automatically succeeded his father as president, thus establishing a pro-American dynasty of political hegemony. In the popular Nicaraguan mind, the Somoza dictatorship had become identified with the United States.

From 1956 to 1979 the Somoza family controlled the destinies of Nicaraguans. When Luis died of a heart attack in 1967, his brother Anastasio Somoza Debayle ascended to the presidency and served as president or chief of the National Guard from 1967 to 1979. Each of the Somozas was successful in persuading Nicaraguan citizens that the United States unconditionally supported the Nicaraguan government. Nicaraguans felt they could do nothing to break the Somoza political stranglehold because of perceived United States backing. As the Somoza dynasty grew more opportunistic and more repressive in the 1970s, Nicaraguans directed their antagonism to both the dictatorship and its ally, the United States. The United States, preoccupied with major problems in China and the Soviet Union, gave Central America little attention and allowed itself, almost inadvertently, to become a factor in intensifying Nicaraguan partisan rivalry.

The traditional conservative opposition, which like the Somoza family, was pro-American and ineffectual during the Somoza period and lost its primary opposition role. Some conservatives were co-opted, some were content to divert their energies to cultural and entrepreneurial pursuits, and others were silenced. With minor exceptions, conservatives chose not to advocate violence. Nicaragua was approaching the point in its history where liberal-conservative arguments were meaningless and neither party seemed aware of a surging anti-Americanism. Into the breach stepped the Frente Sandinista de Liberación Nacional (FSLN) which was organized in 1961 as a Marxist guerrilla movement to overthrow the dictatorship. Small at first and on the verge of extinction at times, the FSLN grew in prestige and in numbers as it carried out successful raids and attacks on the government in the 1970s. The legacy of partisanship and violence was intact but dormant; the Sandinistas revitalized it, linked it with antiimperialism, and rode into office with it.

Sandinista Rule

Having led the insurrection against the Somoza dynasty, the FSLN was rewarded with power in July 1979, when Anastasio Somoza Debayle, his cabinet, and the highest officers of the National Guard fled the country. Of the five-person government junta of national reconstruction, two positions were held by conservatives, while Sandinistas retained the remaining three. No one in Nicaragua at this time doubted, however, that the real authority resided in the hands of the nine commandants who staffed the National Directorate of the FSLN.

The Sandinista government sought, as indicated by statements of the junta and the commandants, to establish a government of political pluralism, mixed enterprise, and non-alignment. Nevertheless, it was clear from the expressed Marxist-Leninist commitment of some of the commandants and from the early steps taken to establish relationships with Cuba and the Soviet Union, and with such other national liberation movements as that of Vietnam that fear of Sovietization among Nicaraguans—especially middle- and upper-class Nicaraguans—was palpable. Outspoken anti-Communist opposition and disillusion with Sandinista leadership appeared in March 1980, giving greater Sandinista representation than before in the junta and in the Council of State. As the Sandinistas tightened their grip on political and economic institutions and revealed the depth of their commitment to redistributive politics, opposition increased. By September 1980, when the Sandinista government announced that it would postpone elections until 1985, the gap between the Sandinistas and the anti-Sandinistas (made up of adherents of all political parties) was sufficiently wide that violent resistance became a distinct possibility.[8]

A nucleus of resistance to the Sandinistas was already in place, in the traditional Central American manner, in border states. Honduras, on the northern border of Nicaragua, gave refuge to a substantial number of ex-National Guardsmen. Costa Rica, the customary haven for middle-class Nicaraguans in exile, already provided sanctuary for many conservative anti-Sandinistas. Guatemala and the United States, especially the State of Florida, housed additional dissidents.

An additional circumstance favoring armed resistance to the Nicaraguan government was the hostility of the Miskitos and other Indian groups toward the Sandinistas. The Indians had lived along Nicaragua's Caribbean shore for centuries and along the Honduran border in relative isolation and had seldom had good relations with any government in Managua, least of all with governments with such centralizing tendencies as the Sandinista government. The Indians were among the first contra recruits.

Throughout Central American history, political exiles have clustered in border countries for the purpose of conspiring against the government in power. Historically, any personal or ideological grievance held by persons in power in one state in Central America against another may have resulted in these individuals taking action to weaken or overthrow a neighboring government. Since several Central American governments, particularly Costa Rica, Panama, and Honduras, had helped the Sandinistas to drive Somoza out, they tended to believe that they had a special responsibility for the behavior of the Sandinistas. When the Sandinista government turned more sharply toward the Soviet bloc than they expected, governments in Costa Rica, Panama, and Honduras—all anti-Communist—began to consider giving aid to the enemies of the Sandinistas. The logical place for staging a counterrevolution was Honduras, which was closest to the Miskito homelands and where most Nicaraguan exiles with military training were located. In addition to being close and available, Honduras also proved to be far more submissive than any other Central American country to United States pressure to assist opponents of the Sandinistas.

A comparison with circumstances respecting Cuba in 1959 and its anti-Castro exiles is instructive. Anti-Castro Cuban refugees in Florida who planned armed resistance were forced to attack by commando-like raids across water. A large-scale invasion was impossible without large-scale United States logistical support and training, and as the Bay of Pigs fiasco shows, even that was inadequate. Once on Cuban soil, there was little possibility of returning to a safe refuge. Anti-

Sandinista refugees, however, could organize and train military resistance with relative impunity in border states, and support from the United States, which they also counted on, could be carried on in friendly countries without exposure to attack. Both the Castro and Sandinista governments enjoyed widespread popular support in the beginning and thus were not likely to be easily toppled by raids or guerrilla warfare. Circumstances favored Castro in Cuba, resulting in a confrontation and a Cuban victory. Circumstances in Central America favored a buildup of forces, clandestine penetration, and guerrilla raids, resulting in a prolonged war.

Role of Ideology and Religion

The intensity of the Nicaraguan political conflict was not unlike previous conflicts in its history. This time, however, ideological differences played a greater role in the division than regional or family ties, and for the first time in Nicaragua's history, Marxism and anti-Marxism replaced conservatism and liberalism as the fulcrum of political debate. Sandinistas minimized but did not conceal their ideology with that of Augusto César Sandino, who had successfully fought against United States Marines in the 1927–33 period. Sandino was neither an ideologue nor a Marxist. In his early years, he was a typical liberal with strong anticlerical views. He became a spiritualist, a nationalist, and an antiimperialist, whose principal motivation was hatred of United States occupation of his country. His sympathies were with the peasants and the workers.[9] Nevertheless, the most determined opponents of the Sandinistas tended to equate the new Sandinismo with Marxism, and the official Washington interpretation supported this contention.

Sandinista leaders of the 1970s and 1980s, who were inspired by Sandino's stand against the United States Marines and who were guided to some extent by his guerrilla strategy, were motivated by Marxism, Leninism, and liberation theology. They cultivated Sandino's nationalism and his sympathy for the masses, but they gave greater attention than he did to the idea of solidarity with liberation causes around the world. In his later years, Sandino did show interest in international issues, but his principal objective was to remove the occupying forces. The Sandinistas planned to decrease Nicaragua's historical dependency on the United States by fostering close relationships with Cuba, Vietnam, East Europe, the Soviet Union, and, it should be added, with Western Europe and the Arab countries. Although Sandinista leadership did not call for public ownership of all enterprises in Cuba and in the Soviet bloc, the commandants did insist that private enterprise be rigidly controlled by the state and that entrepreneurs as a class be excluded from political power.

Contra ideology is not as easy to characterize. For one thing, leadership of the contras changed frequently and different leaders emphasized different programs. For another, the contras concentrated on the war and not on their ideology; they had no specific constituency to respond to and held no significant ideological congresses. Furthermore, the contra leadership, especially in the early phases of the war, consisted primarily of ex-National Guardsmen, who looked to a restoration of a Somoza-like regime and who had little interest in ideology or even in specific political programs.[10] As the FDN organized and civilians took over the role as interpreters of the contra platform of the resistance, expectations of ideological clarity were raised. Changes in leadership orchestrated by the United States and personal rivalries among contra leaders nevertheless prevented a clear, consistent exposition of the contra position. Furthermore, contra military commanders operating from Honduras, Costa Rica, or the United States were isolated from their natural political allies inside Nicaragua. Political opposition to the Sandinistas inside Nicaragua had to minimize contact with the contras lest they expose themselves to the charge of treason. This situation deprived contra leaders of domestic political input. Contras and political opponents of the Sandinistas inside Nicaragua generally agreed on opposition to communism but had little opportunity to carve out a common platform. Some political opponents of the Sandinista government inside Nicaragua, it should be added, were more fearful of a return to Somocismo under the contras than they were of a continuation of

Sandinismo. Political ideology, in short, was not a powerful contra weapon.

Religion played a special role in Sandinista ideology and in the Contra War. Students of the Nicaraguan war generally agree that the Nicaraguan Catholic Church before 1979 was weak institutionally and tended to identify with the oligarchy. Conservatives in Nicaragua, as elsewhere in Latin America, were the principal defenders of the church establishment. Thus, the doctrine of liberation theology, which attempted to turn the hierarchy's attention to the poor and which had growing influence in Latin America in the 1970s and 1980s had a special appeal to those who opposed the Somoza dictatorship. Indeed, of the three branches of the FSLN, the largest branch, the insurrecionistas or the terceristas, was the least Marxist and the most closely identified with liberation theology. It should be pointed out, though, that Christian revolutionaries like Father Ernesto Cardenal, a Trappist monk, not only saw no conflict between Sandinismo and Christianity, but he maintained that being a twentieth-century Christian in Nicaragua made it necessary to be a Sandinista.[11]

To the extent that the Sandinistas sought to capture religious sentiment for their revolution, they constituted a significant threat to traditional Catholic values. The contras successfully courted the Nicaraguan Catholic hierarchy for the purpose of weakening internal support for the Sandinistas. Although their hierarchy never explicitly endorsed the cause of the contras, Catholic refusal to condemn the contras was considered by the Sandinistas as equivalent to an endorsement. Miguel Obando y Bravo, a Nicaraguan priest designated as archbishop of Nicaragua in March 1970 and elevated to the rank of cardinal in June 1985, played a vital role in all major events of Nicaraguan politics during the Sandinista insurrection and the Sandinista government. During the last years of the Somoza dictatorship, he had his own popularity by distancing himself and the church from the government, supporting human rights, and displaying evenhandedness in mediation between the Somoza government and the Sandinistas. As opposition to the Sandinistas grew after 1979, he increasingly identified with this opposition. Although he was never immune to Sandinista criticism and even calumny, his stature as an unimpeachable religious figure rose enormously when he became a cardinal. As in Poland in the same period, the Catholic church became a refuge for opponents of the Sandinista government. As the war came to a close, the Sandinistas and the contras turned to Obando as a mediator. Pope John Paul II's visit to Nicaragua in 1983 was an additional inspiration to Catholic opponents of the Sandinistas, as the Pope made it clear that he disapproved of extremist Sandinista measures.[12]

Antidictatorship was another facet of the ideological confrontation. An intense hatred of dictatorship had united nearly all Nicaraguans behind the Sandinista leadership against Somoza in 1978–79. To the extent that the Sandinistas could identify the contras as somocistas, they won popular support at home. By the same token, the contras attempted to minimize any connection with the former Somoza regime. But as the Sandinista-controlled junta and Council of State marched steadily leftward, courting friendly economic relations with the Soviet Union and Eastern Europe and friendly ideological dialogue with Vietnam and North Korea, and as the Sandinistas began to arrest dissenters, prohibit strikes, censor the press, and in general take arbitrary measures, the advantage of antidictatorial ideology shifted to the contras. In the propaganda war, it was important to be antidictatorial; the contras, despite their past connections to Somoza, shed the mantle of dictatorship, became "freedom fighters," and gained the advantage.

The Role of the Cold War

The cold war fueled the Contra War. The leftist pronouncements of the Sandinistas, their anticapitalist policies, their open harassment of their domestic enemies, their friendship with Fidel Castro, their cultivation of and ultimate dependence on the Soviet bloc, and their attempt to disguise or cloak their affiliation with the Soviet bloc by emphasizing their "nonalignment" overshadowed every moderate policy and action in the minds of contras and Washington policymakers. This was particularly true in the first two years of the Reagan administration, when Gen Alexander Haig served as secretary of state. In reality differences among the nine commandants, who ruled Nicaragua through the directorate of the FSLN and the international pressures of Nicaragua's neighbors, dictated many compromises and concessions to the opposition inside Nicaragua. These concessions slowed the revolutionary process. But Washington's belief in Nicaragua's allegiance to the Cuban model and its fear of the introduction of a substantial Soviet military presence caused it to take the position that tolerance of the Nicaraguan revolution was not possible. President Reagan and his closest advisers were determined to bring down the Sandinistas. The result was all-out support of the contras, even if the American public and the Congress were not as apprehensive about the possibility of a Soviet satellite in Central America as was the Reagan administration.

Polls taken throughout the 1980s revealed conclusively that Americans, while concerned about the Soviet presence on the continent, never supported the policy of removing the Sandinistas from power.[13] Congress' ambivalence is clearly shown in the occasional support of the contras in response to the president's repeated and urgent requests and its contradictory support of the various Boland amendments. These amendments, which were passed by large majorities, prohibited the executive branch from using funds supplied by Congress to overthrow the Sandinista government. This anomaly forced the Reagan administration, in an effort to placate Congress, to resort to clandestine and private measures on behalf of the contras and to moderate its publicly stated aims in the war. While few observers doubted that the Reagan administration sought to destroy the Sandinista government, none of the contra leaders would dare suggest to their own followers that they intended anything less than the destruction of the Sandinistas. To appease Congress, however, Reagan administration spokesmen usually referred to war aims as stopping the Nicaraguan-El Salvadoran arms flow and forcing the Sandinistas to take a more moderate position. Nevertheless, the war aim of the Reagan administration, as carried out in policy and actions, was to remove the Sandinistas from power. This was a position based more on preoccupation with Soviet-United States rivalry than on opposition to the internal dynamics of the Sandinista revolution.

Failure to bring down the Sandinistas eventually drastically modified the war aims of the contras. As the fighting diminished in 1988 and 1989 and as contra hopes for a contra government in Managua evaporated, the Sandinistas began to relax the harsh measures previously taken in response to the Arias Plan and promised to move up the presidential elections originally scheduled in November 1990 to February 1990. The Sandinistas also promised to invite international election observers, including officials from the United Nations and the Organization of American States. Contras, with no hope of replacing the Sandinistas, could congratulate themselves that they had been at least partially responsible for the change.

Conclusion

Civil conflict in Nicaragua has been recurrent and harsh. Nothing, however, in its history has been as destructive as the Contra War of 1981–90, which caused approximately 50,000 deaths and brought economic devastation to most of the country. While bitter ideological differences and political rivalries contributed to the intensity of the fighting, it seems clear that this civil conflict was particularly profound and prolonged because of the decision by the United States to support one side, that of the contras. Alone, the contras could not seriously have challenged Sandinista power. Allied to the United States, they had a chance. From 1981 to 1989 the policy of the United States, although not always expressed openly in these terms, was the use of contra forces to remove the Sandinistas from power and replace them with a government friendly to Washington. For that reason, no diplomatic initiative which fell short of dislodging the Sandinista government was received seriously in Washington during that period. The determination to extract the Sandinistas from Managua was the principal factor in the prolongation of the war. And President Ronald Reagan was the principal reason for this determination.

President Reagan was not alone in his conviction that the Sandinista ideology was communist and therefore unacceptable and that the Sandinista alliance with the Soviet Union posed an inadmissible threat to the security of the United States. But in contrast to the presidents before and after him, namely Jimmy Carter and George Bush, both of whom were willing to accept the Sandinistas in office as long as they did not convert Nicaragua into a Soviet military base, Reagan adamantly insisted on unconditional surrender and resorted to extraordinary, questionable measures to bring the Nicaraguan government down. Despite Reagan's popularity as president and his determination of this issue, he never convinced the United States' public and Congress to give Nicaragua the same priority he gave it. As a consequence, the war dragged on.

Looking at it from the opposite perspective, that of the Sandinistas, there are several determining factors in their commitment to the war, despite the political and financial cost and despite the seeming futility of resisting forces supported by a superpower. First, students of previous revolutions in Latin America knew that massive mobilization was necessary to preserve their victory, and they were willing to risk the economic distress to carry out the mobilization. Second, the nine commandants who governed Nicaragua through their control of the FSLN were experienced guerrilla fighters who had learned patience in the protracted insurrection against the Somoza regime. Third, following the Cuban example, the Sandinistas counted on financial and military support from the Soviet bloc. The Sandinistas did not expect the Soviet Union or Cuba to fight in the trenches in their behalf, but they knew that their Marxist-Leninist outlook, while not a slavish copy of the Soviet or Cuban political model, would force the Soviet bloc to render assistance. Fourth, the Sandinistas came into power in 1979 with an enormously popular following. They had, after all, toppled the hated Somoza dynasty, and their success banked respect and popularity for them, which did not dissipate until the economy was in ruins. And lastly, at least some Sandinistas saw the war and deprivation as an advantage, because it justified the radical measures they favored in the first place. The longer the war, the more radical the revolution. These factors contributed to determined resistance and a prolonged war.

Notes

1. Norma Hamilton et al., *Crisis in Central America: Regional Dynamics and U.S. Policy in the 1980s* (Boulder, Colo.: Westview Press, 1988), contains several articles pertinent to diplomatic negotiations concerning the Contra War. Various Contadora documents are contained in Bruce Bagley et al., eds., *Contadora and the Central American Peace Process: Selected Documents* (Boulder, Colo.: Westview Press, 1985). For the Arias Plan, see Charles L. Stansifer and Michael E. Conroy, eds., *Extraordinary Opportunities and New Risks: Final Report of the LASA Commission on Compliance with the Central American Peace Accord* (Pittsburgh: Latin American Studies Association, 1989); and Marc Edelman and Joanne Kenen, eds., *The Costa Rica Reader* (New York: Grove Press, 1989).

2. Stephen Kinzer, *Blood of Brothers: Life and War in Nicaragua* (New York: G. P. Putnam's Sons, 1991), 341–75.

3. For a full report on the election and its aftermath, see John Booth et al., *Electoral Democracy under International Pressure: The Report of the Latin Americana Studies Association Commission to Observe the 1990 Nicaraguan Election* (Pittsburgh: Latin American Studies Association, 1990).

4. For a summary of President Reagan's statements concerning the overthrow of the Sandinista government and the roles of Central Intelligence Agency director William Casey and others in the formulation of policy, see Roy Gutman, *Banana Diplomacy: The Making of American Policy in Nicaragua, 1981–1987* (New York: Simon and Schuster, 1988), 272–76.

5. Daniel Ortega, "Address by Nicaraguan President Daniel Ortega to the United Nations," 8 October 1987, in Daniel Sheehan and Daniel Ortega, *Assault on Nicaragua: The Untold Story of the U.S. "Secret War"* (San Francisco: Walnut Publishing Company, 1987), 81.

6. See William I. Robinson and Kent Norsworthy, *David and Goliath: The U.S. War against Nicaragua* (New York: Monthly Review Press, 1987); Holly Sklar, *Washington's War on Nicaragua* (Boston: South End Press, 1988); E. Bradford Burns, *At War in Nicaragua: The Reagan Doctrine and the Politics of Nostalgia* (New York: Harper & Row, 1987); Thomas W. Walker, ed., *Reagan versus the Sandinistas: The Undeclared War on Nicaragua* (Boulder, Colo.: Westview Press, 1987); and Stephen Kinzer, *Blood of Brothers: Life and War in Nicaragua* (New York: G. P. Putnam's Sons, 1991). In the many journalistic accounts of Nicaragua in the 1980s, sympathizers with the Sandinista government generally refer to the United States-Nicaragua confrontation as a war. One example is Salman Rushdie, who says, "When the Reagan administration began its war against Nicaragua, I recognized a deeper affinity with that small country." See Salman Rushdie, *The Jaguar Smile: A Nicaraguan Journey* (New York: Penguin Books, 1987), 12. Early in the Contra War Sandinista leaders, accustomed to military metaphors in the guerrilla campaign against Somoza and in the organization of civil matters immediately after their victory in 1979, referred to counterrevolutionary activity as warfare. For example see Daniel Ortega, "An Appeal for Justice

and Peace" (speech before the United Nations General Assembly on 7 October 1981), in *Sandinistas Speak: Speeches, Writings, and Interviews with Leaders of Nicaragua's Revolution* (New York: Pathfinder Press, 1982), 150.

7. Thomas Borge, one of the founding members of the FSLN and minister of interior in the Sandinista government, was the acknowledged head of the FSLN faction known as the Prolonged People's War (Guerra Popular Prolongada or GPP) during the 18-year campaign of the FSLN against the Somoza regime. Relying to a degree on the Chinese and Vietnamese experience, Borge emphasized the need to educate the Nicaraguan peasantry in socialism and gradually to build a powerful rural base before confronting the Somoza regime. However, Borge's faction lost favor at the time of the Sandinista victory in 1979 because the guerrilla strategy of the other two factions, the proletarians (*proletarios*) and the Insurrectionists (*insurrecionistas*), proved to be successful. The Proletarians advocated a showdown with the government in the form of an urban workers' revolt, and the insurrectionists favored capitalizing on the growing anti-Somoza sentiment of all classes on behalf of an immediate confrontation with the dictator. In view of the long counterrevolution of the 1980s and the electoral defeat of the Sandinistas in February 1990, one might expect the GPP faction to recover its ascendancy in the FSLN.

8. The best overall account of the fall of the Somoza dynasty and the first few years of the Sandinista government is John A. Booth, *The End and the Beginning: The Nicaraguan Revolution*, 2d ed. (Boulder, Colo.: Westview Press, 1985).

9. The best source on the ideology of Sandino is Donald C. Hodges, *Intellectual Foundations of the Nicaraguan Revolution* (Austin, Tex.: University of Texas Press, 1986). See also Robert E. Conrad, ed., *Sandino: The Testimony of a Nicaraguan Patriot, 1921–1934* (Princeton, N.J.: Princeton University Press, 1990).

10. *Who Are the Contras? An Analysis of the Makeup of the Military Leadership of the Rebel Forces, and of the Nature of the Private American Groups Providing them Financial and Material Support* (Washington, D.C.: Arms Control and Foreign Policy Caucus, 1985).

11. Ernesto Cardenal, *La santidad de la revolucion* (Salamanca: Ediciones Sigueme, 1976), 20. See also Alvaro Arguello, ed., *Fe cristiana y revolucion sandinista en Nifaragua* (Managua: Instituto Historico Centroamericana, 1980).

12. For a comprehensive analysis of the role of religion in the Sandinista revolution, see Michael Dodson and Laura Nuzzi O'Shaugnessy, *Nicaragua's other Revolution: Religious Faith and Political Struggle* (Chapel Hill, N.C.: University of North Carolina Press, 1990).

13. Washington Office on Latin America Staff and William M. LeoGrande, *Central America and the Polls* (Washington, D.C.: Washington Office on Latin America, 1987).

The Prolongation of the United States in Vietnam

Earl H. Tilford, Jr.

War is the most complex of human undertakings and defies definition. To some observers, war is a condition of formally and legally declared interstate violence. To others, the word *war* can be used to describe a condition of intense economic competition, such as trade war. Modifiers cloud its meaning as they allude to *limited* war and *total* war. While wars manifest themselves in many ways, they do not just happen.

How the United States and the communist Vietnamese approached their war is illustrative. For the communists, their fight with the United States and the Saigon regime was purposeful. Their objectives were constant, achievable, and better defined. Their political and military leaders, in working to achieve those objectives, devised superior strategies which, eventually, produced victory. The communists wanted to make the Americans suffer—over an extended period of time—until they gave up. Washington, on the other hand, constantly changed its ill-defined goals. Three presidential administrations avoided declaring war, and one of the biggest complaints of the generals and admirals who fought—and lost—America's war in Vietnam was that they were never allowed "to win." Whether they could have won or not is, at best, a matter of conjecture. Without clearly defined political goals, however, the generals found it impossible to devise a coherent strategy. The result was a prolonged conflict.

Indeed, the American war in Vietnam was long. It spanned a generation, from 26 September 1945, when Office of Strategic Services Capt Peter M. Dewey was shot by Viet Minh troops at a roadblock near Tan Son Nhut, until 29 April 1975, when US Marine Cpl Charles McMahon, Jr., was struck by shrapnel from a North Vietnamese rocket during the final American exit from Saigon.[1] In the years between 1945 and 1975, 58,000 Americans died in Indochina—46,000 of them in combat—and 300,000 were wounded.[2] The war tore at the fabric of

American society, directly causing the demise of the Lyndon B. Johnson administration and indirectly that of his successor, Richard M. Nixon. For a decade after the fall of South Vietnam to communist troops in 1975, the American nation seemed at times to suffer from self-doubt or, at best, a lack of self-confidence.

Origins of the Vietnam War

Historians do not agree on when the Vietnam War began. Those who examine the war from the Vietnamese perspective trace its beginning to 1885, when Emperor Ham Nghi joined the resistance against the French, which had been sputtering in the countryside since 1859.[3] For the French, their Indochina War began on 19 December 1946, when Viet Minh guerrillas attacked their installations throughout Vietnam and then melted into the jungles as a prelude to an eight-year-long protracted struggle.[4] For the United States, the official casualty count—the one reflected on the Vietnam War Memorial in Washington, D.C.—begins with the names Maj Dale Buis and MSgt Chester Ovnard. They were killed by Vietcong snipers at Bien Hoa on 8 July 1959.[5] Historians do not agree when the war began, or for that matter, when—or if—it has ended.

Another ambiguity is that what is referred to as the Vietnam War involved a number of related conflicts going on throughout Indochina simultaneously. In Laos, there was a civil war between neutralists and pro-American rightists on the one side and communist Pathet Lao and pro-Communist Neutralists, allied with North Vietnam, on the other. In southeastern Laos, the North Vietnamese occupied the border area, where they operated their extensive infiltration and logistical system known as the Ho Chi Minh Trail. In northern Laos, the North Vietnamese and Pathet Lao struggled among irregulars under the command of Gen Vang Pao, a chieftain supported by the American Central Intelligence Agency (CIA). In northwestern Laos, former Chinese nationalist army units, along with Lao and Burmese warlords, struggled for control of the hills on which opium poppies were grown and the mountains, where jungle laboratories processed opium into heroin. To the south, there was a communist-supported insurgency in northern Thailand. Another separate communist-inspired insurgency simmered in the Thai portion of the Malay Peninsula. In Cambodia, the North Vietnamese and Vietcong occupied sanctuaries along the border of South Vietnam and with the approval of the government of Norodom Sihanouk until 1970. With the approval of the same Sihanouk government, the US Air Force began to bomb those sanctuaries in March 1959. In March 1970, after Sihanouk's generals overthrew him, the Cambodian army turned on the Vietnamese communists in the border region, and before long the whole country was engulfed in war.

These wars notwithstanding, the United States, North Vietnam, and South Vietnam were the principal protagonists. Therefore, as one considers the dynamics of prolonged conflict and protracted war, one must focus on the national objectives and military strategies of the parties involved. Ultimately, the only national objectives that mattered were those of the United States and North Vietnam. Furthermore, the years 1959 to 1975 form a unified period during which the United States was militarily involved.

The United States and Prolonged Conflict: 1959–1968

The United States became involved in the Vietnam War as a part of the cold war struggle against what was perceived as an international communist movement. A cold war mindset led the Eisenhower and Kennedy administrations into the initial commitment, prevented Lyndon Johnson from reversing that commitment, and defined the way Richard Nixon extricated US troops from Vietnam. In 1959 Washington believed the real enemy was a Peking and Moscow cabal bent on world domination, not someone in the rice paddies and jungles of South Vietnam. From the beginning, the United States made its commitment to Vietnam as a part of a larger and *prolonged* ideological struggle against international communism.

Until 1965, American foreign policy tended to be declamatory and morally based. The cold war was a moral crusade for both the Eisenhower and Kennedy administrations. President John F. Kennedy set the tone of American foreign policy in his inaugural address when he vowed to "pay any price, bear any burden, meet any hardship, support any friend, oppose any foe to assure the survival and success of liberty."[6] This was an open-ended commitment to employ the resources of the United States to accomplish whatever needed to be done to support "the success of liberty." Since liberty was assumed to be morally superior to totalitarian communism, this was the equivalent of Woodrow Wilson's pledge to "make the world safe for democracy." Such commitments failed to advance specific goals pertinent to the interests of the United States.

Part of the ideological struggle was to support virtually any regime which depicted itself as being "anticommunist." From 1955 to 1963 Washington's commitment to support the Ngo Dinh Diem regime was part of that effort "to assure the success of liberty." While Diem was a nationalist, his regime—in addition to being the creation of the United States—was hardly democratic.

Another factor leading to the prolongation of the conflict was the way America introduced its forces into combat. A great deal of the initial military commitment was made covertly and in small, seemingly tolerable, if not insignificant, amounts. Even among those American units assigned to Vietnam, there was little understanding as to what their mission was supposed to be. For example, the US Air Force's 4400th Combat Crew Training Squadron, sent to Vietnam in later 1961 in Operation Farm Gate, had as its overt mission the training of pilots for the South Vietnamese Air Force. Its covert mission, however, was to fly close-air-support missions for the Army of the Republic of Vietnam (ARVN). The pilots in Operation Farm Gate had volunteered for this outfit thinking they would be flying into places like China or Cuba to insert agents or saboteurs. One of these pilots remembered that "no one bothered to tell us this wasn't an insurgency operation. It would be counterinsurgency, or maybe the Air Force didn't know the difference."[7]

After the ARVN generals deposed and murdered Ngo Dinh Diem in early November 1963, it became increasingly more difficult to depict United States policy as assuring "the survival and success of liberty." In this crucial period, while the Kennedy administration reexamined its Vietnam policy, an assassin claimed the life of the president. Lyndon B. Johnson inherited the approach of his predecessor's administration and, from 22 November 1963 until the presidential election one year later, he focused mainly on getting elected. America's Vietnam policy went on hold during a critical phase of the political and military conflict in Indochina. During that year, as South Vietnamese generals busily overthrew each others' regimes, the Vietcong grew in power in the countryside. Furthermore, Hanoi increased its aid to the Vietcong, rearming them with such Soviet-bloc weapons as the AK-47 assault rifle.

Domestic political considerations also served to prolong the conflict. Johnson did not want to cope with the Vietnam War. He wanted the Great Society and Vietnam intruded on that. But the two were related, whether he liked it or not. Johnson could not have the Great Society unless he was elected in 1964. He believed that United States' involvement in a large war in Asia might jeopardize that election. On the other hand, if the Communists succeeded, even if he were elected president in 1964, Johnson believed the right wing of the Democratic party would have all the ammunition it needed to thwart his liberal domestic agenda. In early 1964 the president and his cabinet, and by extension the Pentagon and State Department, were prisoners of election-year politics. According to the Pentagon Papers, the decision to bomb North Vietnam was made in the aftermath of the Gulf of Tonkin incident in August 1964. But until he had won the presidential election, Johnson wanted to keep his initiatives in Vietnam as low key as possible. American policy in 1964 revolved around covert actions, which were a prelude to what was to be an expanded commitment in 1965. Throughout 1964, a crucial year in which the North Vietnamese rearmed the Vietcong and increased their own commitment to the struggle, US policy was indecisive and ambiguous, because of domestic political considerations. The result was that the Communists took the initiative and never relinquished it. While Washington's half-hearted efforts did nothing but prolong the conflict, Hanoi was left free to protract the war as a part of its overall strategy.[8]

In 1965, the war went to a new level of intensified fighting, with increased involvement by both sides. Operation Rolling Thunder, the bombing of North Vietnam, got underway on 2 March. It became the longest bombing campaign ever conducted by the US Air Force, lasting some three years and five months. During Rolling Thunder, Air Force, Navy, and Marine Corps planes bombed roads and bridges, military barracks and radar installations, railyards and petroleum storage areas, as well as airfields and military headquarters. The bombing had three objectives:

1. Strategic persuasion. To persuade Hanoi to stop supporting the Vietcong and to negotiate an end to the conflict.

2. Interdiction. After July 1965, the focus switched to cutting the flow of men and supplies moving south toward battlefields inside South Vietnam.

3. Morale. To boost the morale of South Vietnamese military elites who were too busy overthrowing each others' governments to effectively prosecute the war.[9]

Three years and five months after the first strikes, Rolling Thunder failed to achieve its objective, even though the United States had flown a million sorties, had dropped nearly 800,000 tons of bombs, and had lost 900 American aircraft.[10] In addition, Rolling Thunder was counterproductive in that it prolonged the war. The objectives of Rolling Thunder were based on misperceptions concerning the sociocultural values of the North Vietnamese and some erroneous assumptions about Hanoi's war goals.

First, there was the assumption that since North Vietnam had only a handful of industries, the threatened destruction of these factories would deter the politburo from supporting the Vietcong. Industries were not targeted immediately. Second, it was assumed that the bombing would affect the will of the people of North Vietnam by depriving them of electricity and by disrupting their transportation system. However, since the bombing was restrained, and its intensity only gradually increased, the people had plenty of time to adjust to it. In fact, the bombing played into the hands of Hanoi's propaganda machine by galvanizing the will of the people behind the war effort.[11] Finally, as for Hanoi's war goals, the United States made what was perhaps its most crucial error of the war by assuming that because Washington's goals were limited, Hanoi's were as well, and that the North

Vietnamese leadership could be dissuaded with limited applications of force.

Further contributing to the prolongation of the conflict, Rolling Thunder escalated, and the target list was expanded over time, giving the North Vietnamese not only the time to get used to the bombing but also time to compensate for its destructive effects. When North Vietnam realized what little impact the bombing was having on their ability to move supplies south, they increased the flow of men and materiel to escalate the ground war in the South. From 1965 to 1968, the flow of men and supplies doubled each year.[12] Furthermore, the North Vietnamese built redundancy into their transportation system throughout the southern panhandle so that by 1968 it was capable of handling three times as much traffic as it did in 1965.[13]

Rolling Thunder ended on 31 October 1968. As British counterinsurgency expert Sir Robert Thompson put it:

> If it is accepted, as on the surface it must be, that the bombing had a minimal effect on infiltration and on the capacity of North Vietnam to wage this type of war, which were the only two advantages the Americans may have got out of it, then all the benefits have been derived by Hanoi.[14]

Rolling Thunder may have been "full of sound and fury," but it truly signified—if not "nothing"—very little. When it began in March 1965, infiltration was at a relatively low level, as was Hanoi's support for the Southern insurgency. As Hanoi saw that it could endure the bombing and increase its support for the Vietcong, they did so, escalating the war accordingly. As historian Larry E. Cable of the University of North Carolina at Wilmington has pointed out, "What Washington did was to apply a remedy to a problem that hardly existed and, in the process, created a problem it could not solve."[15]

The Ground War: The Agony of Prolonged Conflict

Inadvertently, the air war led to and then fed the ground war. First, the deployment of large numbers of combat aircraft prompted Washington to dispatch combat troops to protect airfields from Vietcong attack. Second, the presence of US warplanes on those airfields, and the troops sent to protect them, enticed the Vietcong into action against them. Increased action in and around the airfields caused the change in mission for the troops deployed to protect those installations from static defense to spoiling operations. Each side escalated in reaction to the other, ratcheting up the level of violence and expanding the very notion of "limited war" beyond recognition. In July 1965, when Gen William C. Westmoreland asked Secretary of Defense Robert S. McNamara for the deployment of 44 combat maneuver battalions, the big unit war was on.[16]

Westmoreland oversaw the massive buildup of American forces to support a three-phase attritional strategy using the mobility of the air cavalry bolstered by artillery, helicopter gunships, and the Air Force to ultimately drive the Vietcong into the kind of annihilative, large battles they could not win. Westmoreland envisioned a three-phase buildup, culminating in 1969 when, with over one-half million American troops at his disposal, the United States would find, fix, and annihilate the enemy. In phase one, from 1965 to 1966, Westmoreland planned to build and secure the base network needed to support expanded operations. In phase two, from 1966 to 1967, US troops would begin to conduct offensive sweeps through the countryside to take the initiative from the Vietcong and the People's Army of Vietnam (PAVN). In phase three, beginning in 1968 or 1969, he planned to have the kind of forces in South Vietnam which would force the enemy into battles where mobility and firepower would be decisive.[17]

What evolved was a strategy of attrition in which the Army and Marine Corps went on the tactical offensive. The strategy depended on three factors, all of which prolonged the war past a point of possible victory. First, the attrition part of the strategy depended on building a massive infrastructure to support the high-firepower tactics used in the field. It had to be manned and operated by a force so large that the logistical "tail" stretched far beyond the fighting "teeth" of the American force in Vietnam. Second, the attrition strategy was resource and time dependent. The United States had the resources to support this strategy; what it did not have was the patience to fight a war which would last—at best estimate at the time—into 1969. The third factor upon which the strategy was dependent was beyond the control of Westmoreland or anyone in Washington—the enemy. For this strategy to work, the Vietcong and the North Vietnamese had to make war on the terms defined by the Americans. They did not cooperate.

North Vietnam defeated the United States, because its leadership pursued goals which were both more clearly defined and more encompassing. The Communists had three objectives: two of them nearly "total" and one of them limited. First, they planned to use a combination of military force and political activity to destroy and supplant the government in South Vietnam. To accomplish this goal, they pursued their second "total" objective, which was to destroy the enemy army (i.e., the ARVN). They went about that through a combination of subversion and by mauling it in combat. Taken together, they eventually depleted the ARVN's morale so that it finally collapsed. The third objective was more limited. Hanoi did not have to defeat the United States militarily. It only had to compel the United States to withdraw its forces. To do that the PAVN and Vietcong inflicted casualties which, over time, frustrated the Americans to the point that they disengaged from battle.

While Westmoreland had decided to fight on the tactical offensive to attrit communist forces to the point that they would run out of troops, Gen Vo Nguyen Giap, Hanoi's defense minister and commander of its military forces, decided to fight on the tactical defensive, giving battle only when the prospect of inflicting casualties on the Americans made it worthwhile. Westmoreland believed that if American forces could inflict 10 casualties for every one they suffered, a "crossover point" would be reached in about 1969, at which time the war would become too costly for Hanoi and the National Liberation Front to pursue. General Giap did not have the luxury of American firepower. But he had people who were either committed, or could be forced to endure, the costs and to outlast—and to defeat—the Americans. Giap believed that after the United States had suffered approximately 50,000 dead, the American public would make its government change its policy, and the troops would be withdrawn.

Hanoi's strategy was to use time to prolong the agony of the war and to wear down both the American forces in the field and the will of the people back home. In this view, Douglas Pike quotes Ho Chi Minh:

> Time is the condition to be won to defeat the enemy. In military affairs time is of prime importance. Time ranks among the three factors necessary for victory, coming before terrain and support of the people. Only with time can we defeat the enemy.[18]

If time was the ally of the North Vietnamese, it was the enemy of the Americans. At one level, Westmoreland's strategy had been tied to a timetable culminating in 1969, when he believed he would have the numbers of troops and machines needed to force the enemy into battles of annihilation. His was a time-sensitive strategy in two ways. First, the logistical system became an end unto itself. As the war continued, efficiency counted for as much as, if not more than, effectiveness in determining how the Americans perceived what they were doing. With no enemy cities to be captured or borders to be crossed, the United States saw the war become a numbers game for measuring efficiency. This game ran the gamut from body counts in the field to the number of sorties flown in the air to the number of slides processed for the generals' morning briefings. Over time a myriad of means for measuring success supplanted the need to win on the battlefield. Although the managerial impulse had taken root in the American military in the 1950s, the Vietnam War exacerbated the situation and, in Vietnam, efficiency was more valued than effectiveness. Or maybe too many officials in responsible positions did not know the difference.

Second, at another level, the American public does not like long wars. Gen George C. Marshall, Army chief of staff during the Second World War stated, "A democracy cannot fight a Seven Years War."[19] Vietnam's time-oriented strategy of protracting the war worked against the American tendency to see time as linear. By 1968 time had run out on the Americans, and war weariness was setting in at home as casualties mounted on the battlefield with no end in sight.

The year 1968 was a culmination point in the war. Arguably, the United States lost the war in 1968. It lost it because after seven years of military involvement, the world's greatest power had not accomplished its tenuous, dubious, and constantly changing goals. Meanwhile, nearly 30,000 Americans had been killed in combat and over one-half million men and women were serving in Vietnam each year.[20] The war was hotly debated on college campuses, in the media, and in the halls of government. American society was in turmoil over civil rights, the role of government in fostering social change, and the nearly revolutionary changes in the relationships between the classes, races, and sexes. All this came together in 1968 with the Vietnam War as the focal point of contention.

Meanwhile, throughout 1967 the rhetoric emanating out of Saigon and Washington held that the United States was winning the war. If the military could not win the war on the battlefield, it seemed determined to win the war on the briefing charts, where numbers relating the body count, the sortie count, bomb damage assessments, and countless lists measuring every kind of statistical indicator seemed to show that the strategy of attrition was working. The phrase commonly heard around the Department of Defense was that there was "light at the end of the tunnel."

That light, some pundits have quipped, was on a communist locomotive on a collision course with United States policy. The locomotive hit with full impact in late January 1968 with the start of the Tet offensive. The Vietcong, with considerable support from the North Vietnamese, struck throughout South Vietnam. They attacked in 36 of 44 provincial capitals and a Vietcong suicide squad got inside the compound of the US Embassy in Saigon. Although American and South Vietnamese forces were surprised by the Tet offensive, their military units recovered rapidly to limit the extent of the attacks and, within 48 hours, had taken the initiative from the Communists.[21]

Tet was the turning point in the war. Despite their battlefield losses, the Communists won their most important and significant victory prior to those of 1975. In the sense of winning the war, this was a strategic rather than a tactical victory. It was a strategic victory because the Tet offensive changed the course of United States policy from Americanizing the war to Vietnamizing the war; from buildup to withdrawal.

Shifting Goals and Shifting Gears

The battles of 1968 also served to prolong the conflict. The Vietcong had been decimated to the point of tactical impotence by the 85,000 or so estimated casualties they suffered at Tet.[22] North Vietnamese troops began taking over the fighting inside South Vietnam. This should have vindicated Westmoreland's strategy because by 1969, as intended, the US Army was on the offensive in the countryside, seeking to fight the kind of battles for which it was organized and equipped: the large unit engagements where firepower can be decisive. But, again, the enemy was not obliging.

In the aftermath of Tet, the United States began its withdrawal from Vietnam. Retreat, however, proved as difficult there as it has in every conflict, and disengagement from the Vietnam War was a long and torturous process, which took four years and claimed 20,000 additional American lives.

President Lyndon Johnson's announcement on 31 March 1968 that he would not seek a second term paved the way for Richard M. Nixon's successful bid for the White House. Nixon ran for the presidency by implying he had a "secret plan" to end the war. That plan, announced after the Guam Conference in July 1969, was Vietnamization.

Vietnamization, which actually had its origins in the last days of the Johnson administration, served neither to prolong nor protract the conflict. It simply played out the dynamics of a sad situation which had evolved as a result of two decades of the ambiguity and naivete that guided American policy. The inability of the United States to define goals and the incompetence of the military commanders who could not devise a strategy applicable to the war presented to them by their civilian leaders, combined with the purposeful protraction of the war by the communist Vietnamese, had already led to America's longest war. Getting out was not going to be quick or easy. Whether Vietnamization succeeded depended on two factors.

First, South Vietnam had to accept the policy of Vietnamization. With over 500,000 US military personnel and several thousand other Americans inside South Vietnam, there was no simple way to go home. The Saigon regime controlled the harbors and the airports, and the South Vietnamese were not about to allow the United States to "cut and run." By 1969 the United States was hostage to its own failed policy in Vietnam.

Second, for Vietnamization to work the North Vietnamese and what was left of the Vietcong could not be allowed to indulge themselves in another offensive of the magnitude of Tet 1968. Such a military debacle would derail Vietnamization. In its aftermath, the Saigon government would insist that US combat troops remain to prosecute the war.

Additionally, there were both domestic and international political reasons arguing against any withdrawal which might be construed as defeat. The Republican right wing, which had been powerful within the party since the early 1950s, might turn on Richard Nixon, crippling his presidency. Overseas, Washington's credibility among its European allies and other friends would be severely shaken if the United States sold out Saigon in the face of military defeat.

For the United States, its extrication from Vietnam involved a prolonged agony. There were, quite simply, a great many Americans to be withdrawn. Force levels in South Vietnam peaked at 549,000 in April 1969. Withdrawals began in June of that year and by December the force level had dropped to 434,000.[23] As forces withdrew, the killing and dying continued. Of the 45,929 US soldiers, sailors, marines, coast guardsmen, and airmen killed in Vietnam, 29,907 died between 1968 and 1972, with the pivotal year, 1968, being the bloodiest for Americans, with 14,592 individuals being killed.[24]

After 1969, although the war was lost, Washington at last had a clearly defined goal toward which a military strategy might be devised and applied. Withdrawal of all US combat forces could, however, proceed only if the North Vietnamese army was kept at bay. Air power covered the retreat.

A week before the 1968 presidential election, on 31 October 1968, Johnson ended Rolling Thunder. Two weeks later, on 15 November 1968, Operation Commando Hunt began. Commando Hunt campaigns involved attacks along the Ho Chi Minh Trail to disrupt the flow of men and supplies moving toward the battlefields of South Vietnam. During Commando

Hunt, the air forces of the United States dropped nearly 3,000,000 tons of bombs on Laos, most of it along the 250-mile long infiltration corridor.[25] Air Force AC-130 gunships roamed up and down the Trail at night searching for trucks with infrared sensors and attacking them with computer-aimed 40-mm cannons and, on some models, 105-mm howitzers. Each day, up to 30 B-52 sorties dumped bombs into the pass areas leading from North Vietnam into Laos and from Laos into South Vietnam and Cambodia. The Air Force seeded the trail with sensitive electronic sensors to analyze the truck traffic and to target strike aircraft against truck parks and storage areas.[26]

For all its sound and fury, Commando Hunt was only the facsimile of a strategy. But it did cover the retreat, and fundamental to the process of Vietnamization was the perception that if the United States and the South Vietnamese were not "winning," they were not exactly "losing." To prevent the North Vietnamese and the Vietcong from launching another Tet-style offensive, the United States bombed Laos and Cambodia secretly in 1969 and 1970, joined the ARVN in invading Cambodia in May and June 1970, and supported an ARVN invasion of Laos in the spring of 1971, with helicopters and fighter-bomber sorties. The secret bombings of Laos and Cambodia, when revealed, combined with the perception that the Nixon administration was widening the war by invading and supporting invasions of those nations, undermined the credibility of the administration and worked against its political viability.

Dissension over the war grew, along with frustrations over the seeming duplicity of the Nixon administration. To many it seemed that to extricate itself from a war in one country, the United States had invaded and secretly bombed Cambodia while also secretly bombing Laos. While these actions were all part of "buying time" to accomplish Vietnamization and the withdrawal of American forces, the collateral damage was to the credibility of the Nixon administration. These actions, however necessary to the continued withdrawal, put the Nixon administration on somewhat of the defensive in that the president could not take concerted action to end the war quickly, except in reaction to communist initiatives.

The Year 1972: The Prolonged Agony of Withdrawal Ends

North Vietnam's strategy since 1968 had been to protract the war, while shifting gears from what had been an unconventional war, albeit one with occasional conventional aspects, to a conventional war, which relied heavily on unconventional tactics. Small unit actions and attacks on fire support bases kept the pressure on the Americans without prompting the kind of military responses that might have led to defeat for the Communists. After the ARVN's invasion of Laos in the spring of 1971—Lam Son 719—failed, Hanoi figured the United States had supported its last significant ground operation.

Vo Nguyen Giap and Le Duan, the former head of the Central Office for South Vietnam (COSVN), made the case in the politburo for a large-scale offensive in 1972. The time seemed propitious. The American ground combat presence was negligible, as withdrawals continued to draw down US forces. Also, there would be a presidential election in 1972, and American peace activists visiting Hanoi had convinced the North Vietnamese leadership that the antiwar movement was much more influential than it in fact proved to be. The politburo figured Nixon would be reluctant to risk a major military venture in an election year. Finally, as Sir Robert Thompson has suggested, the politburo was comprised of old men in a hurry. All were over 60 years of age. Most had been fighting since the 1930s, and they wanted to see their life's work completed before joining Ho Chi Minh in some eternal communist pantheon of heroes.[27]

The North Vietnamese "Nguyen Hue Offensive" began on 31 March 1972. Fourteen PAVN divisions and 26 independent regiments invaded South Vietnam. Initially, the North Vietnamese were successful in capturing most of Quang Tri Province and in pushing ARVN units back from the Cambodian border to the town of An Loc, west of Saigon. With massive American air support, the ARVN bent but did not break, and the North Vietnamese ground assault was contained.

American air power was absolutely fundamental in defeating the North Vietnamese spring offensive of 1972. In a series of redeployments, known as Operation Bullet Shot, fighter-bombers and B-52s were sent back to Southeast Asia to reinforce the 15 squadrons of fighters and the B-52s already in the theater. They were on hand to begin an air campaign which got under way in April. Throughout the first month of increased air operations, the US Air Force and Navy planes supported the hard-pressed ARVN units inside South Vietnam. The bombing, which brought Hanoi to a politically amenable position, however, began on 8 May 1972 and was called "Linebacker." Later, it was known as Linebacker I. This campaign lasted from 8 May to 23 October 1972, and it was the most successful use of air power in the Vietnam War. It was successful because the political objective and the strategic military objective came together. The goal was to compel Hanoi to sign a cease-fire agreement acceptable to the United States. The strategy to accomplish this was to isolate North Vietnam from its outside sources of supply by mining its harbors and destroying its internal transportation system while pounding PAVN units inside South Vietnam. The strategy worked because it was appropriate to the military situation. There were 14 PAVN divisions and 26 independent regiments inside South Vietnam, on the offensive, consuming up to 1,000 tons of supplies each day. With Haiphong closed to shipping by mines and with the northwest and northeast rail lines and highways destroyed at key bridges and tunnels in the rugged mountains near the Chinese border, the flow of supplies needed to sustain the southern offensive diminished. Furthermore, after some initial reversals, the ARVN fought doggedly. While they lost some territory, the ARVN contained the PAVN offensive with massive American air support. By June Hanoi's politburo knew it was not going to win the war in 1972.

The United States had won an important battle, one which contributed to ending its prolonged conflict. But containing the North Vietnamese Nguyen Hue offensive was hardly a "clean" kill. Part of "winning" was that Washington had changed vastly its political goals since first becoming involved in South Vietnam a decade earlier. From 1965 until May 1972, the American position was that it would withdraw its troops provided Hanoi remove its forces from South Vietnam. In effect, that policy changed in 1969, when Washington undertook Vietnamization and unilaterally began to withdraw its forces. All the while, Hanoi insisted that it had no troops in South Vietnam, and even if some volunteers had gone south to fight, Vietnam was all one country in any event. After 31 March 1972 Hanoi could not deny that its units were inside South Vietnam. In Moscow, in May 1972, Nixon's national security advisor, Dr Henry Kissinger, told Soviet Premier Leonid Brezhnev that the United States would accept a cease-fire in place in South Vietnam in exchange for the removal of only those PAVN forces that had entered South Vietnam since the start of the offensive.[28] The peace talks in Paris proceeded quickly through the summer of 1972, until 26 October, when Kissinger announced, "Peace is at hand." It wasn't.

Saigon balked at the peace agreement Kissinger had reached with Le Duc Tho, Hanoi's chief negotiator. This forced Hanoi and Washington to return to the negotiation table. From this point, it was to Hanoi's advantage to prolong the fighting through January 1973. Although Nixon had won a resounding victory over Democratic party candidate George McGovern, a more liberal, Democratic-controlled US Congress would be seated immediately after his inauguration. Hanoi was convinced (probably rightly so) that this Congress would curtail or terminate spending on the Vietnam War contingent only on the return of American prisoners of war. On 13 December Le Duc Tho went back to Hanoi and Kissinger returned to Washington with the peace talks stalled.

What Nixon offered to break the impasse was Linebacker II. It was an act of political urgency. He wanted the peace agreement signed before the new Congress convened. The political objective was to secure a peace agreement; the military strategy was to convince Hanoi that it was in its interest to sign sooner than later.

In the 11 days of Linebacker II, 739 B-52 sorties bombed rail yards, supply depots, airfields, and petroleum storage facilities. Some 1,200 fighter-bomber sorties struck at the North Vietnamese air defense system, cratering runways, bombing surface-to-air missile sites, and blasting the Hanoi power plant with laser-guided bombs. The North Vietnamese launched virtually every SA-2 missile in their inventory to shoot down 15 B-52s. Nine fighter-bombers, a reconnaissance jet, and a rescue helicopter were also lost. But on 26 December 1972, the air forces of the United States destroyed what was left of North Vietnam's air defense command and control capability and then, that night, put 120 B-52s over 10 targets within a 15-minute period.[29] The North Vietnamese informed Washington that they wanted to reopen negotiations. On 29 December Nixon limited the bombing to the area below the 19th parallel and the peace talks in Paris resumed. During the 11 days of Linebacker II, B-52s dropped 15,237 tons of bombs and fighter-bombers added another 5,000 tons.[30] The targets struck were not themselves fundamental to Hanoi's decision to come to terms. Tactically and operationally, Linebacker II did little more than rearrange the rubble caused by Linebacker I. Hanoi agreed to negotiate seriously to end the war quickly for three reasons. First, their air defense system was in shambles; they were out of SA-2 missiles; and B-52s could roam over North Vietnam virtually unopposed after 26 December. Second, virtually all legitimate targets had been destroyed. Left were only the dikes containing the Red River and the neighborhoods in the major cities. Only Richard Nixon knows if he would have bombed these targets, but Hanoi did not need to find out. They had already agreed to sign the cease-fire documents—it was Saigon that had balked, prompting everyone back to Paris with additional demands. Hanoi had secured a good peace agreement in October and the changes required to bring it into effect were minimal and not worth the risk of further destruction. Third, after Hanoi agreed to return to the peace negotiations, Nixon did not stop the bombing; he just curtailed it and changed its focus to the southern panhandle of North Vietnam and to South Vietnam, where the planes focused on PAVN units still battling the ARVN. Hanoi knew that if it was ever going to win the war, it had to retain a viable army. The peace talks in Paris moved ahead quickly until 27 January 1973, when the United States, the Republic of Vietnam, and the National Liberation Front signed a cease-fire agreement which did little more than conclude America's combat role in the Vietnam War.

Concluding Thoughts on America's Prolonged Conflict and Vietnam's Protracted War

The United States lost its prolonged conflict in Vietnam because until 1969, its political goals—and what passed for military strategy—were uncoordinated. When political goals were finally—and clearly—established, they revolved around withdrawal. Ultimately, it was the inability of the political leadership to define national goals and the ineptitude of military commanders in dealing with the war at hand that resulted in the prolongation of the war past the possible point of victory if there ever was one. Incompetence wore pin-striped suits as well as green, blue, and brown uniforms.

Richard Nixon and Henry Kissinger changed US goals in Vietnam to limit the damage from defeat. By focusing on Vietnamization and "peace with honor," Nixon and Kissinger defined goals which could be achieved. When the United States changed its negotiation position to agree to North Vietnam retaining over 100,000 PAVN troops inside South Vietnam, the cease-fire could be achieved. Additionally, by making an issue of the return of American prisoners of war—something normally done when hostilities are concluded— Nixon set not only another achievable goal, but also exploited an emotional issue. By comparison to the goal of "going anywhere to assure the survival and success of liberty," Nixon's goals were modest; but they were also more realistic.

Hanoi won the Vietnam War. That fact, so often overlooked by American historians and political scientists (and ignored or denied by American military leaders of the Vietnam era) intent as they were on investigating the sociopolitical, diplomatic, military, and cultural reasons for the American debacle in Vietnam, is important to students of war. The Vietnamese Communists understood war as a sociopolitical and cultural phenomenon to be prosecuted at a number of levels simultaneously. In May 1959 the Hanoi politburo set its goal as the unification of Vietnam under a single Communist system. To attain that goal, they were willing to engage in total war—using every resource at their disposal—to achieve victory. After the United States entered the conflict in 1961, Hanoi adopted an additional limited goal: convincing the United States to withdraw its military and to curtail its political support for the regime it created in Saigon. Then, as the war escalated, Hanoi pursued that goal with every military, political, and diplomatic resource at its disposal.

Clausewitz wrote in chapter two of book eight, "No one starts a war—or rather, no one in his senses ought to do so—without first being clear in his mind what he intends to achieve by that war and how he intends to conduct it. The former is its political purpose; the latter its operational objective."[31] The nation that knows what it wants to accomplish in a war can better regulate the violence to protract the war if it so desires or to conclude the war quickly if it sees fit. The nation that begins a war without a clear sense of what it intends to achieve, or how it plans to fight that war, invites the agony of a prolonged conflict.

Notes

1. See Archimedes L. A. Patti, *Why Viet Nam? Prelude to America's Albatross* (Berkeley: University of California Press, 1980), 320; Harry G. Summers, Jr., *Vietnam War Almanac* (New York: Facts on File Publications, 1985), 59. President Gerald R. Ford declared the Vietnam War "over" for the United States the day before Saigon fell on 30 April 1975. The killing and dying did not stop, even for Americans. A little over two weeks later, on 15 May, 15 Americans were killed and three listed as missing during efforts to rescue the crew of the *Mayaguez* from the Khmer Rouge. See John Guilmartin, Jr., and Michael O'Leary, *The Illustrated History of Helicopters: The Vietnam War* (Toronto: Bantam Books, 1988), 151.

2. Thomas C. Thayer, *War Without Fronts: The American Experience in Vietnam* (Boulder, Colo.: Westview Press, 1985), 107.

3. Stanley Karnow, *Vietnam: A History* (New York: Viking Press, 1983), 107–8.

4. See Bernard B. Fall, *Street Without Joy* (New York: Schoken Books, 1972), 27; Edward Dole, Samuel Lipsman, and Stephen Weiss, *Passing the Torch* (Boston: Boston Publishing Company, 1981), 26.

5. Summers, 29.

6. Theodore C. Sorensen, *Kennedy* (New York: Harper and Row Publishers, 1965), 246.

7. Lt Gen Phillip B. Davidson, *Vietnam at War: The History, 1946–1975* (Novato, Calif.: Presidio Press, 1988), 312.

8. Covert actions in 1964 included Operations Plan 34A (OPLAN 34A). These actions entailed raids by ARVN units along the coast of North Vietnam and into Laos as well as increased bombing along the Ho Chi Minh Trail by American-piloted, usually unmarked, single-engine T-28 fighter-bombers. See Cablegram from the United States mission in Saigon to the State

Department, 18 August 1964, in *The Pentagon Papers, New York Times* edition (New York: Quadrangel Books, 1971), 358–59.

9. Mark Clodfelter, *The Limits of Air Power: The American Bombing of North Vietnam* (New York: The Free Press, 1989), 61–65.

10. See Raphael Littauer and Norman Uphoff, eds., *The Air War in Indochina* (Boston: Beacon Press, 1972), 44; John Morrocco, *Rain of Fire: Air War, 1969–1973*, The Vietnam Experience (Boston: Boston Publishing Company, 1985), 9.

11. William S. Turley, *The Second Indochina War: A Short Political and Military History, 1954–1975* (Boulder, Colo.: Westview Press, 1986), 90.

12. Guenter Lewy, *America in Vietnam* (Oxford: Oxford University Press, 1978), 84.

13. *Pentagon Papers,* Sen Mike Gravel edition, vol. 4 (Boston: Beacon Press, 1971), 222.

14. Sir Robert Thompson, *No Exit from Vietnam* (New York: David McKay Co., 1969), 139.

15. This argument is well developed in Larry E. Cable, *Unholy Grail: The US and the Wars in Vietnam, 1965–68* (London: Routledge, 1991), 109–18.

16. Memorandum from the Secretary of Defense Robert S. McNamara to Lyndon Johnson, "Recommendation of Additional Deployments to Vietnam," 20 July 1965 in *Pentagon Papers,* Gravel edition, vol. 4, 619–22.

17. See William C. Westmoreland, *A Soldier Reports* (New York: Doubleday & Company, Inc., 1976), 172; Andrew F. Krepinevich, *The Army in Vietnam* (Baltimore: Johns Hopkins University Press, 1986), 164–68.

18. Douglas Pike, *PAVN: People's Army of Vietnam* (Novato, Calif.: Presidio Press, 1986), 219.

19. Quoted in Russell F. Weigley, *The American Way of War* (New York: Macmillan Publishing Co., Inc., 1973), 5.

20. Thayer, 107.

21. The Tet offensive became the great paradox of the Vietnam War. United States Army Col Harry G. Summers, Jr., raised the issue in the introduction to *On Strategy: A Critical Analysis of the Vietnam War* (Novato, Calif.: Presidio Press, 1982), 1, where he related his story of a conversation which took place between himself and a North Vietnamese colonel just as Saigon was falling in April 1975. According to Summers, he stated, "You know you never defeated us on the battlefield." The PAVN colonel answered, "That may be so, but it is also irrelevant." The Vietcong carried the brunt of the Tet attack, throwing themselves on American and ARVN bayonets to suffer devastating casualties. The United States lost 1,100, and 2,300 South Vietnamese troops died. See George C. Herring, Jr., *America's Longest War: The United States and Vietnam, 1950–1975* (New York: Alfred A. Knopf, 1979), 188.

22. Dave Richard Palmer, *Summons of the Trumpet: U.S.-Vietnam in Perspective* (San Rafael, Calif.: Presidio Press, 1977), 200.

23. Thayer, 38.

24. Ibid., 107.

25. Littauer and Uphoff, 67–72.

26. John C. Frisbee, "Igloo White," *Air Force Magazine,* June 1971, 48–53.

27. Sir Robert Thompson, *Peace is Not at Hand* (New York: David McKay, 1974), 82–84.

28. Nguyen Tien Hung and Jerrold L. Schector, *The Place File: Vietnam Secret Documents* (New York: Harper and Row, 1986), 58.

29. See Air Force Intelligence Service briefing script, "Linebacker Two," 18–29 December 1972," in *History, Air Force Intelligence Service,* fiscal year 1973, 11–12; Clodfelter, 188.

30. House Committee on Appropriations, Subcommittee on the Department of Defense, *DOD Appropriations: Bombings of North Vietnam, Hearings before the Subcommittee on the Department of Defense,* 93d Cong., 1st sess., testimony of Adm Thomas H. Moorer, chairman of the Joint Chiefs of Staff, 18 January 1973, 40–41.

31. Carl von Clausewitz, *On War,* ed. and trans. Michael Howard and Peter Paret (Princeton, N.J.: Princeton University Press, 1976), 579.

The Soviet War in Afghanistan

Stephen Blank

When Soviet forces invaded Afghanistan at the end of 1979, they hardly expected to leave there ignominiously in 1989. A prolonged war ending in defeat was the last thing they expected. Yet that is exactly what ensued and what they experienced despite their expectations. Moscow intervened to retrieve a Marxist-Leninist revolution that went off course due to the tribalism, erratic radicalism, and incompetence of local radical revolutionaries allied to it. The misplaced native radicalism of the Afghan Communists under conditions of profound socioeconomic backwardness and intense religious and tribal identification had triggered an uncontrollable rebellion against the government. In turn, the spread of the rebellion and the Afghan government's refusal to accept complete Soviet control and a less inflammatory domestic course threatened to bring Islamic religious national elites to power in another country on the Soviet border, and overturn a government that was previously allied with Moscow against the West. To avert this outcome and to replace the Amin regime with a more compliant one, the Soviet Union invaded Afghanistan.

As in all wars, the reasons for the initial invasion and for the ensuring defeat are many. While many observers have analyzed the background to the invasion and the course of the war, more operational analyses have been few. There have been few, if any, published analyses in the former USSR and an apparent inclination to use Afghanistan to settle political scores rather than for public analytical discussions until now, though this tendency may now be changing. Our efforts focus on the major causes that brought about—against all Soviet planning—a prolonged war.

Prolonged war results from a failure in strategic planning. That is, prolonged war emanates from a failure on one or both sides to read the nature of the theater correctly and thus accurately to forecast the course and outcome of the war.

Prolonged war can be conventional or unconventional in nature (e.g., World War I, Operation Barbarosa in 1941, or Vietnam from the American standpoint). When it occurs, prolonged war signifies a massive or gross failure of the military-political leadership at the strategic level of war with effects that resonate down to operational art and tactics.

This analysis highlights four linked structural causes for this strategic-level failure to understand both the nature of the theater and the nature of the peculiar military challenge posed in Afghanistan. They are (1) the corruption of the Soviet intelligence process in Afghanistan that was linked to; (2) a sclerotic, narrowly ideological, excessively secretive, decision-making process; (3) a fundamental misreading of the nature of warfare in the third world and Afghanistan's requirements in particular that was expressed in the doctrine of local war; and (4) a force structure, operational doctrine, and tactics that also were fundamentally maladapted to the theater.

It was the Soviet invasion that made a local civil war an international crisis that drew in outside forces to a degree much greater than was the case before 1979. Throughout this paper I focus on the Soviet side of the war and argue that it was exclusively in Moscow's hands to initiate what turned out to be a prolonged war in Afghanistan. In other words, Moscow alone retained the strategic initiative to decide for or against military intervention that would expand its participation in the war to the strategic level. The Mujahideen were quite incapable of organized strategic action. In the absence of a high degree of outside support for the Afghan belligerents, they could not have mounted more than a series of uncoordinated insurgent actions from a variety of centers. These certainly constitute hostilities and hostile action, but they do not make up the strategic direction needed for a truly protracted war by conscious decision. The splintered nature of the Mujahideen precluded any strategic decision for coherent protracted war other than traditional rebellions and tribal conflicts. Without Soviet and US (and the latter's allies) intervention, this would have never gone beyond the level of a long-term guerrilla war involving many rival insurgents and a traditional Afghan form of conflict and one lacking a clear strategic direction or goals. Moreover, we can demonstrate that the Soviet decision to intervene and the way in which it was carried out embodied profound structural barriers to coherent strategic planning for any local war after 1979. This means that the structural factors underlying the intelligence, decision making, and strategic processes leading to intervention were the factors that prejudged the intervention's failure and hence the war's prolongation.

The Intelligence Failure

Writing in the open about Soviet intelligence activities is, for obvious reasons, a hazardous venture. Nevertheless, both before and after the invasion of Afghanistan, there were signal failures in the way in which the KGB and GRU went about their main mission of intelligence gathering and other activities related to installing pro-Soviet regimes in the third world. Those failures corrupted the process of intelligence gathering and assessment in Afghanistan and contributed substantially to the defeat of the Soviet army.

Intelligence agencies figured prominently as instruments of Soviet third world policy during the period 1964–85.[1] These agencies went beyond intelligence gathering and traditional forms of espionage and recruitment of agents to encompass aid to revolutionary movements, terrorism, and actual involvement in coups, revolutions, and insurgencies.[2] KGB and GRU agents' involvement in such activities, frequently in defiance of or without the knowledge of the foreign ministry personnel in the host country, reflected both the tight compartmentalization of intelligence and the lack of horizontal coordination in policymaking for the third world. Only at the very top was there a means of coordinating the intelligence agencies, military, foreign ministry, International Department of the Central Committee, and Politburo and its agencies. Effective policymaking necessarily depended on vigorous oversight and discussion in the Defense Council or Politburo, and by 1979 under Leonid I. Brezhnev this was out of the question.[3]

The point transcends Brezhnev's and the other leaders' illnesses. Were they physically alert and vigorous, they still might have fallen into this trap because structural and institutional-political factors were also present in the decision to invade. In assessing the role of Andropov and the intelligence agencies, several of these aspects must be considered. First, in 1968 the KGB in Czechoslovakia manufactured evidence of an American "conspiracy" that swayed the Politburo and played on its members' ingrained convictions regarding imperialism's conspiracies. Thus, the KGB went beyond being an instrument of policy to become an advocate and maker, and even fabricator, of policy. It and Ambassador Chervonenko reported that the Czechoslovakian people eagerly awaited Soviet intervention and would support it. The fact that intervention revealed the opposite to be true did not have any repercussions and encouraged a repeat of such behavior. Some analysts also believe that the Soviets deliberately fabricated the crisis leading to the 1967 Six-Day War as a disinformation exercise targeted at the Arabs and Israel, a fact, which if true, makes Czechoslovakia a continuation of an already established trend.[4]

Two other potentially harmful effects of this pattern occurred. One, it reinforced the belief that no internal opposition could exist in countries under communist or Sovietizing leaderships. Such manifestation were only conspiracies manufactured by Washington or other Western agencies. This perspective reinforced the disposition of some observers to view conflicts abroad mainly through the prism of the superpower conflict and not on their own merits. Second, agents likely concluded that successful intervention atoned for reporting what the bosses wanted to hear rather than the truth. And successful intervention abroad carried its own reward as Moscow would undoubtedly favor those who helped expand its power and influence abroad.

The second contextual consideration of the decision to invade Afghanistan was that it took place when Andropov had clearly begun to make his move to succeed Brezhnev. Exactly how this consideration figured in the KGB's entire threat assessment and analysis process up to Andropov may never be fully known. But when combined with the third contextual factor, its importance is seen as a major factor.

The third factor is the fact that in the years since 1967, policy conflicts over intervention in the third world had been the pretext, if not the cause, of major shakeups at the top. In 1967 Shelepin and his clique went down to defeat and Andropov took over the KGB over the issue of aid to the Arabs and the fomenting of the Six-Day War.[5] In 1977 there is reason to believe that failures in African policy were used against President Podgorny even to the degree of using Castro to show up Podgorny's failures in Africa.[6] If one considers that the period after 1973 was one of almost constant detailed Soviet involvement in third world crises and that the issues involved there stimulated high-level political discord that often played out in analysts' commentaries, the importance of the succession factor in connection with a potential military intervention in the third world becomes visible. More recently the rival policy postures on third world issues of contending leaders have become much clearer.[7]

In the late-seventies controversies involving all aspects of Soviet "national security policy" spread throughout that elite. Already in 1967 these issues had fused with the struggle for power. The "wrong position" on crucial issues involving the use of Soviet forces abroad was not just a losing effort comparable to supporting the wrong region or sector in investment policy. A "mistake" here, pace Talleyrand was worse than a crime: it could cost one his political life. Since détente declined after 1975 and Soviet interventions succeeded, there was probably little political capital to be gained from counselling restraint, especially when the leadership perceived that a Western conspiracy coupled with an erratic Afghan leader were leading to a "rollback" of socialism on Moscow's border. If an elite consensus leaned towards intervention, any contender for the top job would join it. It also bears noting that at no time in 1980–82 was there any attack on the KGB for getting Moscow into this mess as there would have been if Andropov's rivals seriously felt that he had helped shape this policy debacle.

The subsequent trenchant critiques of the restrictive ideological perspectives about proletarian internationalism, vanguard parties, and states of socialist orientation that dominated professional and political discourse under

Brezhnev confirm that this cognitive map among key policymakers undoubtedly shaped the intelligence process.[8] If one conjoins the political factors and atmosphere of ubiquitous conspiracies that pervaded the intelligence agencies with those notions, ideas about local war and the external mission of Soviet forces, and the sclerotic nature of the decision-making process at the top, the entire context of the failure in strategic intelligence assessment becomes clearer.

This sclerosis at the top reflected not only age and debility and factors such as the policy of intervention in the third world and the succession, it also stemmed from the a priori belief in ideological notions that could easily be corroborated by fabricated intelligence, as in Czechoslovakia, and the narrowing of the leadership that resulted from the power struggles at the top.

We now know that the Defense Council, the main organ that decided crucial aspects of Soviet national policy, neither enjoyed a regular composition nor scheduled meetings. It met at Brezhnev's convenience and evidently with a composition (and presumably agenda) that reflected his choices. Hence, it could be rigged in terms of agenda, membership, and preceding staff work to come with a predetermined outcome. In this case, only five men, Brezhnev, Andropov, Suslov, Gromyko, and Ustinov (perhaps the already ailing Kosygin as well), made the decision. We also know that the decision to invade flew in the face of professional opinion of the general staff and experts on Afghanistan, who were either disregarded or seldom, if at all, consulted.

The relationships among the institutions these men represented (i.e., KGB, GRU, Ministries of Foreign Affairs and Defense, and the International Department of the Central Committee) were both tangled and uncoordinated. These relationships were structured to prevent the open advocacy within and among them of dissenting views and their mutual coordination. The International Department (ID), led by the veteran Stalinist, Boris Ponomar'ev under Suslov's watchful guidance, monitored CPSU relationships with foreign communist parties. Interviews with officials in the Middle East Department of the MFA have informed us that the MFA's proposals had to go through the ID, which functioned as a gatekeeper. The ID could thus easily frustrate the MFA's initiatives. In view of its much more Stalinist and anti-American ideological bent and commitment to an expansive third world policy than the MFA had, that relationship set the stage for unresolved conflict.[9]

Precedents for Afghanistan previously appeared in Soviet African policy, where Soviet policymaking tended to favor the military and ID working in tandem. Brezhnev's regime preferred to work in Africa through the MOD and viewed Africa mainly in terms of East-West conflicts. All questions of arms transfers and military strategy in Africa were exclusively reserved for the general staff and MOD. The ID directed the ideological and political relationships, often appointing veteran party hacks with no expertise as ambassadors. Generally, as in Afghanistan, these men proved to be poor analysts. Ideology, personal ambition, and politics dictated their opinions. They were intolerant of dissent or of facts at variance with whatever Moscow wanted to hear.[10] In Africa, and probably Asia, too, these alliances, and the autonomy of intelligence agents from ambassadorial scrutiny, led to such fiascos as the abortive coups in Sudan in 1971 and in Zambia in 1977–78.[11]

Whatever Andropov's and the KGB's actual role was, it appears that KGB relationships with the MFA were bad. This was evidently a general condition that was fully operative in Afghanistan. KGB cadres there believed the MFA made several strategic errors in policy, and coordination between them was poor. Certainly Andrei Gromyko was unalterably opposed to close working ties of his men with the KGB.[12]

The ID was driven by its congenital and visceral anti-Americanism and an ideologically based concept of expanding socialism abroad that made a state where communists or pro-Moscow socialists ruled a state of socialist orientation or one having a vanguard party. These terms meant that these states and ruling parties had thrown in their lot with Moscow and were building along Soviet lines towards replicating Soviet sociopolitical and economic structures. ID spokesmen clearly were inclined to label Afghanistan according to such criteria and thus created the notion of an ideological commitment, reinforced by the Soviet-Afghan Treaty of 1978 (i.e., to extend

Soviet "protection" should Afghanistan be threatened from the West or by its own "counterrevolutionaries" acting with Western help). This commitment, embodied in the treaty, also was a critical part of the Brezhnev doctrine. This doctrine did more than just limit the sovereignty of satellite states. Moscow, under its terms, obligated itself to protect its clients in the last resort or else call into question the reliability of its other commitments to satellites. Defense of the socialist bloc was indivisible, and no breaches could be allowed in the bloc. The hidden side of the Brezhnev doctrine was that it conceptualized the bloc as if it consisted of potentially falling dominos and assumed that any one state falling victim to communist (or anticommunist) insurgency immediately threatened its neighbors and all "friendly states." That doctrine fostered a disposition to view intervention as a last, but necessary and justifiable, resort because it explicitly stated that socialist revolutions were irreversible. No political leader, certainly not Brezhnev in 1979, could then turn around and expose himself to "having lost Afghanistan." Finally, in binding clients to itself Moscow had bound itself to guaranteeing their continuation in office. Moscow raised the stakes of failing to intervene against threats to its clients. The prospective threat of doing nothing and letting the situation develop was, therefore, never a real option, particularly after Moscow had started an incremental intervention (like ours in Vietnam but more telescopic in time) of its forces in Afghanistan in April 1979, Moscow was hoisted on the petard of its "commitment."

The lack of structural coordination and cohesion in policy process, the rivalries among the players and their constituencies, and the deliberate narrowing of both the institutional focus of decision making and of possible policy options contributed substantially to the ensuing intelligence and strategic debacle of a prolonged war. But, within the individual institutions; especially the military, MFA, and intelligence agencies, the specific situation in Afghanistan was still worse, making it virtually impossible for anyone either to transmit or receive (i.e., relatively) unbiased or accurate intelligence assessments of internal political situations, the proper strategic mission of any military action, and the right forces for the job, if that was necessary.

In Afghanistan itself the crucial burden of misrepresentation fell to the MFA's ambassadors and the intelligence operatives there. Studies of Soviet policies in the Muslim world show a high quality of Soviet ambassadorial appointments there. These envoys played important and useful roles in contacts among those states, Moscow, and Washington.[13] In Kabul, however, Ambassador Safronchuk, who was sent there as a demotion from his UN post, was unable to refrain from meddling in internal Afghan policies. His successor, Ambassador Puzanov, fared no better. Puzanov's plottings against the Amin government resulted in his being sent home as persona non grata, a fact which undoubtedly led him to counsel the coup against Amin.[14] The narrowness of official decision making also contributed to the failure of the MFA to argue against the invasion. Other ambassadors, who could be expected to report accurately on the impact of invasion in the states where they served (e.g., Dobrynin in Washington), were not informed about what was coming. Predictably, one of the most grievous failures of calculation was precisely this failure to gauge accurately international reaction and capacity to make invasion cost much more than had been assumed.

Just as MFA officials corrupted their reporting due to their involvement in factional infighting and became partisans for rival Afghan factions, so too did KGB and GRU men tend to assume the viewpoints of the faction to which they were accredited. Strong institutional pressures to report what Moscow wanted to hear existed. And there was also a corruption of the responsibility to report objectively that was due to their partisan involvement in Afghan politics.[15] Whatever Andropov's views really were or however they evolved over time, it is by no means clear that he or his colleagues ever received accurate intelligence assessments.

The role of the KGB and GRU has come in for controversy both between the two agencies and among defecting or dissident ex-KGB men. An account in 1989 claims that the KGB general in Kabul, General Ivanov, strongly advocated intervention, while the GRU man, General Gorelov counseled against it.[16] A dissident KGB agent, V. Iu. Korolev, argues that the KGB, virtually alone, instigated the crisis that led to war. He states:

> It has now become clear from reports in the press and on television, from reports in both Soviet and foreign publications, how great was the role of our country, particularly our intelligence service, in the organization of local wars, in Africa, Latin America, Asia, and so forth. As soon as the CPSU changed its policy in the international arena, as it has done now, all these wars stopped—almost all of them, at any rate, all those that were initiated by the Soviet Union. And the initiative came via the KGB. In other words, the KGB executed this policy.[17]

Andropov's role is harder to determine. At a meeting in August 1979, where Andropov was said to oppose intervention the chief of the GRU, General Invahutin, however, insisted on intervention. Andropov also supposedly opposed assassinations in general and of Amin in particular.[18] Yet in Asia and Africa at that time Soviet officials and agents were deeply involved in coups and the like, as well as in the assassination attempts made against East European dissidents and even perhaps Pope John Paul II.[19] Gordievsky also confirms that within a month of this August meeting (i.e., once Amin had launched a successful countercoup against the one plotted by Taraki in Moscow against him) Moscow did indeed plan to assassinate Amin and replace him with a more pliable government led by the reputed KGB agent, Babrak Karmal.[20]

Any component analysis should have told Andropov and other leaders that such action could only deepen Soviet involvement and commitment to Afghanistan's new government and weaken that regime at the very same time. Popular support for a Soviet-made government in Afghanistan would collapse, foreign reaction would be intense, and the regime itself, as any expert on Afghan history could foretell, would not survive without constant Soviet support that entailed heavy and rising costs.

This September 1979 decision meant throwing good money after bad to retrieve the original investment in Afghanistan. And this decision was evidently made in spite of reports from KGB agents advising against intervention. Former KGB agents claim that intelligence from Afghanistan said that any intervention on this scale would only lead to disaster.[21] One of these dissenting KGB men went so far as to assert that Andropov deliberately "disinformed" the Politburo, as he had done in 1968 in the Czech case, feeding it with false reports that the population would greet Soviet troops with open arms while his agents reported that war would break out in that event.[22]

From the foregoing it certainly seems that the KGB leadership failed to appraise the situation correctly and may have even deliberately misled the political leadership to bring about a desired command decision. The KGB failed to gauge Afghan reaction to an invasion, the capacity of Karmal's faction to gain legitimacy, the reaction of the Afghan army and foreign states, or the viability of the "surgical" short coup de main as a military option. These failures evidently lend support to the observations of Artem Borovik that the preinvasion operatives in Afghanistan lacked expertise in the area.[23]

Intelligence failures in other organizations and the fact that the commander in chief and decision makers were civilians do not, however, exonerate the military and the general staff. Much blame must fall upon both the MOD and the general staff. Members of the general staff claimed that it, led by Marshal Ogarkov, opposed the invasion and proposed instead a kind of garrison strategy and that dissent risked Ogarkov's position. This garrison strategy entailed a limited invasion and seizure of key urban areas and bases, a change in government and the stabilization of a situation from which the Afghan armed forces could then either overcome the opposition or the government could bring about an armistice or a negotiated settlement.[24] Supposedly, this took them off the hook for being responsible for the invasion. Since the public debate has begun, the military press has sought to glorify those soldiers who fought in Afghanistan as fulfilling their internationalist duty and stress that they were acting under civilian command and orders which they had no choice but to obey.

Although there is some truth in this line, the point is that even such a limited intervention as proposed by the general staff was based on incredible assumptions that could only have led to much the same outcome. These assumptions are (1) that Afghans would see this as a limited operation just as the general staff did, and not as an invasion or coup from abroad; (2) the new government, clearly dependent on Soviet forces, could command legitimacy in a society that is fiercely Islamic and intolerant of outside invaders; and (3) that the

Soviets could then simply pick up and leave, having accomplished their mission.

In other words, the notion that Moscow could completely control the outcome of a limited intervention abroad in the tense international situation of 1979 and run no strategic risks or unforeseeable liabilities governed this aspect of the general staff's plan. No allowance was made for the fog or friction of war, let alone resistance or intensified foreign support for the rebels. Like many other military plans there and elsewhere, this one lost sight of its entire political dimension. Inasmuch as it is generally agreed that Soviet forces only intervened abroad when the government was confident that no Western or US response sufficiently deterrent in its effect would occur; this plan, like the ultimate invasion plan, disregarded what could have been assumed to be a predictable strategic American and Western (Muslim, too) response. Failure to consider sufficiently the international and local risks of continued resistance and foreign support for it along with the possible collapse of Afghan willingness to support Moscow signify a strategic-level political failure, a failure to weigh sufficiently the likelihood of surprises, "fog," and "friction."[25]

The general staff and the MOD are as guilty of this failure as are Andropov, the KGB, and the other players in this drama. The plan for a limited intervention lost sight of the fact that for Afghanistan this signified a total war. Moscow thereby triggered an asymmetric conflict, a total war for Afghanistan and one that, in keeping with past history, would be fought by guerrilla means for which limited intervention by a conventional Soviet force was maladapted. Not only did the military misjudge the political dimension, but even the general staff plan failed to reckon with the prospect of continuing military resistance. We may phrase this point in another way because it is of profound importance for Soviet military doctrine as a whole, not just the war in Afghanistan.

Until Gorbachev, it was an axiom of Soviet military thinking that any country occupied by the Soviet armed forces must undergo a socialist revolution in its rear for the Red Army to secure the front. Or, in other words, wherever the Red Army was engaged, it had to revolutionize the country, and this was a paramount strategic objective or outcome of that war. This meant that for the invaded country, the war was a total war. There could be no question of limited objectives or of limited war. Of necessity, wherever Soviet forces were engaged, the war became total, and it duly became incumbent on the Soviet command to achieve a total victory. In practice there were no limited objectives in such campaigns. Accordingly, the effort to induce total changes as in Afghanistan by purely limited means, either the garrison strategy or the invasion strategy, was a non sequitur since the war immediately became total (i.e., a war to revolutionize Afghanistan).

Since guerrilla warfare was the only option open to the Mujahideen and the Afghan people, resistance had to become a total war for them. And since guerrilla warfare is in its essence one that trades space for time, the war preforce became a long one. Soviet sources now admit that Moscow was unprepared to fight this kind of war. In effect the effort to secure total ideological objectives by limited means helped cede the strategic initiative to the Mujahideen. Even though they could not fully exploit this gift by virtue of their own defects, that fact alone made Soviet victory unlikely, if not impossible. The Mujahideen determined the nature of the war and much of its tempo. This can be seen in the fact that Moscow never achieved a single strategic objective throughout its occupation. The inability to resolve the unlimited nature of war aims with the limited forces available made it virtually a certainty that the war would be protracted—the last thing Moscow expected.

Many factors contributed to this military-political blindness. First, the Soviets were contemptuous of the Afghans and their capability to mount a successful resistance. Second, this contempt was strengthened by the utter disdain of both Soviet military men and doctrine for guerrilla warfare, whose utility they consistently disparaged and derided. Hence, Afghan guerrillas were doubly regarded as being of little consequence. Third, Soviet experience of recent "local wars" in the third world, where the "revolutionary forces" had won, misled them into thinking that such campaigns would inevitably lead to their allies victory if Moscow backed them up suitably. And these wars also showed that Soviet conventional doctrine, strategy, operational art, and tactics were universally correct in such theaters of war; this conclusion also reinforced the disdain for guerrilla war. A fourth factor was the specific branch of military thinking dealing with local wars (i.e., those in the third world) and embracing both purely conventional and guerrilla wars. This branch of Soviet military thinking also contributed to the misperception of the Afghan theater and its requirements. All those factors blended with the institutional and strategic ones described above to bring about the catastrophic decision to invade.

In applying these factors into the decision-making process, those forces, in and out of the military, who favored an aggressive policy in the third world, allowed ideologically based notions to dominate strategic thinking.[26] Undoubtedly, there were divisions within the military and other centers about the utility of intervention abroad in general and in Afghanistan in particular.[27] But those who prevailed here allowed ideology to become a force subtractor, not a force multiplier, that masked real strategic considerations. They believed that what was needed was a brief "stability operation," an economy of forces operation with little or no risks abroad. This is what they believed doctrine had told them. Not only did it become apparent that they had been wrong, after 1985 it became policy to recognize that the doctrine itself had been wrong. And even before that, it became clear that the Soviet army could not fulfill its own doctrine and requirements in Afghanistan or elsewhere.

Military Failure

One ideological factor that consistently blinded Soviet elites and commanders was their contempt for Afghans and their belief that they could not fight a true "people's war" of national resistance out of any but mercenary motives. Afghans were viewed as violent, backward, corrupt, and ready to sell out their country to the highest bidder.[28] And the conduct of Soviet troops in Afghanistan, their ultimate resort to indiscriminate terror, testified to their belief in the essential inferiority of the Afghans. This refusal to accept their Asian opponents' humanity showed that the Soviet troops in Afghanistan were ready descendants of the imperial Russian tradition.[29] And the resort to terror certainly provided more reasons for Afghans to resist, thus prolonging the war.

Such chauvinism and professional disdain for the enemy clearly contributed to the misplaced optimism of those generals who reportedly first told Brezhnev and his colleagues that the operation would be over by the time the 1980 Summer Olympics began. When that failed, they then revised their estimate to the Sixteenth Party Congress in February 1981, only to demand 300,000 more soldiers in the fall and winter of 1980–81.[30] Another outcome of the war that certainly has its roots in this contempt for the Afghans is the struggle over learning from the war.

There is clearly a struggle over this going on in Soviet military literature. Naturally, many high officers are reluctant to admit that they lost and that therefore doctrine, strategy, operational art, and tactics were at fault. And secondly, they refuse to admit that the despised Afghans beat them. On the other hand, others insist that the armed forces must learn the lessons of this war. One finds this debate today in the comments of military men in the media (or their lack of comments there) and in private opinions they have expressed to Western observers.

Military commanders attempting to glorify the Afghan veterans and retain them in the army for purposes of enhanced professional training and public propaganda, like General Varennikov, former deputy chief of the general staff, strongly advocate disseminating the lessons of the war (of course, this begs the issue of what issues and lessons he has in mind). The aim is to prevent further occurrences of such situations.[31] Similarly, an All-Service Scientific Conference in late 1990 attended by then-Defense Minister Marshal Yazov and Chief of Staff General Moiseyev decided that military historical research must be fundamentally restructured to include recent local wars, specifically Afghanistan.[32] Lastly, some have already begun to draw operational conclusions from the war (e.g., regarding the operation and training of troops in mined terrain).[33] Writers on tactics have demanded wholesale reconsideration of tactics away from the stereotypical, parade-ground, strictly European, and exclusively offensive-minded tactical training of the Brezhnev era.[34] These calls, along with the historical debate opened up by Safonov in 1990, cited below all are campaigns for more, if not all, of the truth about operations there. The opposing forces prefer silence as indicated by former air forces commander in chief, Marshal Efimov. His 1990 "Air Force Day" article recounted every campaign where the Air Force participated, even the ignominious Finnish war, but said nothing about Afghanistan or lessons learned there as if the war never happened for the air force.[35]

This debate goes to the heart of analyzing the military failure because unwillingness to admit the shortcomings of Soviet policy processes or military doctrine, especially about war in the third world, will block prospects for military reform. Military reformers, like Lopatin, have singled out this war as highlighting the need for reform. Lopatin has written,

> The state of the USSR Armed Forces, however, does not correspond to the command of the times. And the measures which are being adopted are of a cosmetic nature and prevent the army (from) being brought to a new quality level, which was reflected in concentrated fashion in the course of the "Afghan campaign." Military reform encompassing all spheres of defense building are essential to achieve this goal.[36]

The other side of this debate was voiced by General Bazhenov, the editor of *Voennala Mysl* (Military Thought). He stated that

> Afghanistan witnessed a limited contingent of Soviet forces, and their operations demonstrated that the method of preparing for war and conduct of operations in mountains were correct. We left Afghanistan not because of defeat but because of political resolution of the Afghan question. I don't know of anywhere that our forces were driven out. Our Airborne performed particularly well. The situation is moving toward stability.[37]

Such attitudes block reconsideration of the concept of local war and the issue of war in the third world that Soviet military men have studied since 1917. An enormous effort went into such studies in the Brezhnev period and since—only to have led to a dead end. Though some, like Vorob'ev and Safonov, have some inkling or understanding of this; even they cannot do more than rationalize the failure of the campaign. Safonov rightly complains that among the major faults were the neglect of unconventional war, partisan activities of Soviet and other forces, and counterpartisan campaigns against such forces. This neglect encompassed doctrine, strategy, operational art, and tactics. Furthermore, the use of Soviet forces in such a scenario was never even considered.[38]

These are telling criticisms, but in the end they are no more than rationalizations. This is because the Soviet military, like ours in Vietnam, learned the wrong lessons already in the 1930s about small-scale war from the Spanish civil war. Whereas correct lessons that were "in the blood" of the generation of commanders that Stalin purged were either forgotten or cast aside.[39] Since 1940, despite the experience of partisan warfare in World War II and the Soviet counterinsurgency campaigns in the Baltic after 1944, unconventional wars or operations were neglected or utterly discounted. When it came to warfare in the third world, such examples as the Arab-Israeli wars—classic conventional operations from 1956 to 1982—were stressed, and little was learned about other kinds of war. For example, it appears that even in Vietnam's case, at the operational and strategic levels, the Soviets learned only ideological lessons that socialism would always win. Tactical, operational, and even many strategic issues or considerations of Vietnam's specific characteristics appear to have been discounted except for the need not to tie down large numbers of troops in such a war.[40]

The apparent digression into the issue of lessons learned and the Soviet debate was taken to emphasize that intelligence failure at the strategic level as well as a comparable ideological failure took place in military doctrine, operational art, strategy, policy, and tactics in the Afghan theater. As noted earlier, a basic error of Soviet military strategy was its attempt to envision conflict in third world theaters as strictly a smaller-scale version of the theater strategic conventional option in Europe that was its most abiding military concern. In the Afghan instance, this vision of warfare became a cropper; yet, it has strong roots in Soviet experience that date back to the civil war, when the founders of the Red Army waged an intense struggle against independent commanders of irregular formations precisely because they were independent of Moscow.[41] Subsequent counterinsurgency campaigns on Soviet soil only reinforced Soviet notions that guerrillas could not defeat regular armies. This became an article of faith for Soviet commanders and remained so even after the war in Afghanistan. A leading Soviet military figure quoted Lenin approvingly on this subject in 1989. Lenin wrote his correspondent, "Your appeals to create partisan detachments to combat the regular imperialist army are amusing to every soldier." Lt Gen V. Serebriannikov's comment was, "As you see, he had the same approach as Engles. Only a regular army can oppose another regular army."[42]

In 1979 this notion went even more unchallenged among senior military and political figures. Soviet views on guerrilla warfare and terror as practiced by their friends and clients abroad sanctioned these two types of campaigns only with misgivings. These forms of warfare were really ultimately acceptable only if they led to the growth of fully conventionally trained, equipped, and deployed armies. And wherever the Soviets could exercise influence on training and deployment in Asia and Africa, they emphasized that type of warfare rather than counterinsurgency and unconventional warfare.[43] The net result was that Soviet clients in Mozambique, Angola, and Ethiopia and the Soviet army itself were unable effectively to counter unconventional guerrilla warfare in these countries.

Indeed, as of 1979, Soviet suspicion about guerrilla warfare reached such a level that some writers denounced a positive outlook on it as Maoism—the greatest imaginable heresy then. For instance, G. Mos'ko, writing in *Military-Historical Journal*, stated,

> Mao Tse-Tung and his supporters frequently absolutized a partisan war as practically the only form of people's war. In that way the Maoists ignored such an important question for the national liberation movement as the creation of an army of a new type in order to accomplish the tasks facing it. It is appropriate here to recall V. I. Lenin's attitude toward partisan war. He said "that never can the party of the proletariat consider a partisan war to be the only or even the chief means of struggle; that this means has to be subordinated to others and has to be commensurate with the chief means of struggle, ennobled by the enlightening and organizing influence of socialism."[44]

Another result of this disparagement of and antipathy towards unconventional warfare was that Soviet writers on conflict in the third world increasingly focused on purely military-technical issues as bringing about the conditions for victory rather than on the difficult political tasks required. In this regard, they slighted even the Vietnamese experience, a war that Moscow should have appreciated in all its details. As one Western analyst of Soviet writings on the military lessons of such wars observes,

> The change by Soviet military thinkers in the evaluation of American actions in local wars from an emphasis on sociopolitical factors during the Vietnam war to an emphasis on military ones after it represented an important progression of ideas. No longer was the outcome of local war seen as being determined solely by sociopolitical factors involved in it. Indeed, such wars were won or lost depending on the degree of military skill each side exercized.[45]

As we noted above, military planners completely lost sight of the political dimensions of any potential war in Afghanistan. Inasmuch as many Soviet commanders and even the official view in the Soviet military encyclopedia voiced the view that local war no longer necessarily risked world war, if the USSR entered it, and that the USSR's military force not only deterred American intrigues in the third world but actually was the main force guaranteeing its allies' victory, there were less restraints on Soviet intervention abroad.[46] Viewed in retrospect, it is clear that the decision to commit forces abroad, when the situation warranted it and where the risks seemed manageable, took place around 1967 in conjunction with the Six-Day War and Soviet participation in the Nigerian civil war.

Since 1967 intervention abroad, supported often by powerful military-political patrons, steadily grew in both size and depth until Afghanistan. During that decade-long growth in interventionary capabilities in both airlift and sea lift, airborne forces, and air defense, the experience and force posture of foreign interventions helped to shape Soviet doctrine for conventional war in Europe. Soviet successes seemed to confirm the denigration of unconventional warfare in the third world.[47] To say this is not to ridicule all the Soviet lessons learned from local wars. In principle, they should be a most valuable learning instrument, and in conventional warfare, the wars in the Middle East have had visible impact on Soviet military developments. Indeed, for a time it appeared that the Soviet conventional warfare emphasis was actually winning, as the Ogaden war between Somalia and Ethiopia, backed by Cuban forces and Soviet commanders and equipment in 1977–78, strongly suggested.[48] Accounts of the climatic battle of Jijiga in the Ogaden war and of Ethiopia's subsequent campaigns against the Eritrean rebellion could be taken easily for Soviet operations in Europe.[49] Therefore, such experiences could easily mislead Soviet commanders into conceiving of a lighting fast coup de main for their initial operation in Kabul. That operation looked backward to Prague 1968 and incorporated an operational art and force structure that was wholly reminiscent, albeit on a smaller scale, of what a European theater-strategic operation might have looked like.[50]

Thus, Soviet notions of local war, as much as intelligence failure, led them astray in Afghanistan. Here, too, ideological conceptions overshadowed true strategic thinking. Although some have argued that Afghanistan was not typical of Soviet interventions or involvement in the third world, Soviet military experience there during the invasion and subsequent military writings on that war do typify Soviet military thinking.[51] As Eugene B. Rumer observes, most of the press coverage of the war in military journals examined tactical situations and provided "cookbook" recipes for various tactical engagements, mainly of airborne troops in Afghanistan or theaters similar to it (i.e., mountains and deserts). Articles in *Voennyl Vestnik (Military Herald)* that dealt with airborne troops in other theaters also were generally "how to" writings occurring in generic theaters and examining generic tactical missions.

> Accompanied by all the necessary ideological preambles, these writings have a great deal in common with Soviet coverage of local wars waged by "imperialist" states. It appears that the authors of these two types of articles are concerned with generic tactical experience of local wars which could be just as useful in a large conventional conflict in a continental theater and display little interest in peculiarities of local warfare.[52]

Rumer even maintains that Soviet military writers show little interest in past and present third world conflicts except insofar as they can be used to draw generic conclusions about modern theater operations.[53]

Though his conclusion may seem exaggerated, the evidence does not. Prominent Soviet military authors also now admit that tactics were stereotyped under Brezhnev and unsuited for any but standardized European-theater conditions.[54] In tactics, as in logistics, preparation of the theater, infrastructural construction of transport and communication, and so much else, this "out of area" operation failed to win the careful attention it merited before troops were committed. As Paul Kennedy notes about operations in the Middle East in World War I, "Sideshows, in other words, made their own operational demands, which armies neglected at their peril."[55]

Certainly Moscow neglected these operational demands. Though Soviet military writers did study the experience of wars in the third world, they realized that their findings virtually omitted any mention of tactics for fighting either as insurgents or counterinsurgents. I. E. Shavrov's collected analyses of such wars, published as a manual for Soviet commanders on which to reflect, focused exclusively on classic conventional forces and operations, even though he and his collective fully understood that "partisan" operations, as in Vietnam, did exercise an influence on American tactics, operations, and strategy.[56]

Shavrov and his collective also cannot be accused of ignoring the importance of such wars in world politics. Indeed Shavrov, in 1975, called them "the epicenter of the superpower conflict."[57] So too did the author's collective recognize the centrality of the time factor, observing that often as time went on, it favored the revolutionaries who restored to guerrilla tactics and (it is implied) made it impossible for the "imperialists" to fight their kind of war. Therefore, all these wars started with surprise attacks aimed to achieve a decisive victory in the initial phase.[58] Thus, while Soviet commanders understood the need to win quickly, they employed operational means and tactics that were utterly unsuited to the theater. A typical example is the many sweeps through the Panjshir valley in 1980–85, each of which followed stereotypical manuals for tactics in armored sweeps for conventional warfare, and each of which failed to secure the valley, as any Vietnam war commander could have expected. One feels they believed they simply had to show massive force and the opposition would crumble at once. They believed and acted this way in Czechoslovakia and miscalculated. And, in August 1991 in Moscow, coup plotters made the same mistake, failing to think through their political strategy and allow for popular resistance by believing they only needed to show force to intimidate potential resistance.

It should be obvious from the foregoing that Soviet tactics too were, for a long time, unsuited to Afghanistan or to the requirements of low-intensity conflict.[59] Analyses of their tactics show that tactical adaptation was slow, partial, incomplete, and ultimately a failure.[60] Soviet logistics took a long time to adjust to Afghan requirements and never fully succeeded in keeping troops supplied with adequate water, food, medical, and other supplies. Tactical intelligence was inconsistent and poorly coordinated with actual combat operations. Soviet troops were slow to pursue intelligence reports of enemy forces. Command, control, and communications were far too centralized to allow for rapid mobility. Soviet forces too often did not dismount from road-hugging APCs or infantry fighting vehicles suited for Europe but not for Afghanistan, where maintenance of equipment was a constant problem. For a long time, operations lacked surprise and were generally standard sweeps or hammer-and-anvil vertical envelopment operations. The nights belonged to the resistance.

And one can go on. After 1980 Soviet commanders began to realize they were in for a long war and restructured training and force structure gradually to bring about a more mobile force and more integration of air and land power. But these adaptive responses were too little and too late. Even though Moscow now reconciled itself to fighting a long war, it had lost both time and the strategic initiative. The multiple failures of Soviet forces in Afghanistan in this phase reveal a strategy and tactics that, in the end, basically updated classic Tsarist wars of conquest in Asia and the Transcaucasia. In those wars, Tsarist and Soviet governments resorted to holding the urban centers, devising divide-and-rule tactics—often of a form of class warfare—among the "natives," undermining the rural economy of the area over time, and, first of all, sealing the borders. Having begun those operations, the Russian forces usually projected power outwardly by building fortified encampments or towns behind which came Russian settlers and governments. All of this, including the specific Soviet policy of forced hunger due to agricultural destruction, took place in Afghanistan.[61] The fundamental difference between those wars and previous Soviet counterinsurgency campaigns is that in Afghanistan the Soviets were unable to seal off the country and could not spare or would not spare the troops to do so.

Failure to seal off the borders or politically isolate the rebels from support abroad appears to have been a decisive operational failure that nullified whatever gains and destructiveness the Soviet forces developed. This failure to seal the borders also cancelled out the gains Moscow and Kabul should have harvested from the Mujahideen's spectacular and continuing political failures. Finally, that failure to close the borders, in the international conditions of the early 1980s, would have alone sufficed to keep outside aid coming and prolonged the war beyond original Soviet intentions.

Once the forces, mission, operational art, and tactics of the invading forces were adapted to the theater and moved towards an airborne- and lighter infantry-type force, the process took several years. Unwilling to risk repeating the American debacle in Vietnam, Moscow then sought to fight a war of attrition, substituting mass and fire for mobility. In effect it accepted the resistance's kind of war, adapted itself in historically stereotypical fashion to the challenge and then failed to achieve the most elementary strategic military mission of sealing off the borders. At the same time, the Soviet military showed absolutely no interest in civic affairs, being utterly contemptuous of Afghan officialdom, and thus divorced operational concepts from political and strategic considerations. This line of conduct flouted the insight that "ultimately attrition warfare is likely to shift the focus of military effectiveness from the operational level to the strategic and political, as was the case with the Vietnam war."[62] As we now know, unconventional warfare requires attention first to answering the insurgents' political challenge, not exclusively military replies.

The failure in Afghanistan, therefore, suggests a continuing inability of Soviet military and political commanders to see warfare as specific to a theater and a unique concatenation of political and military conditions, as well as a generic form of social conflict. A recent history of the Russo-Finnish war could, in many respects, be inserted into a history of this war and few would know the difference. This is especially true regarding the Soviet disparagement of enemy resistance and reliance on incredibly primitive and stereotyped tactics.[63] There, too, Moscow believed it could "hustle" the Finns by showing force that would intimidate them and lead to a quick victory. The same belief apparently occurred in Czechoslovakia, Afghanistan, and Moscow in 1991.

In these instances the ultimate costs were ruinous. Whereas Moscow learned something from the Finnish war and the ensuing disasters of 1941–42, they learned little from the subsequent Afghan disasters of 1979–80. Indeed, at the strategic level commanders in both the civilian, intelligence, and military leadership continued to make the same mistakes. Both the Afghanistan and the Moscow coup of 1991 were products of "old thinking." In both cases the leaders of the operation failed to ponder Bismarck's advice mentioned earlier to those who would readily unsheath their swords in crisis periods. It is not clear whether Kipling, as quoted above, meant "hustle" to speed up or force the pace, or in its colloquial usage of swindling. Certainly, in Afghanistan in 1979 Moscow sought to force the pace and propel the country into socialism, even if it was ready tactically to moderate Amin's breakneck and rush efforts in this direction.

But one can argue that in 1980 that the mode of operations that Moscow chose first in 1979 then in 1980 also represented a form of hustling, in this case a swindle. In both phases of the war, Moscow sought to achieve maximum, and even unlimited, aims with only minimum force. In general, economy of forces is to be desired. But an economy of force operation can succeed only where there is "an economy of aims," for the operation and the forces at hand are optimally combined and deployed. The pursuit of unlimited aims in a theater can never be attained with limited forces, especially when their strategic and tactical guidance was as bad as was the case here. In Afghanistan as in Moscow the army sought to make a genuine counterrevolution with only a handful of its forces and miscalculated the entire "correlation of forces." In both cases the failure is ultimately a strategic one based on the no-longer sustainable contradictions of Leninism and its postulates on force and politics. In Afghanistan and the USSR these operations make clear the end of the road for socialism and Leninism.

Lenin introduced a permanent state of siege into both domestic and world politics, where it flourished. Now that Leninism is finally discredited, new thinking of many kinds has a chance to offer both societies respite from internal and external strife. In Afghanistan and the former USSR, civil and external peace are inextricably tied together. As regional conflicts, like the one in Afghanistan, draw to a close, that former "epicenter" of ideological and superpower conflict may become dormant for a long time to come. Ideological conflicts and wars are almost invariably protracted since basic values are at odds, and the social consensus of society is shattered. Hopefully, civil peace in once-conflicted areas will usher in a period of consensus that will last as long, if not longer, than the period of war.

Notes

1. Christopher Andrew and Oleg Gordievsky, *KGB: The Inside Story of its Foreign Operations from Lenin to Gorbachev* (New York: Harper Collins, 1990), 495–510, 543–63, 574–78.

2. Ibid., Stephen Larrabee, "Moscow, Angola, and the Dialectics of Détente," *The World Today*, May 1976, 177; Robert G. Patman, *The Soviet Union and the Horn of Africa: The Diplomacy of Intervention and Disengagement* (Cambridge: Cambridge University Press, 1991), passim; and Peter Vanneman and W. Martin James III, "The Role of Opinion Groups in the Soviet African Policy Process," *Journal of Contemporary African Studies* 2, no. 2 (April 1983): 224–33.

3. Vernon V. Aspaturian, "The Role of the International Department in the Soviet Foreign Policy Process," and David E. Albright, "The CPSU's International Affairs Commission and the Third World in the Gorbachev Era," *The International Department of the CC CPSU Under Dobrynin* (Washington, D.C.: Center for the Study of Foreign Affairs, 1989), 5–41, 143–45; and David E. Albright, "Vanguard Parties in the Third World," Walter Z. Laqueur, ed., *The Pattern of Soviet Conduct in the Third World* (New York: Praeger Publishers, 1983), 208–25.

4. Andrew and Gordievsky, and 483–88; and Michael I. Handel, "A 'Runaway Deception': Soviet Disinformation and the Six-Day War, 1967," in David A. Charters and Maurice A. J. Tugwell, eds., *Deception Operations: Studies in the East-West Context* (London: Brassey's [UK], 1990), 159–70.

5. Galia Golan, *Soviet Policies in the Middle East from World War II to Gorbachev* (Cambridge: Cambridge University Press, 1990), 58–67; Dina Rome Spechler, "The USSR and Third World Conflicts: Domestic Debate and Soviet Policy in the Middle East, 1967–1973," *World Politics* 38, no. 3 (April 1986): 435–61; and Ilana Diamant-Kass, *Soviet Involvement in the Middle East: Policy Debates, 1966–1973* (Boulder, Colo.: Westview Press, 1978).

6. Patman, 243–46.

7. Ibid.; Ilana Diamant-Kass and John W. Parker, *The Kremlin in Transition* (Winchester, Mass.: Unwin Hyman, 1990); David E. Albright, "The USSR and the Third World in the 1980s," *Problems of Communism* 38, nos. 2–3 (March–June 1989): 50–70; and Francis Fukuyama, *Soviet Civil-Military Relations and the Power-Projection Mission* (Santa Monica, Calif.: Rand Corporation, 1978).

8. Stephen Blank, *Afghanistan: Strategic and Operational Lessons Learned* (Carlisle Barracks, Pa.: Strategic Studies Institute, U.S. Army War College, 1991), chap. 1; idem, and "Soviet Forces in Afghanistan: Unlearning the Lessons of Vietnam," in Stephen Blank et al., *Responding to Low-Intensity Conflict Challenges* (Maxwell AFB, Ala.: Air University Press, 1990), 53–79.

9. Karen Dawisha, "Soviet Decision-Making in the Middle East: The 1973 October War and the 1980 Gulf War," *International Affairs* 57, no. 1 (Winter 1980–1981): 43–59.

10. Alexander Rahr and John Richmond, "Wind of Change in South Africa from Unexpected Quarter," Radio Free Europe/Radio Liberty, *Report on the USSR* 1, no. 19 (May 1989): 25–33.

11. Vanneman and James, 224, 233; Ilyda Dzhirkvelov, *Secret Servant: My Life in the KGB* (New York: Harper & Row, 1988).

12. Arkady N. Shevchenko, *Breaking with Moscow* (New York: Alfred A. Knopf, 1985).

13. Dawisha, 57.

14. Jiri Valenta, "Soviet Decision-Making on Afghanistan, 1979," Jiri Valenta and William C. Potter, eds., *Soviet Decision-Making for National Security* (London: George Allen & Unwin, 1984), 226; "Soviet Motives for Invasion Are Beginning to Lose Their Veil," *Insight* 5, June 1989, 30–31; Andrew L. Eiva, "The Russian Invasion of Afghanistan: The Facts Behind the Takeover," *Islamic Defense Review* 6, no. 1 (1981): 9; Lt Jacob P. Wilkins, USN, *The Soviet Invasion of Afghanistan: Trends and Precedent in Soviet Foreign Policy* (Monterey, Calif.: Naval Postgraduate School, 1984), 43; and Andrew and Gordievsky, 574–75.

15. Wilkins; Blank, *Afghanistan*, chap. 1; and Blank, "Soviet Forces," 69–72.

16. "Kak Prinimalos Reshenie," *Krasnaia Zvezda*, 18 November 1989, 3–4.

17. "Former KGB Officers on Need for Reform," *Foreign Broadcast Information Service, Soviet Union* (Hereafter *FBIS SU*), 2 July 1991, 25.

18. "KGB's Kalugin Interviewed on Career," *FBIS SU*, 29 June 1990, 58.

19. Vanneman and James, 224–33; Vladimir Solv'ev and Elena Klepikov, *Yuri Andropov: A Secret Passage Into the Kremlin* (New York: Macmillan, 1983), 180–83; and Galia Golan, "Gorbachev's New Thinking on Terrorism," *The Washington Papers*, no. 141 (New York and Washington, D.C.: Praeger Publishers, 1990), 1933.

20. Andrew and Gordivsky, 574–75.

21. Ibid.; and "Coups and Killings in Kabul: A KGB Defector Tells How Afghanistan Became Brezhnev's Vietnam," *Time*, 22 November 1982, 33–34.

22. "KGB Colonel Condems Andropov Corruption," *FBIS SU*, 14 September 1990, 58.

23. Artem Borovik, remarks, The Seapower Forum of the Center for Naval Analyses, Washington, D.C., and Alexandria, Va., 25–26 September 1990.

24. "Afghanistan-Podovdiia Itog," *Ogonek*, no. 12 (March 1989): 6.

25. Patman, 115.

26. Oliver Roy, *The Lessons of the Soviet/Afghan War*, Adelphi Papers, no. 259 (Summer 1991): 13.

27. Blank, "Soviet Forces," 53–79; and Blank, *Afghanistan*, chap. 1.

28. Boris Petkov, *Afghanistan Today: Impressions of a Journalist* (New Delhi: Sterling Publishers Private, Ltd., 1983), 11; and Aleksandr Prokhanov, "Afghanistan," *International Affairs* (Moscow), August 1988, 20.

29. Alvin Z. Rubinstien, "Speculations on a National Tragedy," *Orbis* 30, no. 4 (Winter 1985): 599.

30. Wolfgang Berner, address to the twenty-second annual convention of the American Association for the Advancement of Slavic Studies, Honolulu, Hawaii, 20 November 1988; and Karen Sawhny, "The New Afghan Military Balance," *Contemporary Review*, November 1988, 233.

31. "Varennikov at Internationalists' Meeting," *Joint Publications Research Service* (Hereafter *JPRS*), *Soviet Military Affairs*, 21 June 1991, 18.

32. Maj Gen I. N. Vorob'ev, Retired, "Pochemu Taktika Okazalas' v Zastoe?" *Voennaia Mysl'* (Military Thought), no. 1 (January 1990): 37–44; and "Defense Under Conditions of Remote Minelaying," *JPRS, Military Thought*, no. 1 (14 June 1991): 1–2.

33. "Defense Conditions," 8.

34. Vorob'ev, 37–44.

35. "AF CIC Rationalizes Offensive Characteristics of Air Force," *JPRS, Arms Control*, 16 July 1990, 1–2.

36. "Draft Concept of Military Reform Proposed," *FBIS SU*, 21 June 1990, 66.

37. Lt Gen V. Serebriannikov, "Armiia-Kakoi ei Byt," *Krasnaia Zvezda*, 12 February 1989, 2.

38. Col V. G. Safonov, "Afghanistan: Itogi i Vyvody Kak Eto Bylo," *Voenno-Istoricheskii Zhurnal*, no. 4 (May 1990): 68.

39. Jacob Kipp, *The Spanish Civil War and the Politics of Future War; The Red Army's Assessment of War Experience and the Fate of the Theory of Deep Operations* (Fort Leavenworth, Kans.: Soviet Army Studies Office, 1990).

40. William Zimmerman and Robert Axelrod, "The Lessons of Vietnam and Soviet Foreign Policy," *World Politics* 34, no. 1 (October 1981): 16.

41. Francesco Benvenuti, *The Bolshevika and the Red Army, 1918–1922* (Cambridge: Cambridge University Press, 1988), passim.

42. Serebriannikov, 2.

43. Golan, "Soviet Policies," 117–19; Golan, "Gorbachev's New Thinking," 8–11, 31.

44. Mark N. Katz, *The Third World in Soviet Military Thinking* (Baltimore, Md.: Johns Hopkins University Press, 1982), 105.

45. Ibid., 112.

46. Ibid., 65–122; and Patman, 247.

47. Lt Col David Eshel, "The Soviet Expeditionary Force, 1970: A Lesson in Organizational Flexibility," *Born in Battle*, no. 9 (1980): 42–48.

48. Patman, 232–33.
49. "Foreign Report," *The Economist*, 16 August 1978, 2.
50. Blank, *Afghanistan*, chap. 2.
51. Eugene B. Rumer, "The Third World in Soviet Professional Military Thinking," Kurt M. Campbell and S. Neil MacFarlane, eds., *Gorbachev's Third World Dilemmas* (London: Routledge, 1989), 122.
52. Ibid., 123.
53. Ibid., 126.
54. Stephen Blank, "Imagining Afghanistan: Lessons From a 'Small War'," *Journal of Soviet Military Studies* 3, no. 3 (September 1990): 468–90.
55. Paul Kennedy, "Military Effectiveness in the First World War," Allan R. Millett, Williamson Murray, eds., *Military Effectiveness: Volume I, The First World War* (Boston: Allen & Unwin, 1988), 336–37.
56. General of the Army I. E. Sharvov, *Lokal'naia Voina: Istorlia i Sovremennost'* (Moscow: Voennoe Izdatel'stvo Ministerstvo Oborony SSSR, 1981), 124–26; "Lokal'nye Voiny i ikh Mesto v Global'noi Strategii Imperializima," *Voenno-Ostoricheskii Zhurnal*, March–April 1975, *JPRS*, translation, September 1975, pt. 2, 21.
57. Ibid., pt. 1, 8–9.
58. Ibid., pt. 2, 20–21.
59. Rod Paschall, *LIC 2010: Special Operations & Unconventional Warfare in the Next Century* (Washington, D.C.: Brassey's [US], 1990), 45.
60. Blank, *Afghanistan*; idem, "Soviet Forces"; Stephen Blank, "Soviet Forces and Low-Intensity Conflict in Central Asia: Three Case Studies," Lewis B. Ware, ed., *Low-Intensity Conflict in the Third World* (Maxwell AFB, Ala.: Air University Press, 1989), 37–39; Roy, *passim*; Robert H. Scales, Jr., *Firepower in Limited War* (Washington, D.C.: National Defense University Press, 1990), 155–97; Anthony H. Cordesman and Abraham R. Wagner, *The Lessons of Modern War: Volume 3: The Afghan and Falklands Conflict* (Boulder, Colo.: Westview Press, 1990), 3–237, 362–433; David C. Isby, *War in a Distant Country: Afghanistan: Invasion and Resistance* (London: Arms and Armour Press, 1989); and Lt Col Scott R. MacMichael, US Army, "Soviet Tactical Performance and Adaptation in Afghanistan," *Journal of Soviet Military Studies* 3, no. 1 (March 1990): 102–7.
61. Roy, 21–22; Blank, *Afghanistan*, chap. 4; and idem, "Soviet Forces," 102–7.
62. Kennedy, 338.
63. William R. Trotter, *A Frozen Hall: The Russo-Finish Winter War of 1939–40* (Chapel Hill, N.C.: Algonquin Books of Chapel Hill, 1991), 7, 35–41, 74–75, 205–6.

Northern Ireland
A Prolonged Conflict

Benjamin Kline

The "troubles" in Northern Ireland erupted in 1968 and continues after more than 20 years of bloodshed. The combatants have been haphazardly labeled Catholic versus Protestant or Republican versus Loyalist, but neither of these generalizations sufficiently helps readers to understand the forces which have motivated the unceasing violence. Yet, the primary protagonists are more readily identified as the Irish Republican Army (IRA), with its numerous factions, and the British army, with its Loyalist allies. In this particular case, the struggle clearly illustrates that the differences between an intentionally "protracted" conflict and one which is unexpectedly "prolonged" is often determined by the success of the former. Britain had no intention of becoming involved in a lengthy war in Northern Ireland, but its failure to recognize that the causes of unrest were social rather than political has forced Britain into what it terms a prolonged conflict. In contrast, the IRA has deliberately chosen an "armed struggle," one designed to be protracted, to win a war of attrition. That warfare still torments Northern Ireland, despite British efforts to halt it, indicates that the IRA's protracted strategy has become the most appropriate definition of the conflict.

The ability of the IRA to persevere is remarkable, if not surprising. Above all other aims, the IRA is determined to maintain a protracted conflict, however destructive it may be. A protracted conflict holds the key to the organization's existence and is the core of its philosophy. In the words of Terence MacSwiney, an Irish Republican who died during a hunger strike in 1920, "It is not those who can inflict the most, but those that can suffer the most who will conquer."[1] According to his belief, an exhausted Britain will eventually retreat from Northern Ireland when the cost of maintaining its presence becomes prohibitive. When the enemy is thus vanquished, Ireland will be reunited, and the problems of the persecuted Catholics of Northern Ireland will be rectified. It is a simple argument, based on the historical myths and truths of British imperialism and designed to elicit the support of Catholics and Irish Republicans. According to this scenario, the British army represents the power of past and present rulers, and only a war of attrition will dislodge it. However, targets and strategies are secondary issues compared to the necessity of keeping the struggle alive. In such a conflict, military aims are based on winning the propaganda war and demonstrating that no solution will succeed except that of the armed struggle. In the end the IRA is confident that history will judge it right in spite of its faults.

Despite numerous setbacks, the IRA has successfully won the public relations battle among the Catholic population in Northern Ireland. Much of this success emanates from the vacillating nature of Britain's policy. The British government has no interest in a protracted conflict and consequently can only interpret the unceasing opposition to its rule as regrettably prolonged. However, in searching for a solution to the troubles, the British government has become a major cause of its own problem. British troops have become the ideal target for IRA resentment, which depends on traditional Catholic and Republican hatred of Britain for its support. Once embroiled in an armed conflict with the IRA, the British have found it impossible to withdraw without appearing to be surrendering a region of Britain to a minority paramilitary group. In its defiance the British government and its army have, at times willingly, associated themselves with the cause of the Loyalists, thus validating IRA claims.

While justifying its presence by citing the democratic will of the people and the need to prevent civil war, the British government has striven to construct a political solution which would establish the best conditions for an honorable withdrawal. This apparent willingness to leave only convinces the IRA that its policy of protracted conflict is working. Furthermore, in attempting to find a political answer to their problem, British authorities have worked to prevent the alienation of their Loyalist supporters and consequently have refused to become completely committed to the real issue—social reform. Thus, without an effective policy to solve the problems of unemployment, housing, and Catholic grievances, Britain has prolonged the conflict by contributing to the social instability which sustains the IRA. Conor Cruise O'Brien, a leading moderate, has stated the case clearly:

> The real options in Northern Ireland are just two: either continue direct rule or withdrawal. By constantly seeking some attractive but nonexistent option in between, successive governments have given the impression that they crave withdrawal even while they reluctantly stay. That is an important part of the political culture in which the Provisional IRA has flourished.[2]

Background to the Troubles

Although the IRA traces its beginnings to the Irish Republican Brotherhood, a nineteenth century revolutionary movement which sought an independent Irish republic, it was not officially established until 1919. When members of the Republican Sinn Fein party were returned to power in the general election of 1918, they declared Ireland's independence from Britain, and the IRA was formed to defend the newly created Irish republic. As the military wing of Sinn Fein, the IRA fought British forces in the Anglo-Irish War, a conflict which would ravish the land for two years. The war finally came to an end in December 1921, when a controversial agreement, the Anglo-Irish Treaty, was signed. This treaty allowed the loyal northern six counties to remain a part of Britain on a temporary basis and granted dominion status to the southern counties. Opposition within Ireland appeared when Eamon de Valera led the antitreaty delegates out of the Dail and declared their commitment to a united Irish republic. The IRA then split between those who accepted or rejected the treaty. The latter group, keeping the name IRA, continued to demand the unification of all Ireland in one republic. The protreaty advocates formed the Irish Free State and for two years fought a bloody civil war. Finally, the IRA was defeated in the south in 1923, was later outlawed, and was generally inactive until the present troubles began in 1968.

The government in Dublin refused to recognize the legitimacy of Northern Ireland, despite the apparent permanency of the border. In 1949 Ireland became a republic and included in its constitution the right to govern the entire island. Yet, this ideal remained unfulfilled by a government which was too weak to assert its authority against Protestant and British opposition. Furthermore, although the Protestant majority of Northern Ireland had established a state which discriminated against the Catholic minority in government, employment, and housing, Britain continued to support them. Britain assisted the Northern Ireland parliament at Stormont by subsidizing northern industries and maintaining a staunch recognition of the existing partition. This support became particularly apparent after the Second World War, when an infusion of British finance and favor rewarded Northern Ireland's loyalty, as opposed to Ireland's neutrality. The continued denunciations of British imperialism and colonialism, which emanated from Sinn Fein headquarters in Dublin, could do little against the power of the London government. The use of force to champion Republican aims was beyond the means of the small Irish military while the IRA remained ineffectual. This deficiency was made clear by the IRA's failed military campaign against the border from 1956 to 1962. During these years it had demonstrated an inability to win battles or to enlist the support of northern Catholics.

The failure of the IRA to gain Catholic support in Northern Ireland reflected its general ignorance of the discrimination and prejudice which Catholics suffered during this period. These tendencies were certainly issues which could have been exploited. Unemployment was a major problem in postwar Northern Ireland. It hovered, from 1945 to 1963, between 5 and 10 percent, outdistancing the numbers in the depressed regions of Scotland and Wales. By 1971 male unemployment among Catholics was even higher, reaching 17.3 percent, as compared to 6.6 percent among Protestants.[3] The Protestant monopoly of political power used gerrymandering of voting districts and a housing shortage, which often left the Catholics with little recourse but to live in overcrowded slums, to add to this problem. These social conditions were not incorporated into the IRA's strategy and were thus left to be taken up by other organizations.

The campaign for social justice (CSJ) initiated the first significant effort to improve Catholic living conditions. Organized by a group of middle-class Catholic activists in 1964, the CSJ began to publicize and agitate against sectarian discrimination in Northern Ireland. It had some minor successes and became the forerunner of the Northern Ireland Civil Rights Association (NICRA). In February 1967 NICRA, building on the experiences of the CSJ, chose to demonstrate against the poor housing conditions of Catholics and to draw attention to the abuses of local authorities. It further demanded one man-one vote in local elections, no gerrymandering of constituency boundaries, and fair distribution of local council houses. Its criticism of the autocratic nature of the Stormont government was expressed by calling for the repeal of the Civil Authorities Act, which gave the government the power to "take all such steps and issue orders . . . to preserve the peace," and by disbanding the B-Specials, an almost exclusively Protestant and notoriously anti-Catholic part-time police force.[4] These demands were soon followed by public meetings sponsored by NICRA.

Stormont's reaction to the marches and demonstrations of the civil rights movement was uncompromising. The Royal Ulster Constabulary (RUC) and the B-Specials were mobilized to confront the agitators forcefully. It attributed the causes of the disturbances to the IRA and communist instigators. Stormont deplored the uncontrollable sectarian violence which resulted but considered it unavoidable. The violence was a clear and concise strategy which, if kept within the confines of Northern Ireland, might have succeeded. Unfortunately for the Stormont government, the scenes of riots and violence in the streets could not be kept out of the news. Each night television transmitted the images of police brutality, Protestant mobs, and Catholic marchers into the homes of Britain and the rest of the world. As the drama unfolded and continued unabated, the media publicized each event. Background reports on the discrimination of Catholics, the religious segregation in housing and jobs, and the cases of detention without trial were presented in detail. Northern Ireland had been put under the magnifying glass of worldwide observation. The streets of Northern Ireland were uncomfortable places for the Stormont government as well as for the British government. For the civil rights movement, the exposure created support and momentum.

The effectiveness of global awareness of the civil rights movement took the IRA by surprise. It had not initially been involved, but it soon found that many of its followers were participating in the marches. As the inspector general of the RUC, Anthony J. Peacocke, stated, "I do not think the [IRA] are organizing it, but it fits in with their long-term plans for uniting Ireland forcibly."[5] In fact, the rank and file of the IRA had chosen to participate in a movement which seemed to be succeeding where its own organization had continually failed, the undermining of Protestant power. This was an outburst of discontent, deep and resolute, emanating from the core of Northern Ireland society. In essence, the early stages of a prolonged conflict had been reached.

The Intervention of the British Army 1969–1972

None of the antagonists could claim an effective strategy for achieving their goals. Decades of religious, political, and economic divisions had erupted in communities throughout Northern Ireland, resulting in uncontrolled sectarian strife. With few lines of communication, efforts to find solutions were hampered by fear and mistrust on both sides. Instead, an environment was being created in which small violent factions could come to the forefront and be perceived by many as protectors. The emergence of these protectors depended on the escalation of a conflict in which they would be sustained by an atmosphere of insecurity and pessimism. Unfortunately, the British government unintentionally but effectively became the catalyst to establish this condition and thus contributed to its prolongation.

On 20 April 1969 a pipeline at the Silent Valley reservoir was destroyed and seven post offices were burned in Belfast. Responsibility for the bombings remained a mystery—the IRA and the Loyalists blamed each other—and Stormont's inability to quell this new stage of unrest was apparent. Sectarian clashes continued unchecked, and riots in Derry resulted in an estimated 100 casualties. In August Protestants in west Belfast joined the police in tearing down Catholic barricades. The situation quickly deteriorated, and the Catholic Falls area of Belfast was invaded by Protestant civilians, who joined the B-Specials in destroying the homes of Catholics and in beating residents. On the night of 14 August alone, six people died and 150 families were burned out.[6]

Unable to deal with the growing conflict and fearful that it would spread to other communities, Stormont called to Britain for assistance. Subsequently, 10,000 British troops were dispatched to the north, arriving in Derry on 14 August and in Belfast on 15 August. Two companies of the Third Light Infantry arrived with fixed bayonets, loaded machine guns, and armored cars. Wire barriers were soon set up to divide the Protestants from the Catholics. Ironically, in view of later developments, Catholics widely welcomed the move.

> It was greeted with profound relief on the Catholic side, where community leaders had been attempting all day to communicate their plight after last night's widespread house-burning and shooting by Protestant extremists and police. But the troops were met with a cold and hostile reaction from many on the Protestant side.[7]

Richard Crossman, a minister in the British cabinet, was less optimistic and wrote in his diary an insightful prophecy, "I fear that once the Catholics and Protestants get used to our presence they will hate us more than they hate each other."[8] Few leaders in Northern Ireland or Britain shared Crossman's foresight and thus were unaware that this latest event would be an escalation rather than a curtailment of the conflict. The British government had failed to realize it could not maintain its position of neutrality in a society in which a large percentage of the population viewed it as the traditional imperialist enemy.

These events significantly influenced elements within the high command of the IRA. In the past the IRA had held staunchly to its principle of not recognizing the Stormont government or British rule. Subsequently, it abstained from political involvement. Yet, by the late 1960s, a strong Marxist/Socialist following had emerged among the IRA leadership, which was anxious to exploit the power in the mass demonstrations. As the civil rights movement grew in power, the IRA altered its tactics to emphasize a socialist agenda and a willingness to become involved in the social movement.[9]

To become more politically active, the IRA recognized the de facto existence of two Irish governments and Westminster in December 1969. Abstentionism thus was eliminated as a basic principle of the IRA. It subsequently became more closely associated with Sinn Fein and its political agenda. This apparent repudiation of traditional republican ideals appalled many of the old guard. The dissenters withdrew and formed the Provisional Army Council (provisionals). Devoid of socialist aims and committed to abstentionism, the provisionals called for a simpler goal: "the eventual achievement of the full political, social, economic and cultural freedom of Ireland."[10] It renewed its commitment to the protracted armed struggle. The division created the officials, who primarily associated with Sinn Fein, and the provisionals, who committed to the armed struggle, but the term *IRA* is more commonly used to identify the latter.

By August 1969 the British government was prepared to take steps to head off the growing social unrest. As a result, it obtained from Stormont a promise to institute a program to reform those institutions criticized by the Catholic civil rights movement. The stipulations of this program included equality in employment and housing, the prevention of religious discrimination and incitement to religious hatred, the development of the Northern Ireland economy, the fostering of better community relations, and the examination of the roles and functions of the police. By 1971 an ombudsman had been established to investigate the central government, and a similar official had been appointed to do the same in local affairs; electoral changes had brought the franchise at last into line with that of Britain; a points system based on need had been adopted for the allocation of houses; and a ministry of community relations had been created and a community relations commission set in motion.[11]

However, reforms took time and there were signs that their implementation was not popular within the Stormont government. Other areas where there had been little or no movement included the appointment of a public prosecutor distinct from the police, the introduction of antidiscrimination clauses into government contracts, and, according to a British government report, the unemployment rate, whereby Catholics were two and one-half times more likely to be unemployed than Protestants.[12]

In October 1971 Northern Ireland Prime Minister Brian Faulkner published proposals for proportional representation and for a larger House of Commons and Senate at Stormont. He also appointed to a cabinet post Dr Gerard Newe, the first Catholic to hold such a post in the 50-year history of the state. But the effect of both gestures was muted by the fact that, in putting forward his suggestions, Faulkner made it absolutely clear that his administration was still committed to the maintenance of Northern Ireland as an integral part of the United Kingdom and also that it regarded as fundamentally unrealistic the idea of a mixed government of Catholics and Protestants, which was being touted at the time. Consequently, the Catholic community, impatient at the lack of visible progress and deeply confused about the direction of that progress, inclined more and more to the view that reforms received in response to the threat or the actuality of violence were insufficient substitutes for the recognition they sought as full citizens in the province.[13] During this period Northern Ireland suffered from the anxiety of expectations as Catholics sought changes which could not come soon enough, while Protestants dreaded the slippery slope of social reform.

As attempts to resolve the problems of bias and discrimination in Northern Ireland society had been meeting with difficulty, the violence continued to escalate. By the summer of 1970, the relationship between the British army and the Catholic community had eroded into chronic street fighting. British efforts to halt the growth of the IRA and to separate the religious communities had involved crude tactics of mass evictions, dawn raids, and blockading of streets. The *Sunday Times* reported that "the idea, we have been told, was to cut the Provisionals down to size by demonstrating that the army could invade their home territory whenever it wished."[14] Catholics felt particularly targeted, since their communities were identified as IRA havens, and claimed that the British army ignored the illegal activities of Protestants. Whether there was favoritism or not, the strategy to cordon off areas resulted in further polarization and added to the siege mentality, which engulfed many neighborhoods. The British strategy thus had contributed to the prolongation of the conflict by further destroying its credibility as a neutral participant.

Catholic bitterness produced an atmosphere in which the IRA could represent themselves as the defenders of the community in a way that had not been possible since 1922. The situation was quickly exploited by the IRA in January 1970 when it decided to set in motion a three-part strategy, which it hoped would initiate the protracted armed struggle it sought. As nationalist defender it would organize and arm a "militia" secret army. Retaliation would be made possible by gradually provoking the British security forces so that they would alienate the nationalists. This would allow the IRA, as a secret urban-rural guerrilla army, to engage the British whenever and wherever possible with the accumulated arms and the new and trained volunteers. Thus, a constituency would be created from which an army would be equipped to provoke the British army into committing to a serious and costly armed struggle.[15]

The first step in instituting this plan was to attract support from the Catholic population. The IRA's success in achieving this goal can be significantly attributed to the actions of the British army. The army's behavior was often crude, considering the complexity of the society in which it fought, and seldom responded to the sensitive nature of a sectarian conflict. Brig Gen Frank Kitson, who supervised the British army's counterinsurgency strategy in 1971, described the army's approach as "a sort of game, based on intense mental activity allied to a determination to find things out, and an ability to regard everything on its merits without regard to customs, doctrine or drill." However impartial this plan might attempt to be, it ignored the fact that the British would almost always be associated with the Loyalists. Catholic suspicions were further incensed by Kitson's assertion that the legal system should be manipulated "for the disposal of unwanted members of the public. . . . For this to happen efficiently, the activities of the legal services have to be tied into the war effort in as discreet a way as possible."[16]

The result of such tactics, and a turning point in Catholic attitudes towards the army, was the 34-hour curfew imposed on the Falls in July 1970 to facilitate arms searches. The IRA quickly assumed the position as the sole protector of the Catholic community, while the British army, forced to carry out the searches, was associated with the unionist side. IRA membership mushroomed from about 100 to about 800 in the second part of 1970. The IRA became so formidable that when a sniper killed the first British soldier, Gunner Curtis, on 6 February 1971, the Stormont prime minister, Maj James Chichester-Clark, announced, "Northern Ireland is at war with the Provisional IRA."[17]

Such a situation, where public authorities had neither the machinery nor the information to act effectively, was tailor-made for the intervention of paramilitary groups, who came to win control of, and therefore allocate, housing in many areas. The Ulster Volunteer Force (UVF) and the Ulster Defense Association represented Loyalist efforts to counter the IRA. These organizations began to control neighborhoods and enforce their particular versions of law and order. Their effectiveness led many people, in both Catholic and Protestant areas, to praise these actions as valuable community services, while the bitterness of their victims provided fertile ground for extremist politicians. Thus, criminals, gangsters, and extremists were tolerated as the only effective means of protecting the sanctity of the community. Accordingly, the IRA continually claimed to be the only defense force that the Catholics of Belfast could rely on to protect their lives and homes. The bulk of the Catholic population accepted this claim, and although many of them thoroughly disliked the bombing and murdering, they were not prepared to cooperate with the forces of Britain to destroy the IRA. Loyalist extremists were successful at gaining new recruits by making similar claims.

The arms searches had further damaged Catholic perceptions of the British army and enhanced the position of the IRA. Only 47 of the 1,183 houses searched between 30 November 1971 and 30 January 1972 produced any arms. However, the disruptions caused by early morning raids, the kicking down of doors, and instances of vandalism had turned many moderate Catholics against the British army. Northern Ireland's prime minister, Brian Faulkner, and the British army, vastly aggravated their problems by introducing internment in August 1971. When the Stormont government was finally allowed to implement large-scale internment without trial, on 9 August 1971, over 300 IRA suspects were lifted in one operation, and rioting once more broke out on a major scale.[18] The effect on the Catholic community was devastating. Family members and friends of the arrested men turned against the British army, criticizing their methods as unwarranted and barbarian, and looked to the IRA for protection. An atmosphere in which the IRA could attract support for its protracted conflict had thus been intensified by British tactics.

Internment also increased IRA support abroad, particularly among Irish Americans, thus augmenting Provo's financial resources. American politicians of Irish descent have often exploited the plight of Northern Ireland to attract Irish-American voters. Sen Edward Kennedy, certainly the most well-known Irish-American politician, expressed the feelings of many Irish-Americans when he condemned internment, called for the dissolution of the Northern Ireland parliament, the immediate withdrawal of British troops, and the reunification of Ireland. He warned that the Ulster government's rule "by bayonet and bloodshed" would rapidly turn the situation into "Britain's Vietnam."[19] Such anti-British rhetoric has often coincided with the Irish-American support for the use of force as a tactic in achieving reunification. Accordingly, such groups as Irish Northern Aid (NORAID) have continually supplied the financial backing for the IRA and other Republican paramilitary groups. The amount of such contributions varies yearly but can rise significantly following a high-profile IRA success. A clear example of this occurred in 1979, following the assassination of Lord Mountbatten. NORAID received $350,000 in donations following this incident, $100,000 more than it obtained in 1978.[20]

As the situation deteriorated, the British government found itself in an uncomfortable predicament. It was forced to defend its presence in Northern Ireland by arguing that its army was needed to defend democracy and prevent civil war. In support of this claim were Loyalists who argued that Britons could not abandon those who wished to remain British and claimed the democratic will of the people of Northern Ireland. British domestic support for such a cause had to be heeded by the government. Yet, the British army represented the ideal target for those committed to destroying the state. It could not stand by idly or the IRA would eventually achieve the strength to mount an aggressive campaign against it. However, by seizing the initiative through arms searches, it only contributed to the further alienation of the Catholic community. By 1971 there had already been 17,262 houses searched with the inevitable cries of outrage.[21] It was a war in which traditional tactics of warfare could not be engaged if Britain were to win the hearts and minds of the Catholics. Yet, supporting loyal Protestants would antagonize the Catholics, thus increasing the appeal of the IRA, and neutrality would do the same thing. A prolonged conflict seemed inevitable in a situation in which neither withdrawal nor engagement would achieve Britain's objectives.

In 1972, 468 people died from the "troubles." Numerous incidents, in which innocent people were injured and maimed, caused many deaths. One example occurred in March 1972 when an explosion in a crowded cafe claimed 143 casualties. Continuing initiatives were buried completely by Bloody Sunday, 30 January 1972, when paratroopers opened fire on demonstrators, killing 13 civilians. A few weeks later a bomb attempt by the official IRA to take revenge on the battalion responsible resulted instead in the deaths of six civilian women cleaners and a Catholic priest at a barracks in England. On Bloody Friday, in July, the Provisional IRA set off 22 bombs during one afternoon in Belfast, killing nine civilians and two soldiers. The IRA campaign in the north continued to escalate, as international and domestic support increased, following the reactions to Bloody Sunday. The power of the Republicans had grown to such a degree that large sections of Derry and west Belfast were declared ungovernable "no-go" areas for the security forces.[22] The IRA's tactics for a protracted conflict appeared to be succeeding as the British government failed to halt the prolongation of its troubles.

Struggling to Find a Solution

Conditions were deteriorating by the beginning of 1973, and the British government discovered that it was being trapped by the social quagmire of Northern Ireland. Attempting to avoid a prolonged conflict, it made the fatal error of initiating a process in which it would become a more active participant in the complexities of Irish politics. It would seek a political compromise by sponsoring talks between Northern Ireland and the Republic of Ireland in which it would act as mediator. The strategy was ill-conceived at best, since the British government would hardly be viewed as an unbiased spectator. More importantly, it gave the impression that it was weakening in its resolve and thus encouraged the IRA to continue its armed struggle. In this particular case the nonmilitary approach significantly contributed to the prolongation of the conflict. Perhaps a truly neutral agency, like the United Nations, would have had more success in forwarding a political solution. Instead, in a white paper published on 20 March 1973, the British government proposed the abolition of Stormont and its replacement by an assembly which would govern Northern Ireland through a power-sharing executive.

In pursuance of its diplomatic strategy, the British government achieved the signing of the Sunningdale agreement on 9 December 1973. Here, the British and Irish prime ministers and representatives of the Northern Ireland government agreed to the establishment of a council of Ireland. To Brian Faulkner, prime minister of Northern Ireland, this was purely a token concession designed to ensure that the Irish government "fully accepted and solemnly declared that there could be no change in the status of Northern Ireland until the majority of the people of Northern Ireland desired a change in that status."[23] However, Faulkner was immediately repudiated by the Ulster Unionist Council, which opposed any negotiations with the republic, and was forced to resign as leader in January 1974. The Sunningdale agreement was a breakthrough as far as relations between the Irish Republic and Northern Ireland were concerned, but on 5 January 1974, when the executive formally took office, the

Unionist party rejected Sunningdale and power-sharing. In any case, the effort had done little to deal with the complaints of Catholics and only served to demonstrate that Britain was unwilling to deal directly with this element of society.

Sectarian antagonism had not lessened despite British efforts to find a diplomatic solution. The war between the IRA and the loyalist paramilitaries and the British army continued, marked by great ferocity, bombings, and assassinations. The situation had polarized and, by its refusal to disassociate itself from the cause of the Loyalists, the British government seemed trapped by the uncompromising conditions. In November 1974, immediately after the deaths of 19 people by IRA pub bombings in Birmingham, the government passed the Prevention of Terrorism Act, making the IRA illegal and granting the home secretary powers to exclude from Britain, but surprisingly not Northern Ireland, persons suspected of terrorist involvement.[24] This act further evidenced Britain's decision to react to the problems of a prolonged conflict rather than combat the IRA's protracted strategy.

By the mid-1970s, the IRA and its supporters had targeted Britain as the source of Irish political evils. Added to this belief was an almost fanatical faith in the power of armed struggle to eventually break the will of the British government and end its occupation of Northern Ireland. However long this struggle might take, the IRA was confident that history was on its side. The IRA had learned from history that successful revolutions often started off badly and looked hopeless but were inevitably victorious if pressure were maintained. Therefore, solutions which involved negotiations, the ballot box, diplomacy, or compromises were criticized as mere diversions, which weakened the needed resolve. The dream, and the dream alone, would sustain the revolution.

The commitment to preserve the impression of a continued armed struggle dictated the tactics of such a revolution. As a relatively large fighting force—from April 1973 to April 1974, 1,292 terrorists were arrested—the IRA had difficulty controlling these actions. Prevented from exercising a strong central command and given the difficulties of communication and British surveillance, the IRA saw the initiative for attacks often delegated to individual units. These groups chose bombing sites and identified persons to be killed or maimed according to local interests and interpretations of IRA policies. While the army commanders may set down general principles and strategy, they could not completely control their followers. The IRA's internal structure therefore hindered its ability to institutionalize doctrines of discipline and command. Its day-to-day decisions are made on the streets, thereby relying on local intelligence or stumbling on some vulnerable target, with little direction from central leaders. Objectives often varied according to specific needs; sometimes, they organized as formal attacks for psychological reasons. The IRA sometimes made mistakes, killed innocent people, and chose inappropriate targets. Yet, they still believe these problems have little importance in the long run. They want the armed struggle to continue at all costs. IRA supporters willingly endure the problems of inefficiency and poor judgment as the price for victory.

The British government had developed its own rationale for continuing the conflict in Northern Ireland. There was the foreign threat. Ted Heath, the former Tory prime minister, while visiting the US late in 1975, argued that Britain could not leave the six counties, because "if we withdrew, there would be a Cuba on the fringe of Europe." Many British leaders shared the fear of Soviet intervention on the side of the Republicans. Yet, it was a question much closer to home which made the strongest impression. Where would the scenes of secession end if Britain were to surrender to an armed force and relinquish its hold on Northern Ireland? Will Scotland and Wales wish to imitate the example some day? There was also the possibility that a victorious IRA would encourage such disgruntled elements in Britain as minorities and the working class, of which the Irish constituted a large percentage, to press for significant social reforms. Roy Mason, Northern Ireland secretary in 1976, cited this possibility.

> We would be fooling ourselves if we thought that the bloodletting that would flow from the precipitate withdrawal of troops would be confined to Northern Ireland. The undoubted violence could easily spread to the mainland [Britain] with its large Irish population.[25]

Yet, the British government was not attracted to the prospect of ruling Northern Ireland indefinitely. While it defended its presence by arguing that it had a responsibility to recognize the democratic rights of the majority and to prevent civil war, Britain believed the cost of a prolonged conflict was daunting. It needed to find a solution which would not leave it with the stigma of defeat and yet free it from something Reginald Maudling, home secretary in 1970, caustically described, as "a bloody awful country."[26] The answer which succeeding British governments would become increasingly attached to was the creation of some sort of cross-community devolved government. Such an arrangement served two purposes: it would undermine Catholic support for the IRA by attracting moderates, and it would give Britain an opportunity to depart gracefully.

Diplomatic efforts had broken down in the face of sectarian mistrust and violence. Terrorist activities continued during the latter half of the 1970s but with less ferocity. The conflict claimed 252 dead in 1973, 221 in 1974, and 244 in 1975. This lull can be attributed to an IRA cease-fire that lasted for most of 1975. Attempts to reconcile Republicans by phasing out internment by the end of 1975—when all 2,000 suspects, 95 percent of them Catholic, were released—also contributed to calming matters. During this period the British government tried to transfer some of its security responsibilities by gradually reducing the British army from 21,000 in 1975 to 12,000 by 1980, while increasing the various police forces from 14,500 to 19,500.[27]

The death counts continued to decline. In 1978 and 1979 only 31 civilians were murdered, compared with 55 in 1977 and 221 in 1976. This decrease can be credited to the renewed efforts of the British government and the army to counter the tactics of both sectarian groups. The IRA suffered severe losses to the security forces in 1976 when they were penetrated increasingly by army intelligence. It responded in 1978 by augmenting the independence of the small military cells, thus decreasing the possibility of spies discovering the identity of the top leadership, but also expanding local initiative and the possibility of civilian victims. The Loyalist paramilitaries suffered a setback in May 1977 when the British authorities refused to back down to the second Loyalist strike, which had attempted to obtain greater self-government for the north.[28]

Despite British successes, by the end of the 1970s the political future of Northern Ireland appeared little clearer than it had been at the fall of the Stormont parliament in 1972. In all, between 1969 and 1979, 1,994 people died, including 573 members of the security forces.[29] During this period the IRA held fast to its doctrine of maintaining the armed struggle against all odds. It reorganized when necessary, recruited new soldiers to replace those who were killed or imprisoned, and strove to stay in the public eye. Maintaining the appearance of a viable force was imperative and probably was the reason for the purely symbolic and high-profile assassination of Lord Mountbatten in August 1979. However, the ambushing of British troops in County Down in the same month resulted in the killing of 18 soldiers and proved that the IRA could still strike military targets.[30] The conflict had been kept alive and after nearly 10 years had enlisted a new generation. For a decade the children of Northern Ireland had existed in a war zone, witnessing killings and growing accustomed to violence. It was their participation that the IRA and Protestant paramilitaries sought. In particular, it was refusal of the British government to effectively address the problems of unemployment, poor housing, and the perception of biased British justice among Catholics that significantly assisted the IRA recruitment program.

The Ebb and Flow of a Protracted Conflict

By 1980 the IRA seemed to be in a state of stagnation. Its numbers remained stable, but its support among the general Catholic population lacked the fervor of the 1970s. Years of sectarian murders, street fighting, failed initiatives, and botched strategies had weakened its mass appeal. The sentiment was creating a feeling among many Republicans that the IRA was simply a group of killers who had lost the higher moral ground. The IRA did not understand that killing more Protestants and British soldiers would not regain this desired public support. Instead, they sought an issue which would portray it as the defender of Irish rights and the British as the oppressors of the weak. Reinforcing the social divisions, and the IRA's position, was a crucial factor in sustaining the protracted conflict. The timely appearance of this issue emerged in 1976, when Merlyn Rees, British secretary of state, ended the special category status for prisoners claiming political motivation for their crimes. As of 1 March 1976, these prisoners would lose such privileges as wearing civilian clothing and living in specially designated compounds and be treated as ordinary criminals instead.

The order resulted in a protest in which prisoners refused to wear clothes, keeping only their blankets. It soon escalated when the British denied them permission to leave their cells unless they were wearing the prison uniform. In reply, the protestors chose to live in the growing filth of their cells rather than comply with the British demand. When this tactic failed, the prisoners decided to stage a hunger strike on 7 October 1980. Eventually, after failed compromises and a further escalation, 10 prisoners would die. The most celebrated being the death of Bobby Sands, the IRA officer-commanding, on 5 May 1981. Pressure from family members and small concessions from the British government finally ended the hunger strike on 3 October 1981 but not before the IRA had won a resounding public relations victory. The hunger strike had been followed daily by the mass media in Northern Ireland, the Republic, and Britain, as well as in the United States. Subsequently, the IRA's strategy for a protracted conflict was reenergized as the drama attracted renewed backing for its cause.

With its support bolstered, the IRA decided to escalate its attacks. In July 1982 it set off two powerful bombs. The first exploded in a car just as the Household Cavalry was trooping past Hyde Park in London en route to Buckingham Palace. The second went off under a bandstand in Regents Park, just a few hours later, as British army musicians were about to give an afternoon concert. The blasts took the lives of 11 soldiers. The bombings had coincided with Britain's successful war with Argentina over the question of who ruled the Falklands/Malvinas Islands. The IRA ridiculed Margaret Thatcher's claim that the war with Argentina was justified under the United Nations charter. In its London bombing communique, the IRA taunted Thatcher by declaring, "Now it is our turn to properly invoke Article I of the UN statute and properly quote all Thatcher's fine phrases on the right to self-determination of a people."[31] In all, a core of British soldiers were killed by the IRA in 1982, along with 12 police.

The British army and its Loyalist allies further intensified the situation by initiating what appeared to many to be a shoot-to-kill policy. On 11 November 1982 a special unit of the RUC killed three IRA volunteers, Gervase McKerr, Sean Burns, and Eugene Toman. A total of 109 shots were fired at the car carrying the unarmed men. Three members of the elite unit were eventually charged with murdering Toman. However, all were acquitted by Lord Justice Maurice Gibson, who declared them to be "absolutely blameless" and praised their "courage and determination in bringing the three deceased to justice—in this case the final court of justice."[32] Gibson and his wife, Cecily, were later killed by a car bomb by the IRA in April 1987. An RUC special unit was soon involved in a similarly controversial incident when it killed a 17-year-old farm boy named Michael Tighe. He had no IRA connections, but he and a friend had been caught handling illegal weapons in a barn.

Tensions heightened when the IRA attempted to kill Margaret Thatcher and members of her Tory party during the October 1984 Conservative party conference in Brighton. Thatcher survived the 100-pound bomb planted on the fifth floor of the Grand Hotel but five people were killed, including three wives of party officials. The IRA's statement of responsibility, "Today we were unlucky, but remember—we only have to be lucky once; you have to be lucky always," was consistent with its protracted philosophy. However, it did more to bolster Thatcher's public support that to weaken her resolve. Speaking to the press at the conclusion of her November 1984 Chequers summit with Ireland's Taoiseach (prime minister), Garret FitzGerald, Thatcher made reference to suggestions which had been made by the New Ireland Forum, declaring, "I have made it clear that a unified Ireland is out. A second solution was confederation—that is out. A third solution, joint authority, that is out."[33]

Thatcher's bravado did little to deter the IRA, which had been encouraged by the publicity gained during the hunger strikes and the recent attack on the Tory party. Assaults on

Royal Ulster Constabulary installations increased, including the killing of nine members of the police at Newry RUC in February 1985, and resulted in the death of 23 constables by the end of the year. A new phase in the conflict emerged when the IRA warned that they would prevent the rebuilding of these fortifications by civilians. Any and all civilians who aided in the construction of the British war machine were considered legitimate targets. In August 1985 this threat was fulfilled when Seamus McAvoy, a building contractor in Dublin, was murdered. Three months later the German manager of a company which provided catering services to the RUC was killed in Derry. Terrorism is an important factor in maintaining a protracted conflict, and this new phase was specifically designed to accentuate this element of the IRA's strategy.

Considering the emphasis which the IRA put on local initiative, it is not surprising that this policy led to carelessness. In the same month as the McAvoy death, a 65-year-old Catholic was accidentally killed by an IRA hit squad which mistook him for a Protestant building contractor. The resulting negative publicity was compounded by the killing of a man who was driving a car similar to one driven by a known Ulster Defense Regiment (UDR) soldier. Even when the assassins got the right victim, their choice sometimes did little to enhance their popularity. The execution of Catherine Mahon, the first woman killed by the IRA for being an informer, and her husband was widely condemned.

Popular support for a protracted conflict often follows an ebb-and-flow pattern, rising and falling according to the latest atrocity. Subsequently, a reliance on violence, poor judgment, and mismanagement had seriously damaged the positive gains which the IRA had enjoyed following the hunger strike. Added to these setbacks was the diplomatic manueverings of the British government. In the Anglo-Irish agreement (the Hillsborough accords) of 15 November 1985, signed by Thatcher and FitzGerald, the Republic gained a voice in Northern Ireland affairs by establishing an Intergovernmental Council, which would cooperate on such matters as security, administration of justice, border problems, and political questions. Thus, in principle, Britain recognized the right of the Republic to participate in Northern Ireland affairs and that its alliance was crucial to settling the sectarian strife.

However, Britain also achieved a diplomatic victory by including a clause which reaffirmed its policy in Northern Ireland by stating, "Any change in the status of Northern Ireland would only come about with the consent of a majority of the people of Northern Ireland."[34] As many Republicans argued, this stipulation depended on the Irish government's repudiation of its own constitution, which affirmed the essential unity of the country. Furthermore, the influence of the Catholic religion in the legal system (e.g., making divorce illegal) would hardly be welcomed by a majority of the Protestant north.[35] As long as such provisions remained a part of the Irish constitution, it could not fulfill the promise of the Anglo-Irish agreement.

Yet, for the British government, the significance of the Anglo-Irish agreement was more than an effort to increase the participation of Dublin. It represented another attempt to undermine public support for the IRA. Military tactics had failed, and authoritarian methods could be exploited but convincing Catholics, many of whom viewed the IRA as their only defender, that the Irish government could be a viable alternative had been a major goal in the propaganda war. The Anglo-Irish agreement seemed to suggest that the British were closer to achieving this aim.

Despite the diplomatic achievement of the Anglo-Irish agreement, there remained a serious flaw in Britain's strategy to combat the IRA. The British government still failed to realize that the IRA and sectarian strife were only symptoms and not the cause of the war. At the core were centuries of Catholic persecution and segregation. These social problems could not be confused with military or political priorities if the conflict were to end. While the IRA might convince Irish Americans that a united Ireland was the goal, it depended on the perceived and actual cases of social inequality in Northern Ireland for its recruits and public support. Thus, while unemployment had soared to 22 percent in 1987, the Fair Employment Agency, created by the Northern Ireland office in 1976, admitted that male Catholics were still twice as likely to be jobless as Protestant men.

To avoid alienating its Protestant/Loyalist supporters, Britain had refused to effectively enforce its own guarantees for the protection of civil rights in Northern Ireland. In this study of the northern Ireland war, Kevin J. Kelley cited blatant cases of prejudice in employment.

> At three of the North's major industrial firms—Harland & Wolff shipbuilders, Mackie & Sons textiles, and Shorts Brothers aircraft manufacturers—the disproportion between the number of Catholic and Protestant employees is nearly as pronounced in 1987 as it was in 1967. The Catholic share of Shorts' workforce actually declined in 1985, the FEA reported, falling from 17% to 14%. Mackie's 90% Protestant factory can only be expected to become more segregated as it considers a move from the Springfield Road dividing line in Belfast to an exclusively loyalist enclave. Britain meanwhile provides Harland & Wolff with a 68 million pound annual subsidy that helps it retain a 4,000-strong work force containing fewer than 500 Catholics.[36]

Jobless young men, standing on street corners, with nothing to do and little hope, are easily attracted to groups like the IRA and its promises.

By the late 1980s the war of attrition had gained a life of its own. Causes and solutions continued to be debated but had been dimmed by the dedication of both sides to violence. None of the combatants were prepared to compromise or retreat, fearing that any ground given would bring victory to their enemy. In 1988 a chain of events began in March, which symbolized the "tit-for-tat" nature of the conflict. Three unarmed IRA volunteers were killed by the British army in Gibraltar. During their funeral in Belfast, a group of loyalist gunmen opened fire on the crowd, murdering three mourners. This was followed by the deaths of two plainclothes British soldiers who were beaten and shot when their car approached the funeral procession for one of the cemetery victims. The following year witnessed few innovations in the pattern of incessant murders and poor judgment. In 1989 the IRA killed two top RUC officers and accidentally murdered three civilians at an alleged UVF meeting place. They continued to carry the conflict outside of Northern Ireland when a bomb in England killed 10 English army bandsmen and a shooting in West Germany killed a British soldier and his infant daughter. These public relations blunders were soon offset when the IRA won a propaganda victory in October. The impartiality of the

British legal system was brought into question after the Guildford Four were declared innocent by a court of appeal, which found them wrongfully convicted of a 1974 English pub bombing.[37]

By the summer of 1991, the chief constable of the RUC warned that the threat from paramilitary groups in Northern Ireland was at its highest level in two years and that an increase in violence was likely. In particular he was concerned that "if the Loyalists come back into the fray with more random killings, then it will be grim." He considered the IRA to be less of a problem, arguing, "They are not actually going anywhere. . . . Their political mandate is reducing." He predicted that eventually the IRA would be forced to give up its campaign of violence and join constitutional talks.[38] British officials have speculated about the demise of the IRA before and have continually been found to be wrong. It is unfortunate, in view of past history, that the only forecast which may be reliable is the one which sees the continuance of the IRA's protracted conflict.

Conclusion and Future

The IRA's strategy for a protracted conflict is based on a purposeful continuation of the armed struggle while Britain's regrettably prolonged conflict is an admission of Britain's failure to halt the hostilities. After more than 20 years of atrocities, killings, and social unrest, it is therefore evident that Northern Ireland's troubles are most closely defined by the IRA's terminology.

The reasons for the success of this strategy are certainly complex, but perhaps they can best be illustrated by the teachings of one of this century's most successful revolutionaries. Mao Tse-tung, nearly 30 years after the IRA had committed itself to a war of attrition, described the conditions necessary for a successful protracted war. In his metaphor about the fish and the water, he argued that small persistent revolutionary movements could win victory if they existed in an environment which would sustain them. If the IRA is the fish and the approximately one-third of the population which supports it is the water, then the situation in Northern Ireland becomes clearer. As long as circumstances allow the IRA to attract recruits, sympathizers, arms, and finance, they will have the capability to continue their terrorist activities. Added to this is a society which has access to targets, publicity, and safe havens. For over two decades these elements have existed in Northern Ireland. Subsequently, the IRA, and to a lesser degree the Protestant/Loyalist groups, have survived in a supportive environment.

While the environmental conditions maintaining this conflict include objective, traditional, religious, and political divisions, they particularly involve perceptions of actual injustice. The conventional divisions are constantly exploited by antagonists who seek to polarize camps but who do not address the injustices themselves. The concrete social basis for maintaining this environment among Catholics is that they feel they would be better treated as citizens of the Republic of Ireland, while the Protestant majority seeks to protect its culture by preserving its link to Britain.

Whether these injustices could be corrected within the existing state is irrelevant to the IRA. The IRA has incorporated this resentment into *its* program for uniting Ireland and ending British rule, while the actual persecution of Catholics becomes a secondary concern relative to this goal. Britain has done little to undermine this strategy. Far from building a social base for its point of view, the British government has substantiated IRA claims by steadfastly maintaining its presence in Northern Ireland and by refusing to effectively institute the social reforms needed to correct Catholic grievances. As a result it has further alienated Catholics and inevitably allied itself with its Protestant/Loyalist supporters.

A guerrilla war cannot be fought by traditional tactics of combat. The British government finds itself faced by an enemy which it cannot easily identify or flush into the open. Therefore, military methods are often useless and prone to cause more problems than they solve. Realizing this, the British government has turned to a political strategy in their attempt to undermine IRA support by establishing diplomatic compromises involving the Republic of Ireland and Northern Ireland leaders. Unfortunately, this approach has done more to compromise the British government's position by conveying the impression that it would like to find an honorable means of retreating from Northern Ireland. Thus, the IRA and its constituents have been emboldened by this apparent weakening of British resolve. Such beliefs nourish the environment in which the IRA can attract support and can increase the efforts of the Protestant/Loyalists to oppose any negotiations which they might interpret as being "sold down the river" by the British government. Successive generations have become accustomed to such extremist views and have come to accept violence as a way of life. Therefore, the environment remains stagnant, unmoved by an ineffective British government and exploited by the IRA. Until real social reform is instituted or the British government leaves Northern Ireland to struggle with its own devils, the situation will remain the same; namely, a society in which violent agitators can feed and thrive on the fear of the community.

Pessimism about the future of Northern Ireland is easy. In fact, there is little middle ground from which peaceful solutions can be attained. However, speculations about the future of the military factor offer some intriguing possibilities. If the breakdown in the Eastern Bloc continues and the immediate threat to Britain declines, the usefulness of the British army may convince a cost-conscious parliament to deplete its numbers. In Northern Ireland, the effect of changing priorities has been felt already. In the summer of 1991 the UDR, a military force designed to assist the Royal Ulster Constabulary, was disbanded. Yet, this demobilization may make the military solution in Northern Ireland more feasible. As the British army limits its activities, it has one remaining foe in the IRA. Subsequently, while it decreases in size, those troops and resources available could be concentrated within Northern Ireland. A much more aggressive and ambitious military campaign against the IRA would consequently follow. The Irish Republican Army government would have little force to oppose this, even if it wished to, and the Catholics and Republicans in the north would be faced with the choice of either surrendering or turning to the IRA for protection. Still, if the British strategy were quick and decisive, it could eliminate its military enemy and dictate its own solution. The social and economic costs would be high, but if successful, it might allow a British government to force the antagonists to accept a scheme as a precursor to its own departure. The result would hardly be ideal, leaving a volatile society, but if peaceful political solutions continually fail, a future British government might gamble on the military path as an expedient alternative.

The other factor in the equation is the Republic of Ireland. While it has limited resources, it nonetheless is the key to a peaceful resolution to the Northern Ireland question. Many northern Catholics perceive it as a protector. However doubtful its ability to protect Catholics may be, the Republic of Northern Ireland remains the only accepted alternative to the IRA. If it can enhance this perception with the support of the British government and initiate effective social changes, then IRA enlistments will decrease. Simultaneously, it must convince Protestants that Dublin will not interfere with their way of life. If these goals are achieved, the next step will be to unite Ireland, while ensuring that the predominantly Protestant countries are guaranteed a great deal of local autonomy. Such a solution would necessitate an alteration in the Irish constitution, eliminating such religious and social clauses which are abhorrent to Protestants. A type of federation would thus be created between the Irish state and the Protestant counties. It would depend on the flexibility of all the groups involved. At present such a solution seems beyond the capabilities of the existing powers.

Notes

1. J. Bowyer Bell, *IRA Tactics and Targets* (Dublin: Poolbeg Press, 1980), 11.
2. *The London Times*, 6 July 1991.
3. J. J. Lee, *Ireland 1912–1985: Politics and Society* (Cambridge, Mass.: Cambridge University Press, 1990), 412–13.
4. D. J. Hickey and J. E. Doherty, *A Dictionary of Irish History 1800–1980* (Dublin: Gill and Macmillan, 1987), 70.
5. J. Bowyer Bell, *The Secret Army: A History of the IRA 1916–70* (London: Spere Books Ltd., 1972), 420.
6. Lee, 429.
7. *The Observer*, 17 August 1969.
8. R. H. S. Crossman, *The Diaries of a Cabinet Minister* (London: Fontana Press, 1973), 620.
9. Bell, *The Secret Army*, 424.

10. Ibid., 431.
11. F. S. L. Lyons, *Ireland since the Famine* (London: Fontana Press, 1973), 767–73.
12. A. C. Hepburn, *The Conflict of Nationality in Modern Ireland* (London: Edward Arnold Ltd., 1980), 196.
13. Lyons, 776–77.
14. James Kelley, *The Courage of the Brave* (Meath: Kells Publishing Co., 1989), 3.
15. Bell, *IRA Tactics and Targets,* 17.
16. Kelley, 16.
17. Kevin J. Kelley, *The Longest War: Northern Ireland and the IRA* (London: Zed Books, 1988), 150.
18. Lee, 438.
19. *The Guardian,* 21 October 1971.
20. Kelley, *The Longest War,* 309.
21. Lee, 433.
22. Hepburn, 191.
23. Lee, 444.
24. Kelley, *The Longest War,* 227.
25. Ibid., 222.
26. Lee, 434.
27. Ibid., 450.
28. Ibid., 451.
29. Hepburn, 191.
30. Bell, *IRA Tactics and Targets,* 75.
31. Kelley, *The Longest War,* 352.
32. Ibid., 353.
33. Ibid., 368.
34. Kelley, *The Courage of the Brave,* 4.
35. Arthur Mitchell and Padraig O'Snodaigh, *Irish Political Documents 1916–1949* (Dublin: Irish Academic Press, 1985), 216–17.
36. Kelley, *The Longest War,* 366.
37. Ibid., 382.
38. *The London Times,* 6 July 1991.

Tentative Observations and Conclusions

Constantine P. Danopoulos with Rebecca R. Ruelas

Utilizing the analyses and the conclusions reached in the preceding case studies, the concluding essay draws some broad prototheoretical generalizations and patterns regarding the nature and setting of prolonged wars/conflicts and the factors that contribute to their prolongation.

The Nature and Setting of Prolonged Wars

Prolonged wars are not new phenomena. They have occurred throughout human history with considerable frequency and have involved Asian, African, Middle Eastern, Latin American, and European nations. Yet, in the post–World War II period, the majority of such conflicts have taken place on the soil of the developing societies of the third world; they also were influenced by the cold war conditions and the pervasive involvement of outside forces, especially the two superpowers. With few exceptions, most of these wars have involved multiregional and multiethnic societies that were characterized by social and ethnic fragmentation, political instability, and low levels of economic development. These societies experienced a period of political upheaval—much like European states and the United States did over a century ago—which often served as a catalyst to external aggression. This unfortunate situation remains relatively unchanged even to this day. Most developing countries have made little physical progress toward economic development, political stability, or social consensus. By comparison, advanced industrial societies with high standards of living tend to be satisfied and less likely to go to war, which might risk that valued status. For example, it is estimated that since 1945, all but one of the 60 civil wars took place in developing states, and 14 of these conflicts became internationalized. Western European and

North American nations and Japan fit the category of advanced industrial societies.

Colonial domination and specific historical circumstances often provide the background for the emergence and sustenance of prolonged conflict. Analyzing the impact of colonialism on the Ethiopian conflict, Cobie Harris contends that the "Italian role stimulated and nourished Eritrean national consciousness . . . and planted the seeds of Eritrean nationalism which made the fusion between semi-capitalist Eritrean and feudalist Ethiopia inherently unstable." He then concludes that "Italian imperialism and the rise of the Ethiopian nation-state set the stage for the drama of the longest war in post-colonial Africa." Charles L. Stansifer's discussion of the Nicaraguan civil war is also instructive. He states that the country's "political history since attaining independence in the early nineteenth century has been marked by partisan strife." Centered around "family rivalries," this conflict made "peaceful political exchange" impossible and led to the unfortunate practice where "the party in power not only represses the opposition but also makes the out-party pay economically in various ways for its political misfortunes." The Vietnamese people's long struggle against colonial domination also set the stage for the emergence of a mind-set which perceived the presence of the United States as a continuation of foreign imperialism that they had single-mindedly committed themselves to eradicate. Garth Shelton and Karl P. Magyar attribute the emergence of "violent independence movements in Portuguese Africa to the failure of the Lisbon authorities to feel the winds of change sweeping the continent" and the inability of insurgents to comprehend Portugal's "determination to maintain control." Similarly, the ability of the Ian Smith government "to withstand sanctions by developing alternative domestic industries and by circumventing United Nations sanctions," in Herbert Howe's estimation, was responsible for the long and violent conflict in Rhodesia.

In addition, prolonged conflicts seem to occur almost exclusively among authoritarian societies, where loss of human life and the heavy damage that such wars inflict can be concealed and those who speak out can be muzzled or even liquidated. By contrast, prolonged wars are more difficult to sustain in democratic settings. "[C]onstitutionally secure liberal states," says Michael W. Doyle, "have yet to engage in war with each other."[1] The nature of democracy, constitutional restraints, and public opinion makes it difficult for democratic governments to fight prolonged wars. Earl Tilford credits Gen Vo Nguyen Giap of North Vietnam as having correctly reasoned "that after the United States had suffered approximately 50,000 dead, the American public would make its government change its policy, and the troops would be withdrawn."

Yet democratic governments "have reasons to be skeptical of their counterparts that cannot claim to represent their peoples" and may "go to war for crusade reasons—that is, in order to promote democratic values."[2] Discussing the particulars of the conflict in Northern Ireland and British involvement in it, Benjamin Kline states that London has "defended its presence by arguing that it had a responsibility to recognize the democratic rights of the majority." Tilford sees American involvement in Vietnam in similar terms. He argues that Washington's Vietnam policy "tended to be declamatory and morally-based." Its fundamental aim was "to support the success of liberty," which "was assumed to be morally superior to totalitarian communism."

Prolonged war/conflicts (both domestic and international), says Karl P. Magyar, "often degenerate into purposelessness," inflict heavy material, social, political, and psychological damage, and take a heavy toll on human life but rarely accomplish the intended goals of the participants. Frequently, there are no victors or vanquished in the traditional meaning of these terms. In some cases, a participant may realize its goals and objectives not by scoring a decisive military defeat over its opponent but through the other party's decision to cut its losses and to withdraw from the conflict, rather than risk additional loss of life or further economic and social damage.

According to Stephen Blank, when the Soviets entered the conflict in Afghanistan, "[a] prolonged war ending in defeat was the last thing they expected." But when it became apparent that the war had become "a bleeding wound" for his country, Mikhail Gorbachev decided that the humiliation of withdrawal without achieving stated goals was preferable to the continuing agony of prolongation. Though difficult to gauge due to press and other restrictions prevailing at that time, Moscow's decision to terminate the conflict may have been forced by the unpopularity of the war. This unpopularity manifested itself in a number of ways, including draft evasion and military defections. US involvement and final withdrawal from Vietnam displayed remarkable parallels, so did Israel's withdrawal from the Lebanese conflict in the 1980s. In all cases, though tired and frustrated, the "vanquished" party still had sufficient military force and the wherewithal to continue fighting but chose to cut its losses. By comparison, the winning side won by not losing—not because it managed to score a decisive military victory.

The immense human and material toll prolonged wars take on the combatants eventually hampers the ability of combatants to continue the fight. In such cases, the objective of each side remains unrealized, and the disputes between them remain far from resolution. Under such circumstances, the warring parties may seek a truce, accept third party mediation, or agree on different ways of conflict resolution in fratricides. In the Iran-Iraq War, the combatants eventually exhausted themselves, and though none of the disputes that led to war had been resolved, the combatants reluctantly decided to accept a truce. By July 1988 Iran's supreme leader, the Ayatollah Khomeini, became convinced that "further prolongation of the war would . . . destroy the Revolution." The war nearly bankrupted the regime of Saddam Hussein and forced the Iraqi dictator to call for a cease-fire followed by a negotiated settlement when it became apparent that an all-out victory would become less likely. Similar considerations, according to Stewart Reiser, forced Israel and some of its Arab neighbors to seek a negotiated settlement to their disputes. A stalemate ultimately forced the Sandinistas and the Contras in Nicaragua and the government and FMLN forces in El Salvador to seek to resolve their conflict through electoral means.

The Prolongation Factors

Since the original intention to achieve quick military victory does not materialize, why do the warring parties continue fighting? Why do wars or conflicts become prolonged? Clearly, each conflict has its own idiosyncrasies and character, and its own victims. Yet, we can make some generalizations. Generally speaking, we can divide the factors that contribute to the staying power of certain wars or conflicts into three separate but often interrelated categories: general societal, international/regional, and strategic/military. Let us be more specific.

Societal Factors

Societal refers to specific political, social, ideological, and economic factors that provide the means to continue the conflict or create a social milieu that makes termination of the war unrealistic or socially unacceptable. Even though wars are destructive, they benefit some groups. Supplying and smuggling weapons and supplies can be profitable. Finding the resources to pay for these necessities can lead to the cultivation and sale of narcotics and other illegal commodities. The Nicaraguan Contras, the Irish Republican Army (IRA), and numerous other groups have been linked to such activities as have the militaries of El Salvador, Panama, Guatemala, Lebanon, and Afghanistan. Rebel groups and drug cartels in Colombia, Peru, and Bolivia are said to have developed synergic relationships. Colombian farmers benefited more from Pablo Escobar's narcotic-financed philanthropy than the policies of the government in Bogotá. Stopping the drug trade and the low-intensity guerrilla war going on in these countries is clearly antithetical to the economic interests of many Colombian, Peruvian, and Nicaraguan farmers. Right-wing, left-wing as well as Muslim and Christian militias in Lebanon, notes As'ad AbuKhalil, "developed their own economic network that relied for revenues on narcotics and arms smuggling."

Civil conflicts generate mass mobilization and sectarian agitation which further divide the multiethnic and already fragmented political cultures of developing societies. The "demonization of the other sect," argues AbuKhalil, "has helped traditional leaders in presenting themselves as champions of the interests of the community." Demonization, coupled with regional imbalances in economic development, leads people "to attribute their dissatisfaction to the tyranny or misguidedness of the other sect." Under the circumstances, compromise becomes unacceptable and continuation of the conflict is an inevitable outcome. Frédérick Torimiro depicts the Chadian situation in similar terms. The country's lackluster economy and fragmented political culture prompt those groups not in control of the levers of authority to view the "political group in control of the government . . . [with] suspicion," making civilized politics difficult to sustain.

Social mobilization and sectarianism have two additional spin-offs, which contribute to the prolongation of war. One is the multiplicity of dimensions of the conflict, and the other is the emergence of war elites or warlords. The more multi-dimensional the conflict, the more complicated the process of conflict resolution becomes. Meddling by outside powers often sustains or even exacerbates the number of participants and the dimensions of the conflict. The particulars of the Lebanese, Ethiopian, Sudanese, Afghan, and Liberian cases lend support to the argument. Cobie Harris, for example, feels that the collapse of the central government of Haile Selassie and its replacement by the brutal and illegitimate Mengistu regime led to the emergence of competing secessionist movements in Eritrea and Tigre. Having to deal with a number of movements, each of which had its goals and perceived the Ethiopian problem in different terms, hampered Addis Ababa's ability to defeat the rebels and added to the prolongation of the war between the two main protagonists: the Amhara-dominated central government and the Eritrean People's Liberation Front (EPLF). Divisions and splits among the different rebel groups in Liberia and in the Sudanese People's Liberation Army are seen by Karl P. Magyar and Ann Mosely Lesch, respectively, as having played a similar role in the staying power of the Liberian and Sudanese civil wars. AbuKhalil is even more candid. "A main reason for the prolongation of violence in Lebanon which produced the civil war and which could produce more bloodshed and strife in the future, lies in the multiplicity of dimensions of the conflict." He places the blame on the "connection between the various internal and external factors that perpetuated the war."

The emergence of a new war elite, exemplified by the warlords in Chad, Liberia, Lebanon, and Afghanistan, is the other derivative of social mobilization and sectarian agitation. The weakening of the central government, itself a product of strife and social fragmentation, gave rise to a new breed of military leaders who replaced the traditional ethnic or sectarian politicians. In Lebanon this new elite set up "a sophisticated bureaucracy . . . to cope with the rising needs of the thousands of fighters . . . and inherit[ed] the state responsibilities" in different parts of the country. The new leaders who speak "for the war activists" tend to be "youthful and experienced on the battlefield. They are trusted by their fighters for their closeness with the average people on the streets." "This new elite," concludes AbuKhalil, "represented the militancy of the civil war; a militancy that was responsible for the brutality and savagery that characterize it." Prince Johnson and Charles Taylor in Liberia and Hissène Habré and Goukouni Oueddei in Chad played similar roles in the prolongation of the civil war in these two countries. UNITA leader Jonas Savimbi and the different guerrilla leaders in Afghanistan also fit the warlord category.

Political considerations contribute to the prolongation of conflicts as well. Nothing can unite a society more than an external threat. Outside encroachments against a nation's national interests, pride, and physical integrity usually serve as rallying-around-the-flag movements which mobilize the citizenry to close ranks behind the leadership to save the country in times of peril. This reaction can give a weak and tottering regime a chance to buy time and to cloak itself with a mantle of legitimacy and national pride. In addition, by keeping the armed forces preoccupied with the war effort, the government can neutralize military coups aimed against it. Conflict prolongation, then, can serve politically useful purposes for unconsolidated or illegitimate governments and can lead to regime consolidation.

The Iran-Iraq War had significant consolidation benefits for the Baathist Baghdad regime and for the theocratic republic of Ayatollah Khomeini. Shirkhani and Danopoulos note that "the war demonstrated that an external threat could overcome religious loyalties, as the Baath party successfully mobilized the Shiaa majority population in the war effort against the Islamic Republic. Iraqi Shiaa did not switch sides and did not support their Iranian brethren in the war, as expected." The Iranian "clergy's drive to consolidate their rule also benefited from the war [for it] became the vehicle of control and cohesiveness . . . and established the religious fundamentalists as the dominant political group among the heterogeneous, divided forces of the revolution." One could argue also that the bombing of North Vietnam by the United States, instead of forcing Hanoi to submit, strengthened the resolve of the Vietnamese to carry on the war and helped the communist leaders to present themselves as the only legitimate and national-minded force striving to extricate the country from colonialism.

Finally, religious or ideological considerations are important prolongation factors. Ideology, which is often based on religious doctrines, can be a catch-all vehicle that serves as a framework for action, analysis, moral justification, and rationalization, as well as a blueprint for the present and the future. Wars fought on ideological or religious grounds tend to be cast in messianic or apocalyptic terms and promise higher spiritual rewards for those who give up their lives by elevating them to the realm of martyrdom, thus assuring their souls a place of repose, glory, and fulfillment. Such attitudes help to mobilize, galvanize, and maintain popular support for the continuation of the war.

The Sudanese civil war is a good example. The country's dominant political groups (which are Arab and Muslim and live in the North), in Ann Mosely Lesch's words, feel that they have "the right to institute Islamic legal codes concerning not only their own personal matters, but also political, economic, and social life." By contrast, they view the southern-based and predominantly Christian Sudanese People's Liberation Movement (SPLM) as "insubstantial" and its values "threatening" to the wishes of the Muslim majority. The SPLM, on the other hand, wants to create "a united Sudan under a socialist system that affords democracy and human rights to all nationalities and guarantees freedom to all religious beliefs and outlooks." The wide gulf that separates the two communities made "accommodation . . . impossible," thus contributing to the prolongation of the war.

The Anglo-Irish Treaty of the 1920s allowed the northern six counties of Ireland to remain part of Britain while their Catholic southern counterparts were granted Dominion status. This move led to "bitterness" on the part of the Catholics of Northern Ireland and, in Benjamin Kline's estimation, produced an atmosphere which allowed the IRA to recruit sympathizers and to organize a prolonged campaign against the British. Steffen Schmidt, too, believes that the Catholic Church "has always played an important role in [El Salvadoran] politics." Citing an article in which a Jesuit priest sanctions "the legitimation of insurgency prolonged struggle, [and] war of liberation" by elements of the Catholic Church, Schmidt concludes that "this was an important factor in the guerrilla strategy of prolonged war. It gave moral sanction to the process."

Religious or ideological considerations contributed to the prolongation of war in Vietnam, Lebanon, Cambodia, and Nicaragua. Even the Arab-Israeli conflict was in part influenced by differences in religious outlook. Both sides made references to their respective holy books to substantiate their conflicting claims of landownership. However, nowhere was the role of religion more powerful than in the Iran-Iraq War. Shirkhani and Danopoulos contend that "the factor most responsible for the prolongation of the war was the Islamic orientation of the Iranian Revolution with its religious psychology and the messianic zeal of its leaders and their followers." Seeing Saddam as "a puppet agent carrying out foreign satanic orders" and their revolution as a vehicle to "reverse the penetration of Western values into Iran and other Islamic cultures," Khomeini and his colleagues dismissed efforts to bring about cessation of hostilities as "a retreat from religious duties." Rejecting Saddam's cease-fire offer, Khomeini thundered: "We cannot compromise the 'Hussein,' a perpetrator of corruption. . . . We [are] bound by our religion to resist as much as we [can]." Though badly divided the Afghan Mujahedin saw the Soviet "infidels" in remarkably similar terms.

Finally, fear that cessation of hostilities would allow the status quo—against which they rebelled—to return is another consideration that prompts warring factions to continue their fighting. After having suffered heavy casualties and destruction, leaders find it difficult to convince their followers that a return to the status quo ante is desirable. Thus, even though wars may degenerate into purposelessness, leaders believe that a continuation of the struggle may be more justifiable or even preferable than returning to the conditions that caused the conflict in the first place. The Arab-Israeli conflict as well as the Sudanese, El Salvadoran, Chadian, Northern Ireland, and Lebanese fratricides illustrate this thesis. For example, AbuKhalil argues that the Lebanese feared that the peace plans would result in partial or temporary cease-fire but "would not lead to a final resolution of the war."

International/Regional Factors

Decisions or developments beyond the domain of the warring parties often provide the necessary weapons and the logistical, intelligence, diplomatic, and ideological support that contribute to prolongation of the war. This practice was standard during the cold war, when the two superpowers supported opposing sides. Almost all conflicts during this period, domestic or international, were characterized by superpower involvement of one form or another. None of the preceding case studies were exempted from the spin-offs of the Washington-Moscow ideological rivalry and hegemonic tendencies. By supporting opposing sides, the two superpowers strengthened the ability and the wherewithal of each side to continue the fight. In so doing the superpowers promoted their own conflicting national and strategic interests.

The cold war, says Stansifer, fueled the Contra war. Without foreign support, argues Herbert Howe, Zimbabwean "guerrillas could not have prolonged their struggle" against the well-oiled, white Rhodesian regime. Steffen Schmidt's conclusions regarding El Salvador are remarkably similar. Both sides (i.e., the government and the FMLN) "relied heavily for political, ideological, and material support on the Cold War model, and thus on externalities." He agrees with Kate Doyle and Peter S. Duklis's language (which he cites) "that the United States involvement in the war in El Salvador contributed, directly or indirectly, to the extended nature of the conflict." Likewise, in the words of Stansifer, "the Sandinistas counted on financial and military support from the Soviet bloc," while the Contras benefited from President Reagan's "determination to extract the Sandinistas from Managua." Finally, Stewart Reiser attributes the prolongation strategies adopted by the Arabs and Israelis as reflecting "each side's perception of their respective natural advantages as well as their abilities to extract resources from the great powers. . . . External forces helped the conflict," he concludes, "by giving each side the means by which to exercise their protracted strategies."

In addition to the two superpowers, other states, wittingly or unwittingly, have contributed to the prolongation of wars as well. Balance-of-power considerations, national interests, or the desire to play a regional role can be cited as factors that prompt states to intervene in civil wars or conflicts between other states in their region. The Angolan and Namibian civil war offers a case in point. According to Shelton and Magyar, the civil war became a regional war by directly involving South Africa and Cuba, in addition to the two superpowers whose involvement was more indirect. Discussing the civil war in Mozambique, Christopher Gregory believes that the "principal reason for the prolongation of the war [in that country] was and is South African support for the insurgents. Were it not for South African support the RENAMO insurgency might well not have become prolonged." Magyar's findings regarding Liberian fratricide are remarkably similar. He states that Sierra Leonean, Libyan, American, and ECOMOG involvement "ensured . . . a classic tenet of prolonged wars."

In the Sudanese conflict, the SPLM benefited from Ethiopia's "substantial support," Cairo's "credible diplomatic" assistance, and the logistical aid extended by a number of neighboring countries, including Kenya, Zambia, and Uganda. J. Richard Walsh is even more direct. He asserts that "Cambodia's prolonged war was precipitated by the Vietnamese invasion in December 1978." He attributes this to "Vietnam's belief that it must be treated as the dominant player in Indochina"—an attitude that "conflicted with China's own historic interests in the region." Interestingly, decisions by outside powers to remain neutral or to lend support to more than one of the warring parties also contribute to conflict prolongation. French neutrality with respect to the Chadian conflict is a case where a hands-off strategy helped to keep the flames of war alive. Paris' decision deprived the warring factions the crucial support needed to break the stalemate.

On the other hand, by employing the old divide-and-rule tactic of giving simultaneous support to different parties in the conflict, an outside force perpetuates the conflict. Perceiving the Chadian conflict "as part of its anti-Western campaign," the Libyan regime of Colonel Quaddafi lent support to the different warlords, sometimes selectively and other times simultaneously. "The protraction of the conflict," contends Torimiro, "was boosted by Quaddafi's ability to use the principle of 'divide and rule'. . . . [Thus] Libya became one of the patrons whose activities intensified the [Chadian] conflict." Syria's "no victors, no vanquished" policy on Lebanese fratricide was similar to Quaddafi's. By pursuing "a policy that allowed Lebanese factions and militias to combat one another without allowing any side to achieve total victory," Damascus managed to retain its influence and thus became "one of the most important reasons behind the prolongation of the Lebanese civil war." Finally, regional considerations guided Arab policies regarding the Iran-Iraq War. Shirkhani and Danopoulos assert that "Arab leaders did not want either side to win and seemed to have reasoned that prolongation of the war would weaken the warring sides and would force Khomeini and Saddam to shelve their respective hegemonic designs." Ironically, they conclude, "stalemate and bloodshed appeared more desirable to a clear victory by either side" as far as neighboring Arab states were concerned.

Strategic Factors

Strategic decisions made by the political and military leadership regarding the conduct of the war are another pivotal factor that helps to explain prolonged wars. While societal and external considerations provide the supporting reservoir that generate, nourish, and sustain the conflict, strategic or military actions often determine the nature and the outcome of the war. Strategic or military decisions can be deliberate, or they can be forced by existing realities and circumstances. Incompatibility of strategies can lead to prolongation, and so can the use of similar forms of warfare.

The Arab-Israeli conflict is a case in point where the adversaries, having failed to achieve victory "during the two wars (1948–49 and 1956), gradually developed . . . strategies of protraction." In the Vietnam War things were quite different. According to Tilford, US involvement in Vietnam and the conduct of the war suffered from lack of "clear political and military goals." Having made erroneous assumptions regarding Hanoi's war goals and the sociocultural values of the North Vietnamese, Washington pursued a strategy of attrition which relied on bombing enemy supply lines and employing ground forces heavily. By contrast, the Vietcong and the North Vietnamese adopted a strategy designed "to deliberately protract the war." Thus, while Washington's hazy objective was to defeat the enemy, Hanoi "did not have to defeat the United States military. It only had to compel the United States to withdraw its forces." Over time, the Americans became frustrated by their inability to defeat the enemy. The strategy of "tactical offensive" proved too costly and unsuccessful. Gen Vo Nguyen Giap's "tactical defensive" strategy ultimately forced Washington to disengage.

The conflict in Northern Ireland and the Soviet war in Afghanistan display considerable parallels. According to Blank, for the Afghans "guerrilla warfare was the only option open," their objective was to force Moscow to withdraw. Soviet military thinking was based on the axiom that "any country occupied by the Soviet armed forces must undergo a socialist revolution in its rear in order for the Red Army to secure the front." This meant that "there could be no question of limited objectives or limited war." Much like Washington, Moscow failed to understand the true nature of the Mujahedin's objectives and underestimated their capacity to resist and endure. Blank attributes these failures, in part, to the nature of the Soviet policy-making processes. He states that "[t]he lack of structural coordination and cohesion in the policy process, the rivalries among the players and their constituencies, and the deliberate narrowing of both the focus of decision making and of possible policy options contributed substantially to the ensuring intelligence and strategic debacle of a prolonged war."

While military superiority of the enemy and the nature of the conflict forced the IRA, the North Vietnamese, and the Mujahedin to adopt limited, defensive, and hit-and-run tactics, additional considerations have forced insurgents in other settings to resort to similar strategies. These included geographical, economic, and sociopolitical factors. For example, Eritrea's "unaccessible terrain," according to Harris, "allowed the guerrillas to maintain a permanent presence inside the country . . . [which, in turn] . . . created a symbiotic relationship between the armed forces and people." The relative success of RENAMO's "strategy of limited or low intensity" in Christopher Gregory's assessment, can be explained by FRELIMO's "implementation of highly interventionist agrarian policies [that] antagonized the local inhabitants into either active support for RENAMO or passive neutrality." Adopting Maoist "protracted war strategies . . . gave ZANA time to muster political and manpower support from its greatest potential strength—the overwhelmingly, and increasingly politicized [Rhodesia's] black population."

A final factor that affects the combat strategy of insurgent forces is the sharp separation between rural and urban sectors in developing societies. Urban centers, and particularly capital cities, tend to dominate the political and economic landscape while the countryside remains remote, less developed, and often outside the administrative purview of the regime in power. Indeed this gulf is frequently at the center of domestic conflict in African and, to a lesser extent, in Latin American societies. It makes possible the emergence and sustenance of resurgent forces in rural areas who, using low-intensity tactics, often end up dominating the countryside but are unable to capture the capital city, which is controlled by government forces. The ensuing result is "classic protracted struggle." Magyar captures the essence of this dichotomy by stating that

> [a]uthoritative power in Africa is almost totally concentrated in the capital city. . . . In effect, the struggle continues with two governments; one dominating the formal accouterments of state authority in the

> capital . . .; the other gains incremental legitimacy in the countryside. . . . [I]nsurgents rarely topple the government, but by retaining a credible capacity to survive and by launching an occasional terrorist act or attack on a government installation, . . . their fortunes are determined at international negotiations . . . or they fight on sporadically and interminably without resolution.

A Parting Word

This project did not aim to develop a general or comprehensive theory of prolonged war. Its goals were more modest. The introductory chapter sought to define the concept of prolonged war and differentiate it from the classic Maoist theory of protracted war. The case studies dealt with specific conflicts and sought to identify and analyze the factors that contributed to the prolongation of war in different settings. The concluding chapter drew some generalizations and recurring patterns. What we came up with was a number of theoretical propositions that need further empirical testing before a general theory of prolonged war is developed.

Though modest in its aims, we hope our project manages to call attention to a subject that has not been dealt with in the literature on war or international relations. Additional work is needed before we can fully comprehend this rather pervasive-but-little-understood phenomenon. Prolonged wars are neither new nor likely to disappear. As long as poverty, disease, economic disparities, social and ethnic cleavages, unemployment, hopelessness, and the desire for power exist—the factors that generate conflicts—the flames of prolonged wars are unlikely to be submerged, now or in the future.

Notes

1. Michael W. Doyle, "Kant, Liberal Legacies and Foreign Affairs," *Philosophy and Public Affairs* 12, no. 3 (Summer 1983): 205–35; and Steve Chan, "Mirror, Mirror on the Wall . . . Are the Freer Countries More Pacific?" *Journal of Conflict Resolution* 28, no. 4 (December 1984): 617–48.
2. George Sorensen, *Democracy and Democratization* (Boulder, Colo.: Westview Press, 1993), 95.